Biochemical *and* Biophysical
Perspectives
in Marine Biology

Biochemical *and* Biophysical Perspectives *in* Marine Biology

Volume I

Edited by

D. C. MALINS

U.S. National Marine Fisheries Service
Northwest Fisheries Center
Environmental Conservation Division
Seattle, Washington 98122
and
Seattle University
Chemistry Department
Seattle, Washington 98112

and

J. R. SARGENT

Natural Environment Research Council
Institute of Marine Biochemistry, Aberdeen

ACADEMIC PRESS   1974

London   New York   San Francisco

A Subsidiary of Harcourt Brace Jovanovich, Publishers

ACADEMIC PRESS INC. (LONDON) LTD.
24/28 Oval Road,
London NW1

United States Edition Published by
ACADEMIC PRESS INC.
111 Fifth Avenue,
New York, New York 10003

QL
121
.B56
v.1

Library of Congress Catalog Card Number: 74–5651
ISBN: 0–12–466601–9

Text set in 11/12pt. Imprint Ser. 101, printed by letterpress,
and bound in Great Britain at The Whitefriars Press Ltd., London and Tonbridge

Contributors to Volume I

FRED W. ALLENDORF, *Northwest Fisheries Center, National Marine Fisheries Service, National Oceanic and Atmospheric Administration, 2725 Montlake Boulevard East, Seattle, Washington 98112*

E. BILINSKI, *Fisheries Research Board of Canada, Vancouver Laboratory, Vancouver, B.C., Canada*

ARTHUR L. DEVRIES, *Physiological Research Laboratory, Scripps Institution of Oceanography, University of California, La Jolla, California 92037*

DENIS L. FOX, *Division of Marine Biology, Scripps Institution of Oceanography, University of California, San Diego, La Jolla, California*

HAROLD O. HODGINS, *Northwest Fisheries Center, National Marine Fisheries Service, National Oceanic and Atmospheric Administration, 2725 Montlake Boulevard East, Seattle, Washington 98112*

J. MAETZ, *Groupe de biologie marine, Département de biologie du Commissariat à l'Energie Atomique, Station Zoologique, 06230 Villefranche-sur-Mer (France)*

FRED M. UTTER, *Northwest Fisheries Center, National Marine Fisheries Service, National Oceanic and Atmospheric Administration, 2725 Montlake Boulevard East, Seattle, Washington 98112*

Preface

Recent years have witnessed an increasing awareness of the importance of the oceans and their associated life forms in global ecosystems. Reasons for this awareness are not hard to find. The oceans account for two thirds of the earth's surface and contain major supplies of natural resources exploitable by man; their importance in yielding food, fossil fuels and minerals is today appreciated as never before. And yet recent years have also seen much concern over marine pollution. We are now aware that, vast though the oceans are, they can be relatively easily influenced by human activity and seldom for the better.

To those concerned with biology the oceans present many challenges. Multitudes of highly diversified organisms have evolved in and adapted to the marine environment with its marked variations in light intensity, temperature, hydrostatic pressure, salinity and food density. It was in this environment that life started and it is here that nature has conducted, and is still conducting, some of her strangest experiments. Probably many new aspects of the life processes remain to be discovered in the oceans.

We sense the presence of greater numbers of biochemists and biophysicists showing an interest in marine organisms than ever before. Although many motivations exist for such an interest, two in particular spring to mind. Firstly, there are those concerned with understanding aspects of marine ecology in biochemical and biophysical terms. How do different organisms adapt to the different marine environments? How do individuals and species communicate one with another? Secondly, there are those who aim to enrich our basic knowledge of biochemistry and biophysics by studying highly specialized and specific life forms that have evolved from the massive gene pools of the oceans. Many of these workers saw in the organism of their choice an ideal model system where a biochemical pathway was more highly developed than in traditional organisms, or where the structures and functions of tissues were more accessible or manifested more intriguing properties than in terrestrial counterparts.

In undertaking to organize and edit the present series of volumes it was our intention to provide a forum for the scholarly and cogent discussion of biochemical and biophysical explorations into the nature of marine life. We adopted the word "perspectives" in our title because we feel that the capacity to judge the significance and proper place of ideas and facts in the overall picture may be especially useful in the very wide field of marine biology. It is because of the large numbers and great variety of different life forms in the oceans that efforts must constantly be made to search for integrating principles. We asked individual contributors to review the literature in their particular fields, but we also encouraged them not to shy away from cogent speculation in areas where knowledge is far from complete. Furthermore, we tried to encourage individualistic styles in our contributors because we feel strongly that personal prejudices, motivations and interests should not be entirely excluded from the scientific literature. In exercising these prerogatives, we sincerely hope that the end result will be stimulating and helpful to research workers as well as to advanced students of biology, biochemistry and biophysics.

We are grateful to many people who have helped us to bring this series into being. We especially acknowledge the advice and help given by Dr Patrick T. Grant of the N.E.R.C. Institute of Marine Biochemistry, Aberdeen, Scotland and Dr Maurice E. Stansby of the Northwest Fisheries Center, National Marine Fisheries Service, Seattle, Washington, U.S.A. Thanks are due above all to our authors who kindly devoted their time and energy to making the series a reality. Finally we thank the staff of Academic Press, London, for their patient guidance and the various authors and publishers who gave permission to reproduce original material.

D. C. MALINS
J. R. SARGENT
March 1974.

Contents of Volume I

Aspects of Adaptation to Hypo-osmotic and Hyper-osmotic Environments

J. MAETZ

Biochromes. Occurrence, Distribution and Comparative Biochemistry of Prominent Natural Pigments in the Marine World

DENIS L. FOX

Biochemical Genetic Studies of Fishes: Potentialities and Limitations

FRED M. UTTER, HAROLD O. HODGINS AND FRED W. ALLENDORF

Biochemical Aspects of Fish Swimming

E. BILINSKI

Survival at Freezing Temperatures

ARTHUR L. DEVRIES

Contents of Volume 2

Volume 3 *in preparation*

To

Mary and Liz

Aspects of Adaptation to Hypo-osmotic and Hyper-osmotic Environments

J. MAETZ

*Groupe de biologie marine, Département de biologie
du Commissariat à l'Energie Atomique
Station Zoologique
06230 Villefranche-sur-Mer (France)*

Introductory remarks: basic facts about salt and water metabolism

The present review is an attempt to compare osmoregulatory mechanisms in vertebrates and invertebrates. To limit the subject, I shall concern myself mainly with primary aquatic organisms. Thus no reference will be made to marine reptiles, birds or mammals. Iso-osmotic regulators will briefly be referred to in comparison with hetero-osmotic regulators.

In both vertebrates and invertebrates, adaptation to hypo-osmotic or hyper-osmotic environments implies the existence of a compartment, the extracellular fluids, playing the role of buffer between the external milieu and the living cells. A major element in this regulatory system appears to be the control of the entry or exit of water and salt across specialized effectors, which in their simplest forms are epithelial membranes located at the boundaries where the organism remains permeable. The gill, the gut and the excretory organ are the major sites of salt and water transfer. My main object will be the study of the structure and function of these membranes especially those of the gill and the gut. The very interesting problem of the osmotic relations between body fluids and tissues, which has been reviewed recently (Lange, 1972; Burton, 1973; Schoffeniels, 1973), will be neglected.

Solute and water transfers across osmoregulatory epithelial membranes obey physico-chemical laws. These will be summarized briefly but the reader is referred to more comprehensive reviews (Ussing, 1960; Potts and Parry, 1964a; Katchalsky and Curran, 1965; Dick, 1966; Stein, 1967; Maetz, 1968; Koch, 1970; Schafer and Andreoli, 1972).

A. Water movement across simple membranes

Liquid water has a semi-crystalline structure due to hydrogen bonding between neighboring water molecules (Horne, 1972). One type of water movement involves breaking of these bonds in the water lattice. This occurs spontaneously because of the thermal activation of molecules. The second type of movement, bulk flow, corresponds to movements of clusters of molecules. It involves an external source of energy such as hydrostatic pressure. There is some disagreement between specialists as to the mechanisms of water transfer across biological membranes.

1. FILTRATION

One possible type of transfer—by "filtration"—has been the subject of many years' study. It corresponds to a movement by bulk flow, the driving force being the osmotic pressure of the fluids separated by the membrane. Theoretical considerations and experiments on artificial membranes show that solutes dissolved in water exert a negative pressure which drives water from the hypo-osmotic compartment towards the hyper-osmotic compartment (Dainty, 1965). A hydrostatic pressure ΔP would have to be applied to the hypertonic fluid to stop the movement across the membrane.

According to Van't Hoff's law:

$$\Delta P = RT\Delta Cs \tag{1}$$

which is derived from the classical equation $PV = nRT$, the solute being regarded as formally equated to a gas occupying the volume of solvent, n/V being its concentration, Cs, in moles/ml. ΔP is given in atmosphere, if R the gas constant is in ml atm. $^0K^{-1}$.

Van't Hoff's law applies to an ideal semi-permeable membrane, permeable to water but not to solutes. Small osmotically active solutes do in fact cross most membranes and thus the effective osmotic pressure exerted by the solutes across the membrane is smaller than the theoretical value. Staverman (1951) introduced a coefficient called the reflexion coefficient (σ) to take into account the degree of interaction between water and solutes. In an artificial membrane, the effective hydrostatic pressure which is to be applied to stop fluid movement (ΔP effective for $\mathcal{J}_{net} = 0$) is compared to the theoretical value calculated from Van't Hoff's law. σ is determined by the following equation:

$$\sigma = \frac{\Delta P \text{ effective}}{RT\Delta Cs} \quad (2)$$

For a biological membrane, this technique is often not applicable and Katchalsky and Curran (1965) suggest a different method. It consists of equilibrating (i.e. $\mathcal{J}_{net} = 0$) a solution of an impermeant solute (for example raffinose or sucrose) with a solution of the permeant solute for which σ is to be measured.

$$\sigma = \frac{Cs \text{ (impermeant)}}{Cs \text{ (permeant)}} \quad (3)$$

σ is a coefficient which varies from 1 (no interaction between solute and water) to 0 (maximal interaction).

The net water flow across the membrane being proportional to the driving force ΔP, the following equation applies:

$$\mathcal{J}_{net} = \sigma P_{os} A\Delta Cs \quad (4)$$

where $\mathcal{J}_{net}$ is in moles sec^{-1}. P_{os} the osmotic permeability coefficient, which includes the constants RT, is generally given in cm sec^{-1}. A is the surface in cm^2 while ΔCs is in moles/ml. The osmotic water flow or filtration rate is measured by following the volume changes of one or the other solution bathing the membrane.

2. DIFFUSION

A second type of transfer by "diffusion" has been studied only since isotopic tracers of H_2O (HTO, HDO or $H_2^{18}O$) have become available.

Diffusion obeys Fick's law:

$$\frac{dn}{dt} = - D.A. \frac{dc}{dx} \quad (5)$$

The diffusional flux across a surface (in moles sec^{-1}) is proportional to the diffusion coefficient D in cm^2 sec^{-1}, to the concentration gradient of the substances in moles ml^{-1}cm^{-1} and to the surface A in cm^2. Fick's law may be extended to diffusion across a membrane:

$$\frac{dn}{dt} = \frac{-D}{\Delta x} A\Delta C \qquad (6)$$

ΔC being the concentration difference of the substance in the two compartments, and $-D/\Delta x$ the diffusional permeability coefficient P_d of the membrane in cm sec^{-1}, Δx being the membrane thickness.

If a tracer is used for measuring the diffusional water flux, Fick's law becomes:

$$\mathcal{J}_{12} = P_d A C_1 \text{ or } \mathcal{J}_{21} = Pd A C_2 \qquad (7)$$

where $\mathcal{J}_{12}$ is the unidirectional flux of water from compartment 1 to 2 and C_1 is the concentration of water in compartment 1, i.e. 55.3 moles/litre for physiological solutions. The net flux of water may also be derived from these equations:

$$\mathcal{J}_{net} = \mathcal{J}_{12} - \mathcal{J}_{21} = P_d A (C_1 - C_2) \text{ or } P_d A \Delta C \qquad (8)$$

where ΔC, being the water concentration difference, is equal to the solute concentration difference ΔCs.
It may be noted that:

$$\frac{\mathcal{J}_{net}}{\mathcal{J}_{12}} = \frac{\Delta Cs}{C_1} \qquad (9)$$

ΔCs for biological fluids and external media being small (1% or less) when compared with the concentration of water, the net flux of water is a small fraction of the unidirectional flux. The precision of the isotopic techniques is insufficient to allow for the determination of $\mathcal{J}_{net}$ by double labelling, e. g. by adding HTO to compartment 1 and HDO to compartment 2 and measuring transfer rates of T and D. Generally $\mathcal{J}_{net}$ is calculated from the values of $\mathcal{J}_{12}$ and $\Delta Cs/C_1$. For most biological membranes, the $\mathcal{J}_{net}$ calculated in this way differs from the net flow measured directly by a factor of two or more.

To account for this lack of agreement, Dainty and House (1966a,b) suggest that the unidirectional flux of water is grossly underestimated because either at the surface, or inside biological membranes in the epithelial cells, or in the connective layer, unstirred layers occur retarding the diffusion of the tracer. Recent investigations by Hays (1968) and Hays and Franki (1970) demonstrate that high speed stirring of an epithelial membrane such as the toad bladder greatly enhances the diffusional flux. They show also that the isolated connective supporting layer is responsible for one half of the overall resistance of the tissue to diffusion. They suggest that if the unstirred layers within the cellular epithelial layer are taken into account, water can be taken to move across the membrane by diffusion in channels with radii as small as 2.75 Å. Hays and Franki (1970) also showed that the osmotic net flow of water across the membrane is practically unaltered by high speed stirring.

The diffusion coefficient, D, or diffusional flux varies with environmental temperature according to Arrhenius' equation:

$$D = D_o \exp \frac{-E}{RT} \tag{10}$$

where D_o is a reference coefficient and E is the activation energy of the water molecules in joules/mole. In bulk water the E for diffusion is 4.6 Kcal mole^{-1} indicating that an average of two hydrogen bonds are broken for the movement of each water molecule (Wang et al., 1953). In channels with large radii it may be assumed that water in the central core assumes the properties of liquid water and E for diffusion should be as low as 4.6 Kcal mole^{-1}. Such a low value was observed in the isolated connective supporting layer of the toad bladder by Hays et al. (1971), and across the outer border of frog skin at temperatures above 25°C (Grigera and Cereijido, 1971).

In general E for water diffusion across cellular membranes is considerably higher than in bulk solution (Grigera and Cereijido, 1971; Hays et al., 1971; Motais and Isaia, 1972a; Isaia, 1972). This can be explained either by extensive bonding of water molecules to components of the narrow aqueous channels in the membrane, or by the tendency of water to assume an ice-like state in the vicinity of the nonpolar groups in the channels.

These thermodynamic considerations suggest that water moves across biological membranes through small molecular pores and not, as previously proposed by Koefoed-Johnsen and Ussing (1953), through pores sufficiently wide to allow for bulk flow of water. Support for the latter view was provided by the "solvent-drag effect" discovered by Andersen and Ussing (1957). When equal concentrations of certain permeant test substances (thiourea, acetamide) are placed on both sides of frog skin, the ratio of the unidirectional fluxes of these substances is equal to 1 when there is no net flux of water. But when an osmotic flow exists, an acceleration of the solute flux in the direction of the water movement and a simultaneous reduction in the opposite direction can be observed. The unidirectional flux ratio exceeds 1, and the bigger the net flow the bigger the deviation. It is probable that the explanation of this phenomenon resides in different interactions between the solute and solvent, the solute and membrane and the solvent and membrane.

3. "UPHILL" WATER TRANSFER

In most biological membranes water moves "downhill", i.e. *along* its concentration gradient. In some cases however, movement of water occurs "uphill", that is, in contradiction to the physico-chemical laws given above. Net movement of water in the absence of, or against a concentration gradient has been observed in many epithelia: those of the proximal tubule of the kidney (Windhager, 1968), frog skin (Maetz, 1968), the gall bladder (Whitlock and Wheeler, 1964; Diamond, 1964a,b), the intestine of fish (Skadhauge, 1969) and insect excretory systems (Phillips, 1970; Ramsay, 1971). In most cases a possible explanation for such apparently abnormal water movement is that solutes, especially during ion transport across the membrane, are able to "drag along" water molecules by a process similar to that of the solvent drag discussed above. Curran (1960) and Curran and Mackintosh (1962) have suggested a model of three compartments in series, separated by two barriers, which would account for the uphill transfer of water in certain cases. Patlak *et al.* (1963) and Diamond (1965) have carried out thermodynamic analyses of the proposed model. The first barrier is assumed to be semi-permeable and the site of an active transport of solutes. The second barrier is non selective ($\sigma \leqslant 0$) and merely retards diffusion of solutes without playing an osmotic role. The active transport of solute across the external barrier will maintain the intermediate compartment in a steady state of slight hypertonicity and water will flow through the external barrier as a result of "local" osmosis. Thus hydrostatic pressure will build up in this compartment and drive water across the internal barrier into the internal compartment. Morphological evidence has been advanced in support of this model. Electron microscopic studies have shown that many epithelia that are sites of uphill water transport show enlarged intercellular spaces filled with fluid. These spaces would be equivalent to the intermediate compartment, the basal membrane playing the role of the internal leaky membrane. In some cases, micropuncture studies have confirmed the hyperosmotic nature of the fluid in this compartment (see review by Schmidt-Nielsen, 1971).

B. Salt movement across simple membranes

This section will deal essentially with the transfer of Na^+ and Cl^- which are the chief electrolytes of the extracellular fluids and of the external environment.

To analyse the movements of ions, several parameters have to be assessed. (1) It is essential to measure the concentration or better the activity of the ionic species on both sides of the membrane. (2) It is next necessary to measure the flow of solute across the membrane. If there is a net flux of solute it is important to determine with the appropriate isotopes, the unidirectional fluxes, i.e. influx and efflux in moles sec^{-1}. Control of salt transfer across a membrane may be effected by changes of either or both of these parameters. (3) As the solutes are electrically charged ions, knowledge of the electric potential difference between the two sides of the membrane is required in order to evaluate the flux-force relationship. (4) Finally, it is necessary to bear in mind the importance of the drag of solutes by solvent flow across the membrane.

Transfer of solute across a biological membrane may be either an *active* or a *passive* process.

1. PASSIVE TRANSFER

Passive transfer can be accounted for by the ordinary physical forces of the concentration gradient, the electrical gradient and the solvent flow.
(a) The flux of a solute in response to the concentration gradient dc/dx is formulated by Fick's law for pure diffusion which has been applied above for the movements of water (see equation 5):

$$\frac{dn}{dt} = \mathcal{J}_{\text{diff}} = -AD\frac{dc}{dx}$$

Thus $\mathcal{J}_{\text{diff}}$, the diffusional net flux in mole. sec^{-1}, is the product of the driving force (the concentration or activity gradient) and a mobility term D, the diffusion coefficient. When diffusion across a membrane is considered, a simple solution for this equation is to be found by assuming that the concentration gradient is constant across the membrane.

$$\text{Thus } \mathcal{J}_{\text{diff}} = A.P(C_1 - C_2) \tag{11}$$

where P, i.e. D/dx, is the permeability of the membrane to the solute given in cm sec^{-1} and C_1, C_2 are the concentrations or activities in moles ml^{-1} of the solutes in compartments 1 and 2, A is the surface area in cm^2.

If the unidirectional flux is measured with an isotope added in compartment 1, so that the specific activity of the solute (in Ci $mole^{-1}$) is S_1, the net *tracer* flux is given by:

$$\mathcal{J}^*_{\text{diff } 1\rightarrow2} = A.P\,C_1S_1 \tag{12}$$

providing that the quantity of isotope appearing in compartment 2 is negligible and $C_2S_2 = 0$. The unidirectional flux $\mathcal{J}_{\text{diff } 1\rightarrow2}$ in mole sec^{-1} is the tracer flux divided by S_1.

$$\text{Thus } \mathcal{J}_{\text{diff } 1\rightarrow2} = A.P\,C_1 \tag{13}$$

(b) The flux of an electrolyte in bulk solution in response to an electric field or gradient is given by the equation:

$$\mathcal{J}_{el} = - A.D. \frac{zF}{RT} C \frac{d\varphi}{dx} \tag{14}$$

where $\mathcal{J}_{el}$ in moles/sec is again the product of the driving force, the electric gradient $d\varphi/dx$ in volt cm^{-1} and a mobility term. The mobility term includes the diffusion coefficient, the concentration (or activity) of the solute and a conversion factor relating chemical to electrical energy zF/RT, z being the charge of the ion. F, the Faraday and RT are the usual constants. zF/RT is in volts^{-1}. When a flux of electrolyte is measured in response to a potential difference across a membrane, simplifying assumptions have again to be made to obtain a simple solution to equation (14).

If one assumes that the potential varies linearly across the membrane (constant field) and that no other driving force such as a concentration gradient is to be taken into account, the solution is:

$$\mathcal{J}_{el} = A.P. \frac{zF}{RT} C (\varphi_1 - \varphi_2) \tag{15}$$

(c) If in addition there is a concentration gradient of solute across the membrane, the differential equation giving the net flux ($\mathcal{J}_{net}$) is:

$$\mathcal{J}_{diff} + \mathcal{J}_{el} = \mathcal{J}_{net} = - A.D \left(\frac{dc}{dx} + \frac{zF}{RT} C \frac{d\varphi}{dx} \right) \tag{16}$$

Assuming a constant field, Goldman (1943) solved this equation for a homogeneous membrane:

$$\mathcal{J}_{net} = A.P. \frac{zF(\varphi_1 - \varphi_2)}{RT} \cdot \frac{C_2 - C_1 \exp \dfrac{zF(\varphi_1 - \varphi_2)}{RT}}{1 - \exp \dfrac{zF(\varphi_1 - \varphi_2)}{RT}} \tag{17}$$

If the unidirectional flux across the membrane is measured, the tracer being added in compartment 1, the equation simplifies to:

$$\mathcal{J}_{1-2} = A.P. \frac{zF(\varphi_1 - \varphi_2)}{RT} \cdot \frac{C_1 \exp \dfrac{zF(\varphi_1 - \varphi_2)}{RT}}{\exp \dfrac{zF(\varphi_1 - \varphi_2)}{RT} - 1} \tag{18}$$

It may be seen that, despite the simplifying assumption, the flux is a complicated function of the external concentration. Such an equation is barely of use to the experimenter. Ussing (1960), therefore, proposed a different approach for defining passive transport. He suggested that the various variables of equation (16) need not be taken into account if the ratio between the unidirectional fluxes is considered rather than the fluxes themselves. Providing ions move across the membrane in both directions along the same path, C and φ vary with x in the same way. Thus if the membrane makes no distinction between the different isotopes, the solution of equation (16), generally called the Ussing criterion, is the following:

$$\frac{\mathcal{J}_{12}}{\mathcal{J}_{21}} = \frac{C_1}{C_2} \exp \frac{zF(\varphi_1 - \varphi_2)}{RT} \tag{19}$$

This equation holds provided that the movement of the ions through the membrane is independent not only of all other species (thus excluding solvent-drag effects) but also of that of ions of the same species (thus excluding exchange-diffusion).

Ussing's equation simplifies further when simpler thermodynamic conditions prevail. For example: (1) if there is no voltage difference or if the solute is uncharged, the ratio of the unidirectional fluxes is equal to the ratio of the concentrations from which the fluxes are derived; (2) if there is no net flux $\mathcal{J}_{12}/\mathcal{J}_{21} = 1$, and the equation obtained is the famous Nernst equation:

$$\frac{C_1}{C_2} = \exp \frac{zF(\varphi_2 - \varphi_1)}{RT} \text{ or } \varphi_2 - \varphi_1 = \frac{RT}{zF} \ln \frac{C_1}{C_2} \tag{20}$$

which corresponds to a situation where the concentration gradient or diffusional flow balances the electrical gradient or electrical flow. When several monovalent ions are present with their respective concentration gradients, and if there is no net current of charges across the membranes because the total charges transferred by the net transport of cations is equal to those transferred by the anions, a generalization of the Nernst relation is possible.

$$\varphi_1 - \varphi_2 = \frac{RT}{F} \ln \left(\frac{\Sigma P_{cat}C_2 + \Sigma P_{an}A_1}{\Sigma P_{cat}C_1 + \Sigma P_{an}A_2} \right) \tag{21}$$

where P_{cat} and P_{an} are the permeabilities and C and A are the concentrations of the various cations and anions in compartments 1 (C_1, A_1) and 2 (C_2, A_2).

This generalization of the Nernst equation was derived by Hodgkin and Katz (1949). If a concentration gradient for various electrolytes is maintained across a cell membrane, the electric potential across the membrane may be calculated by equation (21). Such a concentration gradient may result from a Donnan equilibrium produced by the presence of non-diffusible negatively charged proteins in one compartment. This will maintain an excess concentration of diffusible anions on the *trans* side.

Such a concentration gradient may also result from the activity of a neutral ion pump exchanging one ionic species against another in a one for one fashion, compensating for a diffusive flow of these ions along their respective electrochemical gradients.

(d) Another deviation from the normal passive behaviour of ions occurs when the movement of solvent caused by differences in osmotic pressure exerts a force on the diffusing solute, so that the movement of ions in the direction of flow is accelerated whereas ions moving against the flow are slowed down. According to Koefoed-Johnsen and Ussing (1960) the general flux equations given above (eq. 16) must also include this drag force which is proportional to the osmotic pressure difference and to the ratio of the frictional coefficients between ion and solute and pore wall and water. When there is no concentration gradient across the membrane and the movement of solvent is due to hydrostatic pressure difference, the convection flow of solute is given in its simplest form by:

$$\mathcal{J}_{conv} = A.\mathcal{J}v.C \qquad (22)$$

where $\mathcal{J}_{conv}$ is in moles. sec, $\mathcal{J}_v$ is in ml/cm^2.sec or in cm.sec^{-1} and C is in moles.ml^{-1}.

In many cases, when the net flow of water is small this correction factor may be disregarded. Solvent drag may, however, be important across the membranes of animals that are in contact with strongly hypertonic or hypotonic media. To calculate this factor, Ussing recommends the use of uncharged hydrophilic test molecules such as urea or thiourea, which have approximately the same dimensions as Na$^+$ and Cl$^-$. By double labelling it is possible to evaluate simultaneously $\mathcal{J}_{in}$ and $\mathcal{J}_{out}$ as a function of the respective concentrations in the inner and outer media. In whole animals this method may not be very practicable. A qualitative method of assessing the importance of solvent drag in hyper-osmotic regulators is to study the effects on the unidirectional ion fluxes of the addition of impermeant solutes (e.g. sucrose) to the outside in order to abolish the osmotic gradient.

2. ACTIVE TRANSFER

Active transfer characterizes movement of solutes *against their electrochemical gradient*. Such a departure from electro-chemical equilibrium occurs when the experimental flux ratio deviates from that calculated according to Ussing's criterion (equation 19). Net "uphill" transport requires an expenditure of metabolic energy equal to the product of the rate of net transport and the electrochemical potential difference across the membrane.

The transfer of energy from chemical reactions to a substance moving across a membrane poses an interesting thermodynamic problem, because non-directional scalar flows are converted into vectorial flows. An anisotropic system such as a biological membrane allows for such a conversion. It is certain that chemical reactions involving carriers or enzymes concerned with ion transfer are membrane-bound.

The involvement of carriers or enzymes in ion transport is strongly suggested by the impressive body of evidence showing that active transport systems are saturable. The active flux depends on the concentration of the transported ion at low concentrations, but approaches a maximal upper limit value ($\mathcal{J}_{max}$) at higher concentrations. The relationship is described by:

$$\mathcal{J}_{in} = \frac{\mathcal{J}_{max}(C_{ext})}{K_m + (C_{ext})} \qquad (23)$$

This equation resembles the Michaelis-Menten kinetics for enzyme activity as a function of substrate concentration. Two parameters characterize the system, the maximum pump flux $\mathcal{J}_{max}$ and the K_m which is the concentration at which the flux is half-maximal and which corresponds to the apparent affinity of the carrier or enzyme for the substrate.

The occurrence of such transport kinetics for fresh water and brackish water hyperosmoregulators and the involvement of membrane-bound enzymes, especially Na^+-K^+ and anion-dependent ATPases, in transepithelial Na^+ and Cl^- transport will be discussed below.

3. EXCHANGE DIFFUSION

The use of Ussing's passive transport criterion and Michaelis-Menten kinetics to define active transport is complicated by the occurrence across some cellular and epithelial membranes of an "exchange difffusion" process. This phenomenon, described by Ussing (1960), takes place when the flux ratio is closer to unity than would be the case if the fluxes were due to simple passive leaks. This type of transport corresponds to a one-to-one exchange of the same ion species across the cell border or epithelium without any *net* transfer. This process can only be detected by means of isotopes. A coupling between unidirectional fluxes mediated by a carrier which would cross the membrane in a combined state, irrespective of the electrochemical gradient, would account for the passive nature of such a transfer. A simple test for such a process is to reduce the concentration of the solute on one side of the membrane in order to reduce drastically the corresponding unidirectional flux, and to verify whether the flux from the *trans* side is simultaneously reduced. Moreover both fluxes should display similar saturation kinetics as functions of the solute concentrations on either side of the membrane. According to Britton (1965, 1970), however, this test requires that the membrane potential should remain unchanged if the solute is an electrolyte.

Exchange-diffusion has been attributed by some investigators (Bryan, 1960b, c; Shaw, 1963; Potts and Parry, 1964a) to the presence of a "leaky" ion pump. An alternative suggestion was made by Kirschner (1955) to explain the observation that the Na^+ efflux across an isolated frog skin into a Ringer solution is about 3 times that of the efflux into a Na^+ — free Ringer solution. According to Kirschner, if the transport takes place in association with a carrier, then when the external Na^+ is low, the ions diffusing out have a high probability of meeting an unoccupied carrier and of being returned; however, if the external Na^+ is high, the carrier is likely to be fully occupied, hence the increased efflux.

C. Water and salt movements in vivo

When studying even simple epithelial membranes in vitro, extreme caution is necessary in evaluating the active and passive components of the ion transfer across the tissue, and the water permeability. The situation is, of course, much more complicated when whole animals are under study. Some of the difficulties found in the study of intact animals are summarized below.

The relative importance of the various biological membranes concerned with salt and water exchanges between the organism and the environment is often not given sufficient attention. For example, ionic influx occurs through gills, skin and gut of fishes while the efflux is shared by gills, skin and excretory organ. Additional experiments or techniques are necessary, for example, catheterization of the excretory ducts, independent measurements of the drinking rate, or in vitro skin permeability determinations, to ascertain the relative importance of the various routes of exchanges of solutes and water.

Measurement of the diffusional water permeability of the body surfaces implies the use of tracers such as HTO or HDO. In general the tracer kinetics are relatively simple since the total body water behaves as one homogeneous compartment exchanging with the external medium according to a simple exponential function (see Motais et al., 1969). Analysis of this function allows for the calculation of the rate constant in per cent of the body water per hour and hence of the unidirectional exchange fluxes $\mathcal{J}_{in}$ and $\mathcal{J}_{out}$. In some instances, HTO kinetics were found to be more complicated suggesting a heterogeneity of the body water, and necessitating multicompartment analysis (Rudy, 1967).

Several other problems also arise when diffusional water permeability is measured in vivo.

(1) Unidirectional flux measurements should be made in the absence of an osmotic gradient to avoid involving the tracer in solvent flow. This condition is often not met in studies on intact animals. Such an effect may explain why the rate constants K_{in} and K_{out}, calculated from tracer influx studies, yield slightly different results from those from efflux studies (see for example: Potts et al., 1967; Lotan, 1969).

(2) Diffusional water fluxes may be underestimated because of the presence of unstirred layers at the boundaries. Aquatic animals with gills are provided with some of the best stirred membranes, because of the low solubility of oxygen in the external medium and the respiratory function of these organs. The distances between external and internal media are minimal (0.5 to $5\mu m$) and a high rate of water flow over the respiratory leaflets is observed in both fishes (review by Hughes, 1963) and crustaceans (Hannan and Evans, 1973). Also, in theory, the blood flow inside the gills of crustaceans and of fish is at least 10 times higher than the unidirectional water flux across the epithelium (Hannan and Evans, 1973; Motais et al., 1969). In practice problems arise because blood may shunt away from the respiratory lamellae (Steen and Kruysse, 1964). Internal unstirred layers, characterized by a specific activity in the blood different from that found in peripheral blood, may be responsible for an underestimation of the diffusional water flux.

Referring to ionic exchanges, Kirschner (1970) has recently summarized the difficulties in obtaining reliable thermodynamical data such as uni-directional fluxes and electrochemical potential differences in vivo. The use of tracers to measure unidirectional fluxes is sometimes severely limited by the complexities of compartmental analysis, although this is fortunately not the case for Na^+ and Cl^- exchanges (techniques in Maetz, 1956b; Motais, 1967; Kirschner, 1970).

While the measurement of the chemical activities of the electrolytes in blood or in the outside medium does not present major difficulties, the measurement of electric potentials presents certain complications. Most workers simply insert a catheter, connected via agar-KCl bridges and calomel electrodes to a high impedance millivoltmeter, into the intraperitoneal cavity and record potential difference with a similar electrode kept in the outside medium (Kirschner, 1970). Few studies have been based on micro-electrode impalements directly across the gill epithelium (Tosteson, 1962; Maetz and Campanini, 1966; Kerstetter and Kirschner, 1972; House and Maetz, 1974).

One useful method for assessing the effects of electrochemical gradients on electrolyte movements is to alter experimentally the external and internal electrolyte concentrations. Such experiments however upset the mineral balance. Accordingly the observer should assess the immediate effects of the concentration changes and dissociate them from the long-term effects which merely reflect feed-back regulatory mechanisms intervening to restore mineral balance. Care should be taken to verify whether simultaneous changes occur in the electric potential differences.

One of the major difficulties when dealing with whole animals is to avoid shock effects resulting from handling. Stress not only upsets water balance but also mineral balance. Animals should be adapted to the experimental chamber well in advance of the actual measurements. Injections should be made by indwelling catheters and blood sampling by catheterization of blood vessels.

One last point should be made. As discussed above, isolated membranes or organ preparations are far better suited than intact animals for obtaining thermodynamic information on the mechanisms of ion and water transfer. There is, however, one imperative corollary: one should verify to what extent information obtained in such a way fits into the overall osmoregulatory balance of the intact animal.

II. Adaptation to Hypo-osmotic environments

A major problem which faces an aquatic animal living in a hypo-osmotic medium is the maintenance of an adequate osmotic environment for cellular function. Some primitive unicellular organisms including bacteria, protozoa, certain algae and fungi (Conner, 1967; Kitching, 1967; Slayman, 1970) as well as multicellular acoelomates such as porifera and coelenterata (Lilly, 1955; Hazelwood et al., 1970; Macklin and Josephson, 1971; Macklin et al., 1973) are, however, able to live in fresh water with their cells in direct contact with the hypo-osmotic external medium. It should be mentioned that in some of these organisms a functional specialization of cellular territories or epithelial membranes for electrolyte absorption and water excretion is already observable.

The present review will be limited to the coelomates which are characterized by a well defined internal body cavity and body fluids. The most important groups to be considered are the Annelida, Mollusca, Crustacea, fish and Amphibia. A comparison with aquatic insect larvae will be made.

The osmotic pressure of the blood, which is representative of the body fluids, varies from 50 mOsmolar in some lamellibranchs to 600 mOsmolar in some crustaceans. In vertebrates the osmotic pressure varies within a much narrower range from about 220 mOsmolar in amphibians to 300 mOsmolar in fish. The most important osmotically active osmolytes are the electrolytes Na^+ and Cl^-. Fresh water, whether Ca^{2+} rich or poor, or Na^+ rich or poor, has an osmotic concentration of 1 to 20 mOsmolar.

Reference will, however, be made to brackish water species which, although hyper-osmotic regulators, are unable to penetrate into fresh water. The reader is referred to previous reviews (Shaw and Stobbart, 1963; Stobbart and Shaw, 1964; Lockwood, 1962; 1968; Oglesby, 1969; Schoffeniels and Gilles, 1972; Parry, 1966; Potts and Parry, 1964a; Potts, 1968; Deyrup, 1964) dealing with the ionic composition and osmotic regulation of the body fluids, since this review will be limited to the role of the effector-organs of osmoregulation that separate the external and internal media. In hyper-osmotic regulators, these organs control osmoregulation by two compensatory mechanisms: excretion of water to eliminate water entering osmotically by way of the permeable surfaces, and absorption of salt from the dilute medium against a considerable chemical gradient to compensate for salt loss *via* the excretory organs and along the chemical gradient across the body surfaces.

Thus the main features of adaptation to low salinity media are:
(1) a reduction of the permeability of the body surface to water thus lightening the work imposed on the excretory organs for "free-water" clearance; (2) a reduction of the Na^+ and Cl^- permeability of the body surfaces, limiting salt loss; (3) the development of an efficient Na^+ and Cl^- uptake mechanism across the body surfaces; (4) the reabsorption in some terminal part of the excretory organs of the Na^+ and Cl^- filtered in the proximal part helps to save salt and allow for free-water clearance. This fourth aspect, having been reviewed by Kirschner (1967) for invertebrates, and by Hickman and Trump (1969) and Deyrup (1964) for the lower vertebrates, will be largely ignored in this chapter.

A. Electrolyte transfer in hyper-osmotic organisms

1. PARTITIONING OF THE ELECTROLYTE TRANSFERS: ROLE OF GILLS AND SKIN

In order to equate the movements of Na^+ and Cl^- with the electro-chemical gradient across the body surfaces it is necessary to take into account the various routes of salt entry and exit.

One possible route of salt entry is the gut. In freshwater animals, the drinking rate in non-fed animals is generally small. In freshwater eels, however, drinking rates as high as $165\mu l\ 100g^{-1}/h^{-1}$ have been reported (Maetz and Skadhauge, 1968; Krisch, 1972b). From an external medium of low sodium content the salt taken up by the intestine accounts for less than 1% of the Na^+ influx. In brackish water, however, this route of salt entry cannot be neglected. In some aquatic insect larvae, such as *Sialis lutaria*, the gut may be the sole route of salt entry (Shaw, 1955b).

Partitioning of the salt exchanges across the body surface of animals with part of the ectoderm specialized for respiratory gas exchange is difficult.

Obviously, the gills, because of the thinness of the epithelium and its extended surface area, are the major site of ionic exchanges.

In vitro studies have confirmed that the gills are the main site of salt uptake in crustaceans such as *Eriocheir sinensis* (Koch et al., 1954), *Astacus leptodactylus* (Bielawski, 1964), *Callinectes sapidus* (Mantel, 1967) and *Carcinus maenas* (King and Schoffeniels, 1969). In fish, salt uptake has *not* been demonstrated in the "incubated gill" preparation (Bellamy, 1961) or in the perfused gill preparation of the eel (Shuttelworth, 1972). Salt uptake occurs in the perfused gill of *Salmo gairdneri* only after reduction of the concentration of NaCl in the perfusing Ringer solution (Richards and Fromm, 1970). It is unfortunately difficult to infer from such *in vitro* flux studies the relative importance of the gill *in vivo*. An adequately perfused preparation necessitates efficient stirring of the outside medium and adequate control of the flow of the perfusing medium as well as leaks resulting from difficulties in catheterizing the blood vessels.

Another technique, the isolated heart-gill preparation devised more than 40 years ago by Keys (1931a,b), has also proved inadequate to demonstrate salt uptake in freshwater eels, although it provided the first unequivocal demonstration of ion transport in seawater eels. Kirschner (1969) tried recently to account for this failure. Salt loss is caused by a slow deterioration of the diffusion resistance in the gill rather than by a decrease of the sodium influx. More recently, Kerstetter et al. (1970) devised an "irrigated gill" preparation for the anaesthesized trout *S. gairdneri* that consisted of externally perfusing the mouth cavity with a small volume of fluid enabling rapid, successive flux measurements to be made. These showed definitely that the gills are the main site of ion exchange. Unfortunately, comparison with fluxes obtained recently in the free swimming trout gives ground for suspecting that the "irrigated gill" preparation may be inadequate for two reasons: the gill-chamber perfusion flow is probably insufficient and the cardiac output is slowed down by anaesthesia (Wood and Randall, 1973). A similar gill-chamber perfusion technique allowed Alvarado and Moody (1970) to demonstrate in *Rana catesbiana* tadpoles that a distinct influx of Na^+ occurs when the gill chamber is perfused with tap water containing $^{22}Na^+$, while larval skin *in vitro* proved unable to transport Na^+. Ussing and Zerahn (1951) were the first to demonstrate active Na^+ transport across the isolated adult amphibian skin.

That the skin of fish does not participate to a great extent in ionic exchange has been demonstrated by experiments on isolated skins of *Lampetra fluviatilis* (Bentley, 1962) and of *S. gairdneri* (Fromm, 1968), and recently by experiments *in vivo* on the eel by Kirsch (1972a) with the help of an improved version of the divided chamber technique originally used by Smith (1929) and Krogh (1937). Separation of the anterior part from the remaining body surface in the eel was also obtained by the "eel-tube" preparation (Chester Jones and Bellamy, 1964). The Na^+ uptake from the anterior part compensated for the Na^+ loss, probably entirely urinary, from the posterior part.

Partitioning of the electrolyte exchanges in aquatic insects has been attempted with success in the culicine Diptera larvae which tap atmospheric air for their oxygen supply. These animals have developed a cuticle which is practically impermeable to salt and water except for expansions of the last segment called the anal papillae. Koch (1938) showed by ligaturing these organs as well as by heat destruction that they are responsible for most of the salt uptake. These observations were confirmed later by Treherne (1954) and Phillips and Meredith (1969) with the use of radioactive tracers. Koch (*in* Krogh, 1939) also demonstrated the role of the anal papillae in electrolyte uptake by showing that they absorb and precipitate silver from dilute solutions of silver nitrate. Using this technique he also demonstrated that the rectal gills in Anisoptera nymphs (*Libellula, Aeshna*) are probable sites of ion uptake. Such nymphs depend upon diffusion of dissolved oxygen through specialised regions of their cuticle. Similar studies were made recently on various Trichoptera larvae (caddisflies) and Ephemeroptera nymphs (mayflies) (see morphological aspects of active transport).

In molluscs and annelids it is probable that the general body surface is responsible for ionic uptake. Whether gills, when present, function in ionic uptake is not known. The gill-less aquatic pulmonates, *Lymnaea stagnalis* or *L. limosa*, exhibit an ionic turnover higher than the lamellibranch *Margaritana margaritifera* (Greenaway, 1970; Chaisemartin *et al.*, 1968; Chaisemartin, 1969).

Concerning electrolyte efflux, the major component to be taken into account is urinary salt loss. Table I illustrates some data concerning urine flow, urine Na^+ concentrations and the absolute values of the total Na^+ efflux and renal Na^+ loss. Estimation of urine flow has been done in amphibians, fish, crustaceans and molluscs. In many species only indirect measurements have been obtained: (a) by dividing the rate of excretion of an exogenous substance (e.g. dyes, creatinine, inulin, $^{35}SO_4^{2-}$ sodium (^{131}I)diatrizoate) by the concentration of this substance in a urine sample (Parry, 1955, 1957; Shaw, 1959b, 1961a; Little, 1965; Chaisemartin *et al.*, 1970; Kirschner, 1970; Lockwood and Inman, 1973); (b) by blocking urine excretion by ligatures or by plugs and then measuring the rate of swelling (Wigglesworth, 1933b,c; Nagel, 1934; Webb, 1940; Shaw, 1955b, 1959b; Bryan, 1960a; (c) by measuring the diffusional water permeability and assuming the diffusional net flow (see equation

TABLE I: Relative contribution of the body surface and urine to electrolyte loss.

	Body Wt.	°C	Electrolyte	Urine flow (ml/100g)	(Na⁺) or (Cl⁻) (μEquiv/ml) urine	plasma	J_{out} urine (μEquiv/100g/h)	J_{out} total (μEquiv/100g/h)	Urine total (%)	References
ANNELIDS										
Lumbricus terrestris	3g	15	Cl	0·75	1	75	3	9	30	Dietz and Alvarado (1970) Ramsay (1949)
Nereis diversicolor	250mg	19	Cl (2·5mM)	2·1	13	110	28	356	8	Smith (1970*a–c*)
			Cl (10mM)	2·1	71	135	150	560	27	
MOLLUSCS										
Margaritana margaritifera	20g	22	Na	3·4	1·3	16	4·4	9·5	46·5	Chaisemartin (1968*a*, 1969)
Viviparus viviparus	1–2g	19	Na	1·5–5·4	9	32	13·5–48·5	—	—	Little (1965)
INSECTS										
Aëdes aegypti	1mg	28	Na	1·4	4	90	6·4	125	5	Wigglesworth (1933*c*) Stobbart (1959, 1965) Ramsay (1953)
CRUSTACEANS										
Astacus fluviatilis	35g	20	Na	0·34	6	204	2·0	36	5·5	Bryan (1960*a*)
Potamon niloticus	14–33g	25?	Na	0·002	240	260	0·5	80	0·6	Shaw (1959*b*)
Eriocheir sinensis	—	20	Na	0·78	325	303	250	—	100(?)	De Leersnyder (1967)
	150g	12	Na	0·165	280	280	47	210	22·5	Scholles (1933) Shaw (1961*b*)
Palaemonetes varians		15	Na	2·0	200	204	400	1800		Potts and Parry (1964*b*)

Palaemonetes antennarius		16	Na	2·2	100	177	350	350	100	Parry and Potts (1965)
Gammarus duebeni	50mg	10	Na	2·1	83	250	175	870	20	Lockwood (1961), Lockwood and Inman (1973), Sutcliffe (1967a,b)
Gammarus pulex	50mg	10	Na	1·5	27	150	40	235	17	
Carcinus maenas	50g	12	Na	1·25	300	300	370	1800	21 (40% SW)	Shaw (1961a)
CYCLOSTOMES										
Lampetra fluviatilis	35g	17	Cl	0·65	10	100	6·5	66·5	10	Morris (1956)
TELEOSTS										
Carassius auratus	100g	16	Na	0·73	7	158·5	5·1	17·5	29	Maetz (1972b)
Salmo gairdneri	150g	14	Na	0·45	10	150	4·5	18·8	24	Holmes and Stainer (1966) Kerstetter et al. (1970)
AMPHIBIA										
Ambystoma tigrinum			Na	0·76	7·6	100	5·8	8·4	69	Kirschner et al. (1971)
Rana esculenta	50g	20	Na	2·50	4·6	92	11·5	24	48	Mayer (1969) Ehrenfeld (1972)

(9)) to be a minimal estimate of the osmotic net flow. (Smith, 1970a,c; Oglesby, 1972; Lockwood, 1961; Sutcliffe, 1968).

In some cases the contribution of the renal loss of electrolytes to the total efflux has been measured directly by comparing the rate of Na^+ loss in deionized water in an animal with the urinary opening plugged and unplugged (Krogh, 1938), or by comparing the rate of Na^+ efflux in animals placed alternatively in fresh water, or in fresh water rendered iso-osmotic by adding an impermeant solute to the medium (Lockwood, 1965; Croghan and Lockwood, 1968; Sutcliffe, 1967a,b).

Urinary Na^+ excretion represents a variable portion of the total Na^+ efflux, nearly 100% in *Palaemonetes antennarius* (Parry and Potts, 1965) against 0.6% in *Potamon niloticus* (Shaw, 1959b). The species in which the urine is iso-osmotic with respect to the body fluids are not necessarily those with the highest urinary Na^+ excretion relative to the total Na^+ loss. *P. niloticus* is a typical example. This crustacean is characterized by a considerable reduction of osmotic water permeability of the body surface. Urine is scarce and hence Na^+ loss by this route is very small. Production of a hypo-osmotic urine, that is 'free water clearance', is certainly not an essential feature of hyper-osmotic regulation. Extreme reduction of water permeability obviates expenditure of energy for salt reabsorption against a concentration gradient in the excretory organ.

The extrarenal electrolyte efflux most probably occurs by way of that part of the body surface which is specialized for ion uptake. Stobbart (1959, 1960), comparing the Na^+ efflux of *Aëdes aegypti* larvae with or without anal papillae, showed that about 90% of the efflux occurs across these specialized structures. Similar observations have been obtained in fish with the "dividing chamber" and "irrigated buccal cavity" methods (Kirsch, 1972a; Kerstetter *et al.*, 1970).

2. REDUCTION OF THE IONIC PERMEABILITY OF THE BODY SURFACES IN RELATION TO LIFE IN MEDIA OF LOW SALINITY

Reduction of the Na^+ and Cl^- effluxes is particularly well documented in euryhaline species which exhibit much higher Na^+ and Cl^- turnover rates in sea water than in fresh water.

Thus in the euryhaline flounder, the Na^+ turnover rate represents 45% of the exchangeable Na^+ per hour in sea water and less than 0.2% in fresh water (Motais, 1967). The decrease of the electrolyte effluxes corresponds to a true reduction of the diffusional flows as shown by calculations taking into account the changes in the electrochemical gradient and the relative importance of the various routes of electrolyte loss, the urinary component being negligible at high external salinities (House, 1963; Potts and Evans, 1967; Evans, 1967a,b; 1969a; Potts *et al.*, 1967).

The separation of the electrolyte efflux into simple diffusion and exchange-diffusion components will be discussed in the second part of this review.

Such an adaptive reduction in the electrolyte permeability of the gill is observed not only in euryhaline teleosts which switch from the hypo-osmotic to the hyper-osmotic type of regulation, but also in stenohaline species such as the goldfish. Branchial Na^+ efflux in goldfish kept in salt water more or less iso-osmotic to the internal medium is about 25 times higher than in freshwater adapted animals (ref. Lahlou et al., 1969 and Maetz, 1972b).

Similar observations have been made in amphibians. Kirschner et al. (1971) comparing larval Ambystoma tigrinum, adapted to iso-osmotic Ringer and to fresh water, conclude that the Na^+ permeability of the body surface is higher in Ringer than in fresh water. The 3-fold increase in Na^+ efflux is only partially explained by the 20% increase of the Na^+ concentration of the body fluids, while the transepithelial potential decreases from 15 to 6 mV (inside positive).

Jard (1958) and Maetz (1959) also observed in Rana esculenta that had been transferred from tap water to salt water, a progressive increase in the Na^+ efflux across the skin measured in vitro in the absence of an electro-chemical gradient. This result is at variance with the observations of Jørgensen (1954) on saline-loaded Bufo bufo and R. esculenta which suggest that the excess Cl^- is excreted mainly by the kidneys and that extrarenal elimination of Cl^- is of minor importance.

The same adaptative reduction in the ionic permeability of the body surface with decreasing external salinities is observed in various euryhaline crustaceans. It characterizes species switching from the hypo-osmotic to the hyper-osmotic type of regulation, for example the prawn Palaemonetes varians (Potts and Parry, 1964b). It is also observed in species switching from the iso-osmotic to the hyper-osmotic type of regulation, for example Eriocheir sinensis (Shaw, 1961c) or Gammarus duebeni (Sutcliffe, 1967b, 1968, 1971) and in stenohaline typical freshwater species such as Palaemonetes antennarius (Parry and Potts, 1965) and Astacus fluviatilis (Bryan, 1960a, c).

No comparable studies have been made on freshwater insect larvae or freshwater molluscs. In the lamellibranch Margaritana margaritifera, which adapts only to a restricted range of external salinities, the Na^+ efflux remains constant. In the more euryhaline gastropod Lymnaea stagnalis, the Na^+ efflux decreases with decreasing external Na^+ concentration (Chaisemartin, 1969). Unfortunately the relative importance of the urinary loss of Na^+ was not investigated.

In the annelids, the euryhaline Nereis diversicolor which is an iso-osmotic regulator at higher external salinities and a hyper-osmotic regulator at the lower range of salinities, has been studied recently. Smith (1970a) and Oglesby (1972) clearly demonstrated a progressive decrease of the integumental Cl^- and Na^+ effluxes in relation to a diminution of the salinity of the adaptation medium. In these studies the relative importance of the other components of sodium chloride efflux—exchange diffusion and urine loss—were taken into account.

3. ELECTROLYTE EXCHANGE AND THE ELECTROCHEMICAL GRADIENT ACROSS THE BODY SURFACES

a. *Evidence for an active Na^+ and Cl^- uptake in hyper-osmotic regulators*

In Table II are collected the values for external and internal Na^+ or Cl^- concentrations, unidirectional fluxes and differences of electric potentials measured across the body surfaces of most of the freshwater animals studied so far. For the given values, the theoretical potential calculated by using Ussing's criterion for passive exchanges (equation 19) is also given. For this calculation the possible influence of the osmotic inflow of water occurring across the permeable surfaces (solvent drag) was ignored since C (equation 22), the concentration of the solute in the external medium, is very small.

It may be seen that in all cases, active Na^+ and Cl^- transport prevails. That the body surface of freshwater animals actively transports both Na^+ and Cl^- was suggested by Krogh (1939), although at that time no information about the bioelectric potentials and the isotopic fluxes were available. Krogh based his hypothesis on the observation that salt-depleted animals are able to absorb both electrolytes from extremely dilute solutions.

b. *Origin of the bioelectric potential*

Table II shows that, in most aquatic animals, the body fluids are negatively charged with respect to the external medium. In amphibians, however, the opposite potential difference is observed across the skin studied *in vivo* and *in vitro* (see review by Ussing, 1960). In larval amphibians there are contradictory reports. *Ambystoma gracile* and *A. tigrinum* exhibit skin potentials positive to the body fluids *in vivo* and *in vitro* (Dietz et al., 1967; Alvarado and Stiffler, 1970). *Rana catesbiana* tadpoles exhibit the opposite potential difference shifting to zero values when the external NaCl concentration is increased, an effect also observed in isolated skins of these tadpoles bathed with Ringer solution (Alvarado and Moody, 1970). The potential difference switches to positive (inside values) upon metamorphosis (Taylor and Barker, 1965; Alvarado and Moody, 1970). Active Na^+ and Cl^- transport occurs across the body surfaces of the tadpoles but the site of uptake is most probably the gills rather than the skin, as discussed

Table II: Active transport of Na^+ and Cl^- in aquatic animals.

	ion	C_{out} (μequiv/ml)	C_{in} (μequiv/ml)	J_{in} J_{out} (μequiv/100g/h)	Measured $\phi_{int} - \phi_{ext}$ (mV)	Theoretical $\phi_{int} - \phi_{ext}$ (mV)	References
ANNELIDS							
Nereis diversicolor	Cl	2·5	110	356 328	−17·3	+97·5	Smith (1970a)
	Cl	60	155	730 480	−3·7	+34·5	Smith (1970a)
Lumbricus terrestris	Cl	0·5	47	8·8 3·5	−21	+137·5	Dietz and Alvarado (1970)
MOLLUSCA							
Lymnaea limosa	Na	0·25	55	84 77	−10	−138	Chaisemartin (1969)
Lymnaea stagnalis	Na	0·35	57	12·2 8·0	−16·4	−139	Greenaway (1970)
Margaritana margaritifera	Na	0·25	16	9·8 5·1	−10	−121·5	Chaisemartin (1969)
INSECTA							
Aëdes aegypti	Na	2	105	125 125	+9·2	−100	Stobbart (1959, 1965)
	Na	2	93	935 175	+28·8 (Na deficient)	−140	
Aëdes campestris	Na	0·1	75	(flux ratio taken as 1)	−48	−167	Phillips and Meredith (1969)
CRUSTACEA							
Astacus (Austropotamobius) pallipes	Na	0·4	203	38 35·8	+4·1	−158·5	Bryan (1960a)
	Cl	0·3	184	70 50	−28·2 (Cl deficient)	+170	Shaw (1960c)
Orconectes limosus	Na	0·65	186 ad. high Ca media	(flux ratio taken as 1)	−35	−142·5	Chaisemartin (1967)
	Ca	4·90		—	−35	−142·5	
Astacus fluviatilis	Na	0·27	204 ad. low Ca media	—	−42	−167	
Austropotamobius pallipes	Ca	0·27		—			
Palaemonetes antennarius	Na	0·5	177	345 150*	−33	−169	Parry and Potts (1965)

TABLE II : (*continued*)

	ion	C_{out} (μequiv/ml)	C_{in}	J_{in} (μequiv/100g/h)	J_{out}	Measured $\phi_{int} - \phi_{ext}$ (mV)	Theoretical $\phi_{int} - \phi_{ext}$ (mV)	References
Palaemonetes varians	Na	9·2	204	1800	1400	−32	−84·5	Potts and Parry (1964b)
Gammarus duebeni	Cl	10·8	212	2820	2400	−32	+79·0	Lockwood and Andrews (1969) Sutcliffe (1967b)
	Na	10	250	870	695	−11·8	−86·5	Lockwood (1961)
TELEOSTS								
Carassius auratus	Na	0·1	158·5	17·3	12·4	−20	−194	House and Maetz (unpublished) Maetz (1972b)
	Cl	0·1	129	29·9	14·8	−20	+198	deRenzis and Maetz (1973)
Anguilla anguilla	Na	0·2	120	6·0	4·0	−21	−171	Maetz and Campanini (1966) Maetz (1971)
Salmo gairdneri	Na	1·0	150	17·9	16·0	+8·5 (Ca present)	−129	Kerstetter et al. (1970)
	Na	1·0	150	19·3	14·3	−14·8 (no Ca)	−138	Kerstetter and Kirschner (1972)
	Cl	1·1	120	19·3	11·4	−14·8	+131·5	
AMPHIBIA								
Bufo bufo	Na	3	100	4·9	7·7	+77	−77	Barker Jørgensen et al. (1954)
	Cl	3	80	52	4·6	+77	+86	
Ambystoma tigrinum	Na	1·2	100	8·8	2·6	+15·5	−142	Dietz et al. (1967)
Rana catesbiana (tadpoles)	Na	1·5	85	14	16·5	−7·5	−98	Alvarado and Moody (1970)
	Cl	1·5	65	6	8·5	—	+86	
Ambystoma gracile	Na	1	104	14	16	+14	−113·5	Alvarado and Dietz (1970)
	Cl	1	82	9	7	+14	+117·0	

* Na concentration of urine taken as one half of blood concentration.

above. The relative contributions of the skin and the gills to the generation of the potential is not known.

Concerning the origin of the bioelectric potential across frog skin, the classical study by Koefoed-Johnsen and Ussing (1958) suggests that it results from the sum of two Nernst potentials (equation 21). The apical side of the epithelial cells would behave as a Na^+ electrode and the basal side as a K^+ electrode. Thus

$$[\phi_{int} - \phi_{ext}] = \frac{RT}{F}\left(\ln \frac{Na^+_{ext}}{Na^+_{cells}} + \ln \frac{K^+_{cells}}{K^+_{int}}\right) \qquad (24)$$

where Na^+_{ext}, Na^+_{cells}, K^+_{cells}, K^+_{int} are the activities of Na^+ in the external medium and epithelial cells, and of K^+ in the cells and internal medium.

According to this model the high K^+ and low Na^+ levels in the epithelial cells would be maintained by a non-electrogenic one-for-one Na^+/K^+ exchange pump located at the basal or intercellular membrane of the cells. Such a perfect system of "electrodes" in series is observed only when Cl^- ions do not interfere with the cationic movements. According to this model these ions are passively transported. Flowing along their electro-chemical gradient, they "follow" the actively transported Na^+ ions and thus "short-circuit" the skin.

Motais and Garcia-Romeu (1972) have summarized more recent investigations, the results of which are inconsistent with the model of Koefoed-Johnsen and Ussing. I shall summarize the main points and add more recent references.

(1) It is probable that Na^+ enters the epithelium by a carrier-mediated process and not by simple diffusion. Biber (1971) proposed that Na^+ uptake is made up of two components, a linear component and a saturating component. Using mannitol rather than inulin as an extracellular marker, Erlij and Smith (1973) were able to show that simple saturation kinetics prevail at this entry step. The possibility that this carrier-mediated process is effected by an active transport mechanism has been proposed by Leblanc (1972) on the grounds that metabolic inhibitors (cyanide, dinitrophenol) interfere with Na^+ entry.

(2) The existence of an obligatory Na^+/K^+ exchange linking trans-epithelial Na^+ net flux with K^+ influx into the epithelium from the internal medium has been questioned, but the most recent evidence confirms such an exchange. At the same time there is disagreement concerning the stoichiometric relationship. A 2 or 3 to 1 ratio is observed for the isolated frog skin epithelium, the net Na^+ transport being in excess of the K^+ influx, a relationship which suggests an electrogenic Na^+ pump (Biber et al., 1972).

(3) Finally, the model proposed by Koefoed-Johnsen and Ussing considers Cl⁻ movements as being diffusive on the basis of the earlier investigations by Koefoed-Johnsen *et al.* (1952) on isolated *Rana* skin bathed in Ringer solutions. Yet all observations concerning amphibians studied *in vivo* placed in dilute NaCl solutions demonstrate active Cl⁻ transport without ambiguity (Barker-Jørgensen *et al.*, 1954; see also Table II). Moreover Cl⁻ transport has clearly been shown to occur in the isolated skin of the South American frog *Leptodactylus ocellatus* bathed in Ringer solution (Zadunaisky *et al.*, 1963; Zadunaisky and de Fisch, 1964). More recently similar behaviour of Cl⁻ was observed by Martin and Curran (1966) in isolated skins of *Rana pipiens* and *R. esculenta* when the skins were bathed in Ringer of low Cl⁻ concentration. Huf (1972) and Kristensen (1972) confirm the existence of an active Cl⁻ transport in various other frogs and toads. In *R. temporaria* skins bathed with sodium sulphate—Ringer containing 1 milliequiv/l Cl⁻ the magnitude of this transport is small, about 1/50th of the rate of Na⁺ transport. *In vivo*, at identical Na⁺ and Cl⁻ concentrations, these ions are absorbed at similar rates (Jørgensen *et al.*, 1954; Garcia-Romeu *et al.*, 1969; Garcia-Romeu, 1971).

To what extent does this Cl⁻ pump participate in the generation of the transepithelial potential differences? Dietz *et al.* (1967) have shown that, in the larval salamander *Ambystoma*, the external Na⁺ concentration alone determines the potential difference (positive inside), the Cl⁻ concentration being without effect. The authors suggest that the potential difference (P.D.) is generated by active inward transport of Na⁺ irrespective of whether the accompanying anion is permeant Cl⁻ or impermeant SO_4^{2-}. Uptake of Cl⁻ appears to occur by an independent non-electrogenic transport mechanism. The nature of this pump will be discussed below.

In most of the remaining aquatic vertebrates the body fluids are negative with respect to the external fluid. It is often assumed that the sign of the P.D. reflects the relative efficiencies of the Cl⁻ and Na⁺ pumps, assuming that both are electrogenic. That such a generalization is unwarranted is clearly demonstrated by the case of the eel, *Anguilla anguilla*, a species characterized by a negative inside P.D. in fresh water (Maetz and Campanini, 1966). This fish is remarkably inefficient in pumping Cl⁻. Garcia Romeu and Motais (1966) could not detect any chloride influx. Moreover, Kirsch (1972a) measures an influx of 0.01 μEq/100g/h, a value which is 100 to 200 times smaller than that found for Na⁺ (Maetz *et al.*, 1967b).

An alternative explanation of the P.D. is that it results from a Nernst potential, the sign being accounted for by the relative permeability of the body surfaces to Cl⁻ or to Na⁺ (or other negative or positive ions). The chemical gradients maintaining high Na^+ and Cl^- concentrations in the body fluids would be accounted for by non-electrogenic pumps. Such pumps would operate ionic exchanges allowing for Na^+ and Cl^- uptake against endogenous ions of the same electric charge.

Several observations indicate that external Ca^{2+} plays an important role in potential generation. Increasing the external Ca^{2+} activity brings about a shift from negative to positive (inside) of the transepithelial P.D. in three species: *Salmo gairdneri* (Kerstetter *et al.*, 1970), *Lymnaea stagnalis* (Greenaway, 1971) and *Astacus pallipes* (Greenaway, 1972). In the two last species it appears that within a certain range of Ca^{2+} activities the body surface behaves as a Ca^{2+} electrode, a suggestion also made by Istin and Kirschner (1968) in their study of the isolated clam mantle. It must be pointed out that external Ca^{2+} may also intervene indirectly by changing the relative Na^+ and Cl^- permeabilities of the body surface, thus altering the Nernst potentials.

c. *Saturation kinetics of the uptake processes*

In all hyper-osmotic regulators studied so far, whether freshwater or brackish-water species, a curvilinear relationship between Na^+ and Cl^- influxes and the external electrolyte concentrations is observed. Figure 1 compares the results obtained on three species, *Aëdes aegypti*, *Astacus pallipes* and *Carassius auratus*, for which investigations of Na^+ and Cl^- transport have been carried out. The flux values obtained can be interpreted in terms of Michaelis-Menten kinetics, suggesting the existence of a saturable process with a transport-carrier characterized by a limited number of transport sites. Equation 23 describes the observed relationship. It can be seen in Fig. 1 that the maximal rates for Na^+ and Cl^- may be quite different (*Aëdes*), or that the affinity constants may also differ (*Astacus*, *Carassius*).

Kirschner (1955) was first to demonstrate saturation kinetics in frog skin *in vitro*. The calculated K_m, the apparent affinity constant of the Na^+ carrier, is about 4 milliequiv./l. Brown (1962) confirmed Kirschner's observation and found identical K_m values (between 3 and 10 milliequiv./l) when frog skins were compared *in vivo* and *in vitro*. The problem needs to be re-investigated because in the most recent *in vivo* studies a much higher affinity of the carrier for Na^+ has been found ($K_m = 0.2$ milliequiv./l) in *Rana pipiens* (Greenwald, 1971). Recent measurements of the Na^+ uptake across the outer barrier of the epithelial cells suggest that it is the apical surface which is responsible for the saturation kinetics (Biber and Curran, 1970; Erlij and Smith, 1973). It may be noted, however, that the K_m values reported in these *in vitro* studies are exceedingly high (14 to 24 milliequiv./l).

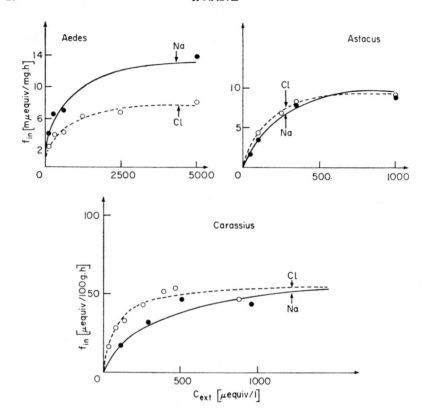

FIG. 1. Relationship between Na^+ or Cl^- influx and external electrolyte concentration.

(a) *Aëdes aegypti* (according to Stobbart, 1967)
(b) *Astacus pallipes* (according to Shaw 1959, 1960c)
(c) *Carassius auratus* (according to Maetz, 1972b; de Renzis and Maetz, 1973).

Ordinates: Na^+ or Cl^- influx (units given on axis).
Abscissa: Na^+ or Cl^- concentration in μequiv./l.

In Table III are summarized most of the relevant data concerning the apparent affinity constants and the maximal transport rates measured *in vivo* for a wide variety of hyper-osmotic regulators, including brackish and freshwater species. For fish, a few hypo-osmotic regulating species, whether euryhaline or stenohaline, have been included for comparison. The maximal transport rates are difficult to compare as the list of species covers a wide range of animal sizes. Smaller animals have in general a relatively higher body surface:body weight ratio, which corresponds to a higher electrolyte turnover (see also Table I for efflux). This point will be

TABLE III: Parameters of Na$^+$ influx in freshwater animals

Species	Body Wt.	Temp. (°C)	ion	K_m μEq/l	J_{max} μEq/100g/h	Habitat	References
ANNELIDA							
Lambricus terrestris	3g	15	Na	1300	10	moist earth (adapted FW)	Dietz and Alvarado (1970)
MOLLUSCA							
Margaritana margaritifera	20g	22	Na	100–250	12·5	FW*	Chaisemartin (1969)
Lymnaea limosa	2g	22	Na	500	120	FW	Chaisemartin (1969)
Lymnaea stagnalis	2–4g	10	Na	250	22·5	FW	Greenaway (1970)
INSECTA							
Aëdes aegypti	1mg	28	Na	550	1200–1300	FW (starved and salt depleted larvae)	Stobbart (1965, 1967)
			Cl	200–500	700–800		
Aëdes campestris	—	—	Na	2000	150	BW* (salt-depleted)	Phillips and Meredith (1969)
CRUSTACEA							
Triops longicaudatus	200mg	24	Na	200	400–650	FW	Horne (1967)
Chirocephalopsis bundyi	20–50mg	24	Na	560	325	FW	Horne (1968)
Astacus pallipes (Austropotamobius)	10g	12	Na	250	95	FW	Shaw (1959, 1960c)
			Cl	100	~120		
Gammarus pulex	50mg	10	Na	100–115	220–310	FW	Sutcliffe (1967a)
Gammarus palustris	50mg	10	Na.	100–150	260	FW	Sutcliffe and Shaw (1967)
Gammarus duebeni	35–85mg	10	Na	400–700	1600–2400	FW	Sutcliffe (1967b)
	75mg	10	Na	2400	2300	BW (ad. to 0·25M)	
Gammarus zaddachi	50mg	10	Na	1250	2000	or 0·5M	Sutcliffe (1968)
Gammarus tigrinus	35mg	10	Na	1250	2300	or 0·5M	
Marinogammarus finmarchicus	55mg	10	Na	8000	2000	SW* (ad. 20% SW)	
Mesidotea entemon	0·5–1g	10	Na	3250	800	FW	Croghan and Lockwood (1968)
Eriocheir sinensis	150g	12	Na	12 000	850	BW	Shaw (1961b)
			Na	1000	215	2%SW	Shaw (1961a)
Carcinus maenas	50g	12?	Na	20 000	300–1050	SW (ad. 20%SW)	

TABLE III: *(continued)*

Species	Body Wt.	Temp. (°C)	ion	K_m μEq/l	J_{max} μEq/100g/h	Habitat	References
CYCLOSTOMES							
Lampetra planeri (ammocoete larvae)	3–5g	10	Na	260	36	FW	Morris and Bull (1970)
TELEOSTS							
Platichthys flesus	150g	16	Na	800	35	FW	Maetz (1971)
	80–330g	16	Na	400 000	4200	SW	Motais et al. (1966)
Poecilia latipinna	1–8g	24	Na	8000	1200	FW	Evans (1973)
Serranus sp.	30–80g	16	Na	17 000	3300	SW	Motais et al. (1966)
Carassius auratus	100g	16	Na	50 000	2950	SW	Maetz (1972b)
		16	Cl	300	65	FW	De Renzis and Maetz (1973)
				75	55	FW	Kerstetter et al. (1970)
Salmo gairdneri	150g	19	Na	450	33·3	FW anaesthetiz.	Kerstetter and Kirschner (1972)
	150g	19	Cl	225	~29		
	250g	13	Na	20	65	FW free swimming	Wood and Randall (1973)
AMPHIBIANS							
Rana pipiens	50–80g	20	Na	3000–10 000	15	FW semi-terrestrial	Brown (1962)
Rana cancrivora	26g	20	Na	200	23	BW–SW	Greenwald (1971)
Xenopus laevis	15·1g	24		400	38	FW	Greenwald (1972)
Amphiuma means	45g	24		50	10	FW	
Bufo americanus	98g	24		200	8	terrestrial	
Ambystoma gracile (larvae)	29g	24		~1100	~70	FW	Alvarado and Dietz (1970)
	10–20g	15	Na	300–500	20		
			Cl	300–500	10		

* FW, fresh water; BW, brackish water; SW, salt water.

discussed in more detail in the section concerning water turnover, which has been more fully investigated in relation to this phenomenon (p. 69).

The K_m values are interesting to compare. It can be seen in Table III that the brackish water species exhibit higher K_m values than typical freshwater species. (Compare *Aëdes aegypti* with *A. campestris*; *Carcinus meanus* and *Eriocheir sinensis* with *Astacus pallipes*; *Rana cancrivora* with *R. pipiens*; *Platichthys flesus* with *Carassius auratus*). The value is also higher in semi-terrestrial species (*Bufo americanus, Lumbricus terrestris*). In euryhaline animals, the salt water-living individuals, whether iso-osmotic or hypo-osmotic regulators, exhibit a much higher K_m value than the individuals adapted to low salinity media (*Mesidotea, Gammarus duebeni, Poecilia, Platichthys*).

Shaw (1959a, 1961b) was the first to point out that acquisition of a high affinity uptake mechanism for Na^+ and Cl^- overcomes the need for an excessive reduction in the permeability of the body surface in order to maintain salt balance at low external electrolyte concentration.

The existence of a Na^+ or Cl^- efflux component dependent upon the external electrolyte concentration, first observed by Kirschner (1955) on frog skin *in vitro*, has been confirmed in *in vivo* studies on mosquito larvae (Stobbart, 1965), crustaceans (Shaw, 1959a) and fish (Kerstetter *et al.*, 1970; Maetz, 1972b). In at least one species, *Salmo gairdneri*, it was shown that the transepithelial potential remains more or less constant despite variations of the external Na^+ and Cl^- concentrations. Thus the changes of the Na^+ efflux cannot be explained in terms of passive movements of Na^+. Kerstetter *et al.* (1970) propose that such changes are mediated by exchange diffusion. It is possible that the "leaky pump" mechanism, which implies that part of the Na^+ efflux occurs through the same pathway as the Na^+ influx (see introductory remarks), is responsible for this component.

4. INDEPENDENCE OF THE SODIUM AND CHLORIDE ION UPTAKE: EXCHANGE WITH ENDOGENOUS IONS

The possibility of an independent uptakes of Na^+ and Cl^- by the body surface of freshwater animals was formulated by Krogh (1939) and confirmed since in almost all the species studied so far. That the mechanism of Na^+ uptake is to some extent different from that of Cl^- is borne out by the following observations: (1) Na^+ and Cl^- exchanges in animals placed in a dilute NaCl solution are frequently of different intensities, resulting in net transfers which may be of the same sign but of different values, or may be of opposite signs, indicating a net loss of one ion and a net gain of the co-ion. This phenomenon is observed in the eel (Garcia-Romeu and Motais, 1966) and occasionally in the goldfish (Garcia-Romeu and Maetz,

1964). It is constantly seen in animals preadapted to artificial media lacking either Na^+ or Cl^-, or both of these ions. Examples are found in *Aëdes aegypti* larvae (Stobbart, 1965) in *Astacus pallipes* (Shaw, 1960*a*, *c*; 1964), in *Carassius auratus* (Garcia-Romeu and Maetz, 1964) and in the frog *Calyptocephalella gayi* (Garcia-Romeu *et al.*, 1969). (2) Either Na^+ or Cl^- is absorbed without an accompanying ion when the latter is impermeant. This is observed in *Aëdes aegypti* (Stobbart, 1965), in various crustaceans including *Astacus* (Shaw, 1960*a*, *c*) and *Triops longicaudatus* (Horne, 1967), in teleosts such as *Carassius auratus* (Krogh, 1938; Garcia-Romeu and Maetz, 1964) and *Salmo gairdneri* (Kerstetter *et al.*, 1970; Kerstetter and Kirschner, 1972) and in amphibians including the larval *Ambystoma* (Dietz *et al.*, 1967; Alvarado and Dietz, 1970), *Rana catesbiana* tadpoles (Alvarado and Moody, 1970) and the adult frogs *Leptodactylus ocellatus*, *Calyptocephalella gayi* and *R. esculenta* (Salibian *et al.*, 1968; Garcia-Romeu *et al.*, 1969; Garcia-Romeu and Ehrenfeld, 1972). (3) A selective inhibition of the absorption of one of either Na^+ or Cl^- can be produced experimentally while the net flux of the other ion remains unchanged. Addition of H^+ or NH_4^+ to the external medium, for example, selectively blocks Na^+ uptake in *Carassius* (Maetz and Garcia-Romeu, 1964; Maetz, 1973*a*) and in *Astacus* (Shaw, 1960*a,b*). Addition of anaesthetics in the anionic or cationic form blocks Cl^- or Na^+ uptake in *Calyptocephalella* (Garcia-Romeu *et al.*, 1969). Addition of amiloride selectively blocks Na^+ uptake in frog skin *in vitro* (Kristensen, 1972; Erlij, 1971) and *in vivo* (Ehrenfeld, 1972; Kirschner *et al.*, 1973) and in trout and crayfish (Kirschner *et al.*, 1973).

In order to preserve electrostatic neutrality in such single ion transfers, the body surfaces must contain a pair of ionic exchange systems capable of operating independently of each other. Recent research has attempted to define the nature and location of these exchange mechanisms.

a. *The Na^+ exchange mechanism: Na^+/H^+ or Na^+/NH_4^+ exchange*

For many years it was thought that absorption of Na^+ involves exchange with endogenous ammonia in the form of NH_4^+. Such an exchange was suggested by Krogh (1939) on the grounds that most of the freshwater species excrete their nitrogenous wastes in the form of ammonia, the major route of excretion being the body surface rather than the excretory organs. Furthermore, in animals in a steady state with the external medium, the rate of NH_4^+ excretion is similar to that of Na^+ net uptake by the body surface (see review by Maetz, 1972*a*). That these two rates should be similar, however, does not prove that such an exchange occurs. Such a proof can only be obtained by measuring the two fluxes simultaneously on the same animals and determining the degree of correlation.

The theory correlating ammonotelism with an aquatic life must be accepted with caution for several reasons. (1) Sea water teleosts, which also excrete ammonia by way of the gills, excrete rather than absorb Na^+ ions by these organs. (2) In most molluscs ammonia appears to form only a small part of the total non-protein nitrogen excreted according to Potts (1967). (3) Many aquatic insects are known to produce ammonia as their main nitrogenous product, but the significance of the ammonium bicarbonate in their rectal fluid remains undetermined (Staddon, 1955, 1959; Shaw, 1955b; Shaw and Stobbart, 1963). (4) While larval or neotenous forms of amphibians excrete their nitrogenous wastes in the form of ammonia (Fanelli and Goldstein, 1964; Alvarado and Moody, 1970; Dietz et al., 1976; Alvarado and Dietz, 1970), adults, in relation to their semi-terrestrial habitat, have switched to ureotelism and very little ammonia is excreted through the skin (Garcia-Romeu and Salibian, 1968; Garcia-Romeu et al., 1969), although the skin is the site of Na^+ uptake. Simultaneous measurements of the transepithelial net fluxes of Na^+ and NH_4^+ show, in the two South American frogs Leptodactylus and Calyptocephalella, a much lower rate of NH_4^+ excretion than that of Na^+ absorption. Thus the Na^+/NH_4^+ exchange proposed by Krogh is ruled out in these species.

Na^+/H^+ exchange.—In two species of frogs, Calyptocephalella gayi and Rana esculenta, the counter-ion of Na^+ has effectively been shown to be H^+ (Garcia-Romeu et al., 1969; Garcia-Romeu and Ehrenfeld, 1973). This was shown with frogs kept in a sodium sulphate solution to avoid interference between the two exchange mechanisms (see below). In all the experiments Na^+ uptake was accompanied by a downward shift of external pH, but the increase of H^+ concentration was small and could not have accounted for the actual addition of H^+ to the bath because of the progressive buffering of the external medium. Only total acidity titration with dilute NaOH allowed for the evaluation of proton excretion.

Figure 2 illustrates typical experiments which compare the ammonia and H^+ excretion rates with the Na^+ absorption rate in Calyptocephalella. It shows that H^+ and not NH_4^+ accounts for the Na^+ uptake. From Fig. 3 it can be seen that an excellent correlation between H^+ excretion and Na^+ excretion occurs, suggesting a one for one relationship. In R. esculenta a similar relationship has been observed (Ehrenfeld, 1972).

In Aëdes aegypti evidence for Na^+/H^+ exchange has been obtained by Stobbart (1971a). A progressive buffering of the Na_2SO_4 solution in which the animals were kept was observed. The H^+ ion excretion was measured by titration. A detailed ionic balance study revealed that of the Na^+ taken up, 50% is accounted for by H^+ excretion, 33% by K^+ loss and the remainder by SO_4^{2-} accompanying Na^+. The NH_4^+ excretion was not

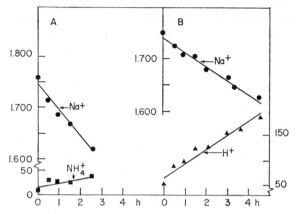

FIG. 2. Comparison between the excretion of endogenous cations and the uptake of Na⁺ in two different frogs, *Calyptocephalella gayi*, kept in sodium sulphate solutions. (A) ammonia excretion *vs* Na⁺ absorption; (B) H⁺ excretion *vs* Na⁺ absorption.

Ordinate: external concentration in μmoles/l.
Abscissa: time in hours.
(according to Garcia-Romeu *et al.*, 1969).

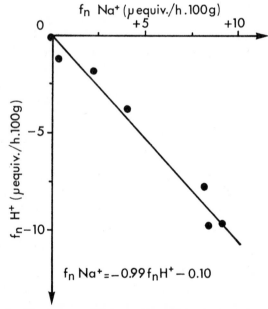

FIG. 3. Regression line between Na⁺ uptake and H⁺ excretion in *Calyptocephalella*.
Ordinate: net excretion rate of H⁺ (μequiv/100g/h).
Abscissa: net uptake flux of Na⁺ in the same units (μequiv/100g/h).
(according to Garcia-Romeu *et al.*, 1969).

measured. The possibility that NH_4^+ is the counter-ion was rejected on the grounds that the addition of ammonia to the external bath did not interfere with Na^+ uptake. In contrast, addition of H^+ ions was followed by a well-marked inhibition of Na^+ uptake which suggests a strong competition for the available sites on the Na^+ carrier (Stobbart, 1967). It should be noted that a thorough demonstration of a Na^+/H^+ exchange in *Aëdes* awaits the determination of the degree of correlation between these two variables.

Na^+/NH_4^+ exchange.—The following indirect evidence suggests that the Na^+/NH_4^+ exchange mechanism operates in ammonotelic aquatic animals: (1) Addition of ammonium salts to the external medium is followed by an inhibition of Na^+ uptake. In the crayfish, Shaw (1960b) showed that this inhibition is observed at a relatively low external ammonia concentration, lower than that recorded in crayfish blood. The inhibition is inversely related to the external ammonia concentration which was thought to inhibit ammonia excretion and hence the coupled Na^+ absorption mechanism (Maetz and Garcia-Romeu, 1964). More recently Maetz (1973a) found inhibition of Na^+ influx at lower ammonia concentrations which do not interfere with the ammonia excretion rate, suggesting that competition between external Na^+ and NH_4^+ for a common carrier occurs. (2) Injection of ammonium salts into the intraperitoneal cavity, a procedure which is known to hasten ammonia clearance by the gill (Wolbach *et al.*, 1959; Pequin, 1967; Maetz, 1972b) was followed by a prompt increase of Na^+ uptake and net absorption in the goldfish (Maetz and Garcia-Romeu, 1964) and the trout (Kerstetter *et al.*, 1970). Recent investigations suggest that the enhancement of the Na^+ uptake results from an increase in the turnover of the available transport sites without any change of their affinity (Maetz, 1972b).

Indirect evidence is, however, inconclusive because other ions, for example H^+ ions, also interfere with Na^+ uptake when added in small amounts to the external medium. This effect occurs in crayfish (Shaw, 1960b), goldfish (Maetz, 1972a, 1973a) and the trout (Packer and Dunson, 1970). Injection of dilute acid solutions, however, has variable effects, presumably because of the buffering capacity of the blood (Maetz, 1972a).

Direct evidence for a Na^+/NH_4^+ exchange, that is attempts to correlate Na^+ uptake and ammonia excretion, have failed to reveal an obligatory exchange between these two ionic species. Thus transfer of crayfish, carp, trout and goldfish into Na^+-free media, a procedure which blocks Na^+ uptake, is without effect on ammonia excretion (Shaw, 1960a, de Vooys, 1968; Kerstetter *et al.*, 1970; Maetz, 1973a). Figure 4 illustrates the absence of a correlation between ammonia excretion and Na^+ uptake in a series of experiments on goldfish transferred to media in which the external Na^+

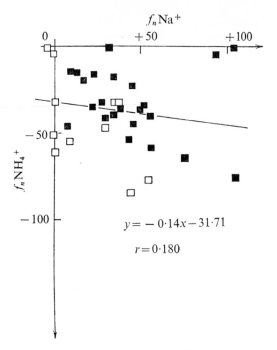

FIG. 4. Absence of a correlation between ammonia excretion and net Na⁺ uptake in *Carassius auratus* kept in sodium sulphate solution.
Ordinate: net (total) ammonia (units as preceding figure).
Abscissa: net sodium uptake in the same units.
(according to Maetz, 1973*a*).

concentration was varied experimentally. Furthermore, pharmacological substances which block Na⁺ uptake either remain without effect on ammonia excretion, as is the case for acetazolamide injection in the trout (Kerstetter *et al.*, 1970), or only inhibit ammonia excretion to a slight extent, as is the case for amiloride added to the medium outside the trout or the crayfish (Kirschner *et al.*, 1973).

Thus, ammonia movements across the gill cannot alone account for the Na⁺ uptake without an accompanying anion. Obviously another counter-ion must intervene. On the grounds that Na⁺ uptake is sensitive to acidification of the external medium in all the aquatic animals studied so far, it may be proposed that H^+ ions fill this role. Shaw (1960*a*, *b*) was the first to make this suggestion, considering it probable that both NH_4^+ and H^+ intervene in the exchange. At very high rates of Na⁺ uptake, when ammonia excretion is insufficient to account for the amount of Na⁺ taken up, H^+ excretion would supplement NH_4^+ excretion. At very low rates of Na^+ uptake, only a fraction of the ammonia excreted would exchange with Na^+ and the remainder would be excreted accompanied by an endogenous anion, probably HCO_3^-. An alternative suggestion (Maetz, 1972a, b, 1973a) is that ammonia is excreted in the molecular form, NH_3, but is then trapped in the external medium by the respiratory CO_2 as ammonium bicarbonate. Kerstetter et al. (1970) and Kirschner et al. (1973) suggest that, as in amphibians, H^+ ion excretion alone accounts for the Na^+ uptake in all ammonotelic animals and ammonia is always excreted in the unionized form. Such hypotheses regarding H^+ ion excretion and the form in which ammonia crosses the gill have been submitted to experimental analysis.

Kerstetter et al. (1970) were the first to demonstrate the excretion of H^+ by the gill of the trout. At the highest rate of Na^+ uptake, when this greatly exceeded ammonia excretion, a significant downward shift of the pH in the external medium was recorded. At low external Na^+ concentrations, the Na^+ balance being negative, a slight although not significant upward shift of the pH was found. Although the authors failed to titrate the amounts of H^+ excreted, they felt justified in proposing an obligatory H^+/Na^+ exchange. In a more recent series of experiments on the effects of amiloride on the exchange mechanism in the frog, trout and crayfish, the amounts of H^+ ion excreted by the various organisms were calculated by taking into account the pH shift and the pK of the buffer solution used in the external bath (Kirschner et al., 1973). In addition, all the ammonia lost in the medium was considered as being excreted in the form of H^+ together with NH_3 in the molecular form. The authors found that amiloride blocked Na^+ uptake in all three species. H^+ ion excretion was simultaneously blocked in the frog and the crayfish, and to a lesser extent in the trout. Ammonia excretion was partially inhibited in both trout and crayfish. These data suggest that a coupled Na^+/H^+ exchange occurs in all freshwater animals, but also allow the possibility of NH_4^+ excretion balancing a fraction of the Na^+ uptake. Nevertheless the methods used by the authors may be criticized on the grounds that the buffering capacity of the external medium is likely to change rapidly, and thus calculations based on a constant pK may not be justified. Maetz (1973a) in a recent series of investigations was able to titrate H^+ excretion by the goldfish. Figure 5 illustrates two examples of such experiments. In the graph on the left-hand of the figure, Na^+ absorption is accounted for by NH_4^+ excretion. In the graph on the right hand side of the figure, Na^+ absorption is accompanied by H^+ ion excretion. These investigations show that in the goldfish both H^+ and NH_4^+ movements have to be taken into account. Figure 6 shows that there is an excellent correlation between net Na^+ uptake and the sum of the H^+ and NH_4^+ movements. At a high Na^+ uptake, an effective

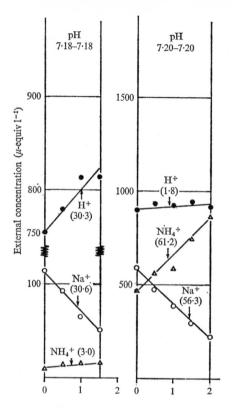

FIG. 5. Changes of the external Na$^+$, total ammonia and titrable acidity as functions of time in *Carassius auratus* (according to Maetz, 1973a).
On the left, a typical case of H$^+$/Na$^+$ exchange in goldfish with poor ammonia excretion.
On the right, a typical case of NH$_4^+$/Na$^+$ exchange in goldfish with poor H$^+$ excretion rate.
Ordinate: concentration in μmoles; abscissa: time in h.
The slopes represent net fluxes (values given in μequiv/100g/h in parentheses).
The changes of external pH are also indicated for each period.

H$^+$ ion excretion is observed, while at a low Na$^+$ uptake, or after partially blocking the Na$^+$ uptake by acidification of the external medium, a progressive alkalinisation of the bath is observed which is due to NH$_3$ excretion resulting in the formation of NH$_4^+$ by proton trapping. It is important to note that NH$_4^+$ ions escape the titration technique.

Concerning the problem of the form in which ammonia crosses the gill, most studies on the toxicity of ammonia in ambient water conclude that ammonia crosses the gills in the molecular form (Downing and Merkens,

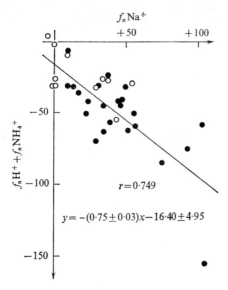

F_IG. 6. Correlation between ammonia excretion and Na$^+$ uptake when H$^+$ movements across the gill are taken into account.
Ordinate: sum of the net fluxes of ammonia (as NH$_4^+$) and H$^+$.
Abcissa: net Na$^+$ uptake in μequiv/100g/h (see Maetz, 1973a).

1955; Wuhrmann and Woker, 1948, 1953; Lloyd and Herbert, 1960; Fromm and Gillette, 1968). Because of its high lipid solubility free ammonia, NH$_3$, readily traverses most biological membranes. Non-ionic diffusion is a passive process conforming to physico-chemical laws. The diffusion of ammonia as NH$_3$ should depend on the partial pressure gradient of NH$_3$, ΔpNH$_3$ across the membrane and should occur from the side of greater tension.

ΔpNH$_3$ is related to the concentration of NH$_3$, (NH$_3$), by the equation:

$$(NH_3) = \frac{\alpha}{22.1} pNH_3 \qquad (25)$$

α being the solubility coefficient in litres STPD per litre solution and 1 mm Hg, pNH$_3$ being in mm Hg, and (NH$_3$) in moles litre^{-1}. Unfortunately the value of α may be uncertain and different in external media and in the body fluids. (NH$_3$) is related to the total ammonia concentration (Am)$_{tot}$ by the following equations.

$$(NH_3) + (NH_4^+) = (Am)_{tot} \qquad (26)$$

and
$$pH = pK + \log \frac{(NH_3)}{(NH_4^+)} \qquad (27)$$

The problem of defining the relative concentrations of (NH_3) and (NH_4^+) in solution is complicated by the uncertainty about the pK values which vary with the pH, temperature and salinity of the medium (Bates and Pinching, 1949; Bank and Schwartz, 1960; Bromberg et al., 1960).

From the three equations given above it may be seen that NH_3 will pass from the side of higher pH to that of lower pH when $(Am)_{tot}$ is identical on both sides of a membrane. It may even move against a concentration gradient of *total* ammonia.

Ionic ammonia, NH_4^+, may cross the membrane actively or passively, that is against or along its electro-chemical gradient. Active transport of NH_4^+ has been shown to occur in some epithelia (Mossberg, 1967). In any case, ionic ammonia crosses the membrane irrespective of the pNH_3 gradient which is responsible for the movement of the unionized form.

Maetz (1973a) has recently reinvestigated the movements of ammonia across the gill of the goldfish. The ΔpNH_3 gradient across the gill was calculated in various experimental situations involving different external pH values after addition of ammonia to the external medium. Ammonia may cross the gill against the ΔpNH_3, and this was taken as proof that ammonia moves in the NH_4^+ form. These experiments are, however, open to criticism because the calculations were made using the pH of the bulk solution rather than the pH of the medium in the vicinity of the branchial epithelium. Lloyd and Herbert (1960) and Holeton and Randall (1967) with the trout and Dejours et al. (1968) with the goldfish indicate pH differences of the order of 0·2 or more, depending on the buffering capacity of the medium, between water taken in the buccal and opercular chambers. Thus the problem of the form, whether molecular or ionic, under which ammonia crosses the gill membrane needs further investigation.

b. *The Cl^- exchange mechanism: Cl^-/HCO_3^- or OH^- exchange*

The hypothesis that HCO_3^- ions are the endogenous ions exchanged against Cl^-, as originally suggested by Krogh (1939), stems from the fact that the gill is the major route of CO_2 excretion and the amount of CO_2 excreted, up to $350 - 500\mu M/100g/h$ in the goldfish or trout for example, is more than sufficient to cover the needs for the postulated Cl^-/HCO_3^- exchange. Furthermore, the presence of carbonic anhydrase in the gills of fish (Leiner, 1938) or crustaceans and in the body wall and gills of molluscs (Maetz, 1946) also suggests such an exchange.

The above hypothesis was submitted to experimental analysis in the goldfish by Maetz and Garcia-Romeu (1964). The addition of HCO_3^- (in the form of the K^+ salt) to the external medium produces a specific inhibition of Cl^- uptake and removal of HCO_3^- by rinsing with a dilute

NaCl solution fully restores the initial pumping activity. HCO_3^- added to the external medium probably acts by competing for the Cl^- pump. An indirect effect mediated by pH changes is, however, also possible. Similar experiments were reported by Stobbart (1967) on the mosquito larvae *Aëdes aegypti*. He showed that addition of OH^- as well as of HCO_3^- resulted in an inhibition of the Cl^- uptake, OH^- being more efficient than HCO_3^-.

Injection of HCO_3^- salt solutions was shown by Maetz and Garcia-Romeu (1964) to produce a stimulation of the Cl^- influx in the goldfish. The same response was observed in the trout by Kerstetter and Kirschner (1972).

Direct demonstration of a Cl^-/HCO_3^- exchange was provided by Garcia-Romeu *et al.* (1969) in *Calyptocephalella gayi* and by Ehrenfeld (1972) in *Rana esculenta*. Both frogs were kept *in vivo* in choline chloride solutions. Titrable alkalinity, measured as a function of time, was observed to increase and this increase was correlated with the net Cl^- uptake. Figure 7 illustrates this correlation for *Calyptocephalella*. Approximately 3 Cl^- ions are absorbed against 4 HCO_3^- or OH^- ions. Stobbart (1971a) also

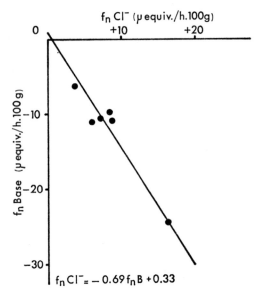

FIG. 7. Regression line between Cl^- uptake and base excreted in *Calyptocephalella gayi* (Garcia-Romeu *et al.*, 1969).
Abscissa: Cl^- uptake.
Ordinate: base excretion in μequiv/100g/h (see Garcia-Romeu *et al.*, 1969).

measured by titration the quantity of base excreted by *Aëdes aegypti* against the Cl^- absorbed from KCl solutions. The OH^- or HCO_3^- excretion accounted for only 40% of the Cl^- taken up, while 20% was accounted for by K^+ ions accompanying Cl^- ions. Thus 40% of the Cl^- uptake is unaccounted for. Complete verification of a Cl^-/OH^- exchange in *Aëdes* would depend on determining the degree of correlation between these two variables. Direct demonstration of a Cl^-/HCO_3^- exchange has also recently been provided for the goldfish by de Renzis and Maetz (1973). In ammonotelic animals, however, the relationship between HCO_3^- or OH^- excretion and Cl^- uptake is complicated by the fact that the ammonia excreted is trapped in the form of NH_4HCO_3 by the respiratory CO_2 in the external medium. When the ammonia excretion rate, simultaneously measured, is taken into account an excellent correlation between the rate of Cl^- disappearance and of OH^- or HCO_3^- appearance in the external medium is observed. Figure 8 illustrates a typical experiment and Fig. 9 shows the regression line representing Cl^- uptake as a function of HCO_3^- excreted. The maximal rate of excretion of bicarbonate ions, presumably in the ionized form, represents one third of the total CO_2 excretion rate.

One important point must be stressed in relation to all the above mentioned experiments in which the alkalinity of the external bath is titrated. Such titrations cannot decide whether it is OH^- or HCO_3^- which is excreted in exchange for Cl^-. In the external bath OH^- would, in any case, be in the form of HCO_3^- since respiratory CO_2 would combine with OH^-. The fact that carbonic anhydrase intervenes in the exchange process rather suggests that HCO_3^- is the counter-ion. This point will be discussed in the section on biochemical aspects of ionic transport.

c. *Simultaneous functioning of the Na$^+$ and Cl$^-$ exchanges*

When Na^+ and Cl^- ions are both present in the external medium, it is difficult to decide whether the exchange mechanisms observed in Na^+-free or Cl^--free salt solutions are functioning simultaneously or whether a completely different mechanism prevails.

Observations on the conductivity of the external medium during simultaneous Na^+ and Cl^- absorption suggest, at least for the ammonotelic crayfish (Shaw, 1960a,c) and the goldfish (Garcia-Romeu and Maetz, 1964) that both ions taken up are replaced by endogenous ions. If a test sample in the bath is submitted to a mild vacuum, its conductivity decreases because of the escape of CO_2 and NH_3. It may be argued, however, that the NH_4HCO_3 of the bath may also originate from simultaneous trapping of CO_2 and NH_3 crossing the gill in their molecular forms.

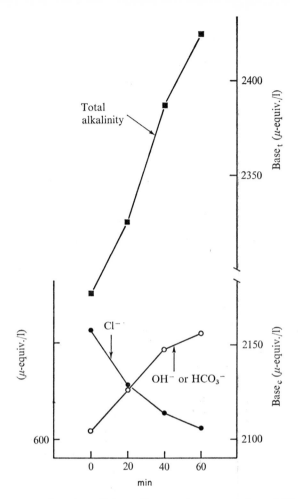

FIG. 8. Changes as a function of time of external concentration of Cl⁻, total base (base$_t$), or base corrected for ammonia present in bath (base$_c$), in a typical experiment with goldfish kept in a choline chloride solution.
Ordinates μequiv./l.
Abscissa: time (min) (according to de Renzis and Maetz, 1973).

Amphibians placed in NaCl solutions take up Na⁺ and Cl⁻ simultaneously, and a decrease in the conductivity of the external medium is observed (Salibian et al., 1968). This is to be expected whether Na⁺ and Cl⁻ are taken up as electrically linked entities by the independent Na⁺/H⁺ or Cl⁻/OH⁻ or HCO_3^- exchanges, or by a different mechanism.

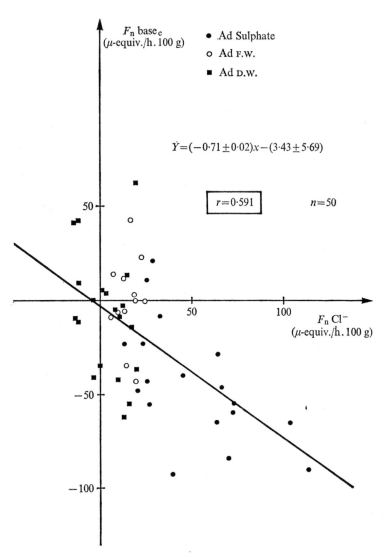

FIG. 9. Correlation between the net flux of base excreted (corrected for the net flux of ammonia) and chloride uptake from a choline chloride solution in *Carassius auratus*.
Coordinates as in Fig. 3.

New experiments have to be devised to verify whether the two exchange mechanisms operate together. Comparison of the influx or net uptake of Na^+ or Cl^- observed in NaCl solutions and in solutions where Na^+ or Cl^- are accompanied by impermeant co-ions, reveals in most cases that the Na^+ or Cl^- uptake is most efficient in NaCl solutions. In *Astacus pallipes*, Shaw (1960a, c) showed that the nature of the anion had no effect on the Na^+ influx, while Na^+ net uptake was slightly less from Na_2SO_4 solutions. Conversely, when the impermeant cation K^+ is replaced by Na^+, the chloride influx remains the same but the efflux decreases suggesting that a Cl^-/Cl^- exchange is replaced by a Cl^-/HCO_3^- exchange. In *Aëdes aegypti* the net uptake of Na^+ from NaCl solutions was greater than from $NaNO_3$, $NaHCO_3$ and Na_2SO_4 solutions. The ions K^+, Ca^{2+}, Mg^{2+} and NH_4^+, when added as chlorides, stimulate the influx of Na^+ from a $100\mu M/l$ NaCl solution. When added as nitrates or sulphates they either have no effect, or they are inhibitory (Stobbart, 1965). Conversely the net uptake of Cl^- from KCl is much slower than from NaCl (Stobbart, 1967). In the trout, *Salmo gairdneri*, Na^+ uptake is independent of the accompanying anion (Kerstetter *et al.*, 1970) while Cl^- influx from NaCl is similar to that from choline chloride and magnesium chloride but not from KCl, although the gill is permeant to K^+ (Kerstetter and Kirschner, 1972). Very recently de Renzis and Maetz (1973) observed in the goldfish that Cl^- uptake from NaCl is higher than from choline chloride. Conversely, in the frog, *Calyptocephalella*, and in the primitive crustacean, *Triops*, Na^+ uptake from NaCl is faster than from Na_2SO_4 (Garcia-Romeu *et al.*, 1969; Horne, 1967).

Thus an interdependence of the uptake mechanisms is observed in almost all the epithelia. Stobbart (1967) derived an explanation for this from the double exchange model. Thus, if Na^+ is taken up from a Na_2SO_4 solution by way of a H^+/Na^+ exchange, HCO_3^- would accumulate in the body fluids and the ionization of H_2CO_3 would be depressed, thus reducing the availability of H^+ and so the efficiency of the Na^+/H^+ exchange. Conversely if Cl^- is exchanged against HCO_3^- from a KCl solution, accumulation of H^+ would again depress the ionization of H_2CO_3, reducing the availability of HCO_3^- and so the efficiency of the Cl^-/HCO_3^- exchange. In a NaCl solution no accumulation of either H^+ or HCO_3^- would occur and ionization of H_2CO_3 would not be depressed.

De Renzis and Maetz (1973) observed accumulation of H^+ and reduction of plasma HCO_3^- in goldfish kept in choline chloride and, conversely, accumulation of plasma HCO_3^- in fish kept in sodium sulphate, but the perturbances develop rather slowly. Yet the enhancing effect of Na^+ upon Cl^- uptake is observed as soon as the fish is transferred from choline chloride to sodium chloride solution. It is suggested that acid-base perturbances take place in a much more limited space than the body fluids, possibly in the epithelial cells responsible for ion transfer.

5. CONTROL OF ELECTROLYTE TRANSFER

Various feed-back control mechanisms intervene to maintain various parameters of the body fluids in a steady-state: the Na^+ and Cl^- concentration, the pH and the volume are all observed to act on the various effector organs of osmoregulation by altering the active Na^+ or Cl^- uptake or the passive losses of these ions.

a. *Control of Na^+ uptake in relation to internal Na levels*

One of the feed-back responses related to adaptation of aquatic animals to water of lower salinites, namely the progressive reduction of the ionic permeability of the body surfaces, has been briefly discussed in a preceding section. This change in permeability is completed by adaptive responses of the excretory organs, the urinary loss of salt being increased in salt-loaded animals and decreased in salt-depleted animals. I shall discuss here the regulatory responses of the Na^+ active uptake mechanism in more detail.

Figure 10 summarizes the effects of salt-loading and salt-depletion on the Na^+ influx of the goldfish plotted as a function of the external Na^+ concentration. The Na^+ influx is considerably increased in the Na^+-depleted fish and decreased in the salt-loaded fish. The curvilinear relationship between influx and Na^+_{ext} is maintained irrespective of the salt status of the animal. Both kinetic parameters, that is the maximal rate of transport and the apparent affinity of the carrier, may be modified. There is a small, probably insignificant, increase in the affinity in salt-depleted animals and a considerable decrease in the affinity in salt-loaded animals compared with the controls. These effects on the active component are completed by an adaptive change of the Na^+ efflux, that is a decrease in the salt-depleted animals (Cuthbert and Maetz, 1972) and an increase in the salt-loaded fish (Maetz, 1964). Thus the net uptake is enhanced after salt depletion and depressed after salt loading. Krogh (1938, 1939) was the first to observe the regulatory response to salt depletion in a great variety of freshwater animals.

The first detailed analysis of the effects of salt depletion on the Na^+ influx-concentration relationship was made by Shaw (1959a) on *Astacus pallipes*. In this species, the K_m remains constant. Shaw schematised the self-balancing mechanism for internal Na^+ concentration in relation to changes of external and internal Na^+ concentration. In this model the

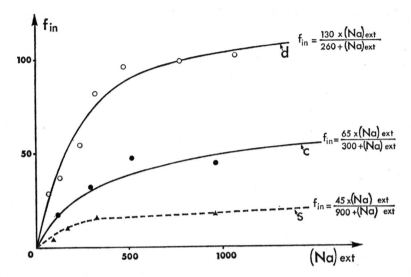

FIG. 10. Relationship between Na^+ influx and external Na^+ concentration in the goldfish. Comparison of control, salt-depleted and salt-loaded fish.
Control: fish kept in freshwater. $\left.\right\}$ according to Maetz
Salt-depleted: fish kept in deionized water for 3 weeks $\int$ (1972b)
Salt-loaded: fish kept in 0·9% NaCl solution for 3 weeks (G. de Renzis and J. Maetz, unpublished).
Coordinates as in Fig. 1.

rate of loss is taken to be constant. Thus, as a fall in blood concentration consequent to transferring crayfish to a low external Na^+ concentration is remedied by an increased efficiency of the Na^+ uptake, so an increase in the blood concentration after transfer to a high external Na^+ concentration is adjusted by a decrease in the uptake rate. Such a system is even more efficient if the rate of loss also varies according to the physiological state of the animals. In a second, more sophisticated model which was suggested to Shaw (1961b) by a comparison of various brackish water and freshwater living crustaceans, the combined effects of an adaptive variation in the affinity constant of the uptake mechanism, and of a change in the passive permeability of the body surface were considered. From this second scheme it appears that an increase in the affinity of the transport carrier overcomes the necessity for an excessive reduction of passive loss in order to maintain salt balance at lower external salinities.

A rapid survey of numerous studies concerning the feed-back responses of the Na^+ uptake mechanism shows that, in most cases, the maximal rate of transport is altered without a concomitant change in the affinity constant. Such a mode of regulation was clearly demonstrated in *Rana pipiens* by Greenwald (1971) and suggested by Alvarado and Dietz (1970) in the larval *Ambystoma gracile* and by Alvarado and Moody (1970) in *R. catesbiana* tadpoles. In invertebrates a similar mode of regulation has been suggested or observed in the mosquito larvae *Aëdes aegypti* (Stobbart, 1967); in numerous crustaceans including *Triops longicaudatus* (Horne, 1967), *Eriocheir sinensis* (Shaw, 1961b), *Potamon niloticus* (Shaw, 1959b), *Asellus aquaticus* (Lockwood, 1959), *Gammarus pulex* (Shaw and Sutcliffe, 1961), *Carcinus maenas* (Shaw, 1961a) and *Palaemonetes antennarius* (Parry and Potts, 1965); in molluscs such as *Lymnaea stagnalis* and *L. limosa* (Greenaway, 1970; Chaisemartin, 1969) and in annelids such as *Lumbricus terrestris* (Dietz and Alvarado, 1970). In *Ambystoma gracile* the various components of the Na^+ influx were calculated using a derivation of the Goldman equation (Alvarado and Dietz, 1970). While the diffusive component may be neglected, the exchange diffusion component is of importance. This component increased with salt depletion.

In other species a change in the affinity constant, accompanied or not by a change in the maximal rate of uptake, is observed; the best example investigated is shown by the isopod *Mesidotea entemon*. Brackish water animals from the Baltic sea and freshwater specimens from inland lakes were compared (Croghan and Lockwood, 1968), and in this study the various components of the Na^+ influx were also calculated. The diffusive influx may again be neglected, but in this species the exchange diffusion component has a greater importance in the salt-loaded (brackish-water living) specimens.

In the ammocoete larvae of *Lampetra fluviatilis*, Morris and Bull (1970) observed that salt depletion is characterized by a 100% increase in the affinity of the branchial carrier and by a 25% increase of the maximal transport rate. The affinity change is considered to be a short-term response while the maximal rate increase is a long-term response. Morris and Bull consider that the reduced Na^+ loss observed in the salt-depleted animals reflects an increased efficiency of the transport carrier for "back-transport".

The situation is less clear when brackish water and freshwater specimens of *Gammarus duebeni* from various Irish or British localities are compared (Shaw and Sutcliffe, 1961; Sutcliffe and Shaw, 1968; Sutcliffe, 1967b, 1971). Some populations, especially from Britain, show an increased maximal rate in Na^+ influx with little change of the affinity. Others, mostly from Ireland, exhibit on the contrary a high affinity in freshwater populations. This is correlated with the ecology of the population, those collected in Britain being found in streams with a higher Na^+ content. An experimental population of British origin, however, was adapted for several years in water of low Na^+ content in Lake Windermere. An increase of the affinity was observed in this very interesting case (Sutcliffe, 1970). Thus physiological differences between geographically separated populations are probably phenotypic in origin. Such distinctive physiological characteristics presumably become invested in the genotype as suggested by Croghan and Lockwood (1968) for *Mesidotea* and by Sutcliffe and Shaw (1968) for *Gammarus duebeni*.

b. *Control of Na^+ and Cl^- transfer in relation to acid-base balance*

The preceding section was concerned with the control of Na^+ uptake from NaCl solutions. The question whether blood Na^+ concentration can be maintained at its normal level irrespective of the Cl^- concentration has been investigated in the crayfish by Shaw (1964). Although Na^+ can be taken up by a salt-depleted animal independently of Cl^-—from a sodium sulphate solution for example—only a fraction (generally about one third) of the Na^+ is regained and this contrasts strikingly with the almost complete recovery of Na^+ from a sodium chloride solution. To regain more Na^+, Shaw showed that a known amount of Cl^- has first to be taken up from a KCl solution independently of Na^+. If the animal is then returned to Na_2SO_4 a net uptake of Na^+ follows once more. Such a cycle can be repeated a number of times and the uptake is thus cumulative, providing alternative pulses are given. Eventually Na^+ uptake ceases when all the Na^+ lost during depletion has been recovered. Complementary experiments showed that the rise in net uptake produced by a "Cl^- pulse" results from an increase in Na^+ influx and not from a decrease of the Na^+ efflux. These experiments clearly demonstrate that Na^+ uptake is regulated by a primary control system triggered by the internal Na^+ level *and* further, by a feed-back system which (i) decreases the rate of uptake as the internal Na^+ level exceeds that of Cl^- and (ii) increases the rate of uptake as the internal Cl^- level exceeds that of Na^+. Thus this second control mechanism seems to depend upon the difference between internal Na^+ and Cl^- levels.

A second series of experiments on the crayfish by Shaw (1964) demonstrated that the feed-back system, which is triggered by the difference between internal Na^+ and Cl^- levels, is of utmost importance for the regulation of Cl^- uptake. For these investigations salt-depleted crayfish were used. Presumably, as a result of differences in the passive permeability of the gill to Na^+ and Cl^-, Na^+ losses exceed Cl^- losses in animals kept in deionized water. Thus salt-depleted crayfish take up Cl^- rather slowly from a dilute KCl solution, but the rate of uptake is greatly increased if the animals have access to Na^+. A series of experiments similar in design to those described in the preceeding paragraph with "Na^+ pulses" instead of "Cl^- pulses" showed that an increased Cl^- influx was responsible for the increased rate of uptake following a Na^+ pulse. Successive pulses produced effects which were progressively damped. The Cl^- influx changes were always the reverse of those found for Na^+, i.e. when the Na^+ influx was low, the Cl^- influx was high and *vice versa*.

Unlike that of Na^+, however, the Cl^- uptake does not appear to cut-off sharply at a critical value of sodium chloride concentration difference. The Cl^- uptake rate decreases as the blood concentration increases, but a *firm control* never appears to be established during independent Cl^- uptake. In conclusion, the animals seem to have no independent control of the Cl^- uptake insofar as there seems to be no limiting blood Cl^- concentration at which Cl^- uptake is maintained at a steady-state level. Near perfect regulation of both Na^+ and Cl^- concentrations can thus be achieved only if both uptake mechanisms operate simultaneously. The operation of the regulatory process may be visualized as follows. First, if recovery starts with an existing blood concentration difference between the two electrolytes, then the rate of uptake of one ion will exceed that of the other owing to the second feed-back mechanism responding to the concentration difference. Then, once the concentration difference is abolished the two uptake rates will be equal and virtually "locked together" since any tendency to drift away from this position will be opposed by an increased uptake of either one of the ions. Finally as the blood sodium chloride concentration rises, the primary control system reduces both uptake rates until they fall to steady state values.

Concerning the second feedback mechanism which appears to depend upon the difference between internal Na^+ and Cl^- levels, or total amounts, Shaw suggests the factor which is involved may be a more general property of the blood which is related to the concentration difference but which may also be influenced by other factors. "This possibility that this property may be the blood pH is worth considering" as "it is known that the independent uptake of Na^+ and Cl^- results from an exchange with weak base-forming cations (H^+, NH_4^+) and weak acid forming anions (HCO_3^-) respectively".

The possibility that a similar, secondary control mechanism operates in other groups of freshwater animals is suggested by various observations. For instance, Alvarado and Dietz (1970) noted in larval *Ambystoma gracile* that less Na^+ accumulated from sodium sulphate solutions than from sodium chloride solutions and, conversely, that less Cl^- accumulated from KCl over a period of time than from NaCl. In *Ambystoma*, however, the initial Cl influx and net uptake is observed to be as high in KCl as in NaCl solutions, a situation which contrasts with Shaw's observation on the crayfish. The mosquito larva *Aëdes aegypti* resembles *Ambystoma* in this respect (Stobbart 1965, 1967). Stobbart suggests that this difference in behaviour results from the fact that in *Aëdes* kept in deionized water, Cl^- loss is greater than Na^+ loss.

Various investigations on freshwater teleosts suggest that ionic balance is intimately linked with acid-base balance. Lloyd and White (1967) showed that acclimation of the trout to external media with high levels of CO_2 is followed by a considerable increase in the plasma HCO_3^- level and a corresponding decrease in the Cl^- level. This change is slow, however, taking about 24 h to be complete. Recovery on removal of CO_2 is relatively faster. Cameron and Randall (1972) confirmed this observation and demonstrated that the increased HCO_3^- level serves to minimize the plasma pH shift resulting from an increased arterial pCO_2, allowing for the maintenance of the blood to water CO_2 gradient. These results are consistent with the hypothesis that blood pH is regulated via a Cl^-/HCO_3^- exchange mechanism. De Renzis and Maetz (1973) demonstrated that, in the goldfish kept in deionized water, Na^+ loss exceeds Cl^- loss as in the crayfish and, correspondingly, the body fluids become acidotic and plasma ammonia increases. Upon return to a sodium chloride solution, Na^+ influx greatly exceeds Cl^- influx. The fish gains Na^+ and loses Cl^-. If the fish is maintained in a choline chloride solution, the pattern is similar to that observed in salt-depleted fish. After an initial Cl^- gain and Na^+ loss, both electrolytes are lost into the external medium, the Na^+ loss exceeding the Cl^- loss. Correspondingly a decrease in blood pH and total CO_2 occurs while plasma ammonia increases. Upon return to a sodium chloride solution, Na^+ is gained and Cl^- is lost (Garcia-Romeu and Maetz, 1964). When the fish are kept in a sodium sulphate solution, however, blood pH and total CO_2 content increase. Both electrolytes are lost into the external medium, the Cl^- loss exceeding the Na^+ loss. Upon return to a sodium chloride solution, branchial Cl^- influx greatly exceeds Na^+ influx. A net uptake of Cl^- is observed contrasting with a Na^+ loss. Dejours (1969) noted earlier that transfer from a sodium chloride to a sodium sulphate solution caused a dramatic reduction of the respiratory CO_2 output for as long as 24 h. When the animal was replaced in a sodium chloride medium, CO_2 excretion was resumed and considerably enhanced.

All these results may be readily interpreted by assuming a second feedback control mechanism related to acid-base balance and to independent Na^+/H^+ or NH_4^+ or Cl^-/HCO_3^- exchange mechanisms. For example, upon transfer from tap water to a sodium sulphate solution, the Cl^-/HCO_3^- exchange mechanism is blocked and HCO_3^- accumulates, hence the tendency towards alkalosis and retention of respiratory CO_2. Both factors presumably alleviate the pH-shift. Upon return to a sodium chloride solution the Cl^- uptake is greatly enhanced by a forced Cl^-/HCO_3^- exchange permitting the elimination of excess plasma HCO_3^-. Figure 11

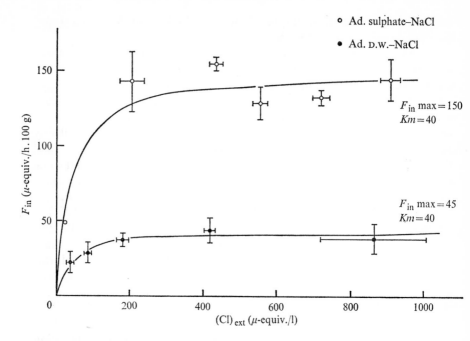

FIG. 11. Cl^- influx from a sodium chloride solution as a function of external Cl^- concentration in goldfish pretreated in deionized water or in sodium sulphate solution. (according to de Renzis and Maetz, 1973).
Ad. sulphate-NaCl: animals pretreated to Na_2SO_4 solution studied in NaCl solution.
Ad. D.W-NaCl: animals pretreated to deionized water studied in NaCl solution.

illustrates the considerable difference in Cl^- and Na^+ uptake efficiencies from NaCl solution in animals preadapted to sodium sulphate solution or to deionized water. It may be seen that the affinity constant (Km) remains unchanged, but the maximal influx is increased about three-fold suggesting an increased number of transport sites or, more probably, an increased turnover of the pre-existing sites.

c. *Control of Na^+ and Cl^- uptake in relation to volume regulation*

The problem of the maintenance of body fluid volume forms an integral part of the overall osmoregulatory processes in both vertebrates and invertebrates. This problem has been the subject of extensive studies in mammals but has been neglected in lower vertebrates and in invertebrates. The most obvious responses of organisms to changes in internal volume are variations in urine flow and in the amount of water taken in either by osmotic influx across the body surface or by drinking. Volume regulation implies as well the maintenance of osmotic and ionic levels of the blood by a modulation of the salt uptake mechanism.

That body volume regulation is tied to ionic content changes in aquatic animals was first observed by Jørgensen and Rosenkilde (1965) in the goldfish. They followed the daily changes in Cl^- content and weight of undisturbed goldfish over several weeks. A statistically significant correlation was found between daily Cl^- content and weight changes. Retention or excretion of 150 to 240 μEquiv of Cl^- was followed by retention or excretion of 1 ml of water. Whether Na^+ and Cl^- play an equivalent role in this process remains to be investigated.

The problem of volume regulation has been extensively studied in the euryhaline gammarid, *Gammarus duebeni*. As will be discussed below, moult is accompanied by a change in osmotic permeability of the body surface which causes an increase in body volume. Lockwood and Andrews (1969) observed that a massive augmentation in the rate of Na^+ uptake accompanies a volume change. The correlation between experimental volume changes and modulation of Na^+ uptake was further examined by Lockwood (1970). Individuals adapted to brackish water were placed for a few hours in isotonic mannitol or sucrose solutions, a procedure which results in a 10% loss in blood volume with only a slight decrease in blood Na^+ concentration. The animals respond to this volume change by a vigorous increase in Na^+ influx, and by a net uptake of Na^+ from a low Na^+-containing medium (2.5% seawater). This increase was shown to result not from an increased drinking rate but from an increased active transport of Na^+. In another series of experiments Lockwood (1970) showed that removal of blood also results in an increased Na^+ uptake.

Conversely, removal of water, accompanied by a rise in blood concentration depresses the Na^+ influx. This is observed when animals acclimated to 50% sea water are placed for a few hours in a sucrose solution iso-osmotic to seawater and thus hyper-osmotic to the body fluids. In this case the primary control mechanism, triggered by changes in the internal Na^+ level, takes precedence over the body volume control mechanism. Thus, four types of stimuli regulate ionic uptake: the absolute internal level of Na^+, the difference between Na^+ and Cl^- internal levels, blood pH and a fourth factor related to body volume changes.

How this fourth factor intervenes in the effector-organs of osmoregulation to restore body volume remains to be investigated in detail. In brackish waters it is probable that activation of salt uptake by the gill results in water movements by a mechanism comparable to Diamond's local osmosis model which was discussed in the introductory remarks.

In media of salinities lower than those in blood, Lockwood (1970) suggests that a similar mechanism intervenes to regulate the relative entries of salt and water. By increasing electrolyte uptake and decreasing water permeability, a fluid hypertonic to the external medium may be absorbed, thus contributing to body fluid volume restoration. The calculations of Jørgensen and Rosenkilde (1956) for the goldfish show precisely that the fluid crossing the gill is more concentrated in Cl⁻ than are the body fluids.

6. MORPHOLOGICAL ASPECTS OF ACTIVE TRANSPORT

It is disappointing that frog skin, the biological membrane on which the most numerous studies on active transport have been made, is a structure that shows few striking morphological features in relation to its function. Recent physiological observations indicate that the dermis can be completely eliminated without affecting the epithelial transporting capacity. Thus, it is clear that transport of electrolytes is effected by the squamous stratified epithelial layer (Aceves and Erlij, 1971; Rawlins et al., 1970).

Experimental investigations combining morphological and bioelectrical studies showed that the outermost layer of the *stratum granulosum* is the site of transport (Voûte and Ussing, 1968). This reactive cell layer swells reversibly in response to experimental changes in active transport. More recent studies making use of ruthenium red and colloidal lanthanum show that these substances cross the *stratum corneum* readily and that the outer border of the *stratum granulosum* is the external high resistance barrier of the skin. The tight juctions between the cells of the *stratum granulosum* are impermeable to these markers (Martinez-Palomo et al., 1971). More recent studies by Ussing (1971) suggest that these "tight seals" open up when there is an osmotic gradient in the outward direction, and close up when the direction of the gradient is reversed. There is a clear cut correlation between the sizes of the intercellular spaces of the various strata below the reactive cell layer, and the transepithelial transport rate between Ringer solutions (Carasso et al., 1971). Histochemically, Na⁺ and Cl⁻ have been shown to be accumulated in frog epidermis on the external surface of the *stratum corneum* cells in an area corresponding to the presence of an extraneous mucous coat; penetration of these ions occurs through the cells of the *stratum corneum* and the *stratum granulosum* and they exit by way of the intercellular space (van Lennep and Komnick, 1971) in accordance with the model suggested by Farquhar and Palade (1964, 1966). Recent studies by Voûte et al. (1972) show that adaptation of the frog to salt water or deionized water, corresponding to considerable variations in the Na⁺ transport rate of the skin, is accompanied by a change in the number and activity of one cell type of the *stratum granulosum*: the mitochondria-rich cells. In skins from salt water animals, these cells appear "unstimulated" in that they contain numerous mitochondria with a moderate amount of dense bodies, but a not very abundant smooth endoplasmic reticulum. In skins from animals in deionized water, these cells contain far more mitochondria. Ribosomes are more abundant and the subcorneal pole of the cell is characterized by a large amount of vesicles and an increased number of microvilli.

In the teleostean gill, the presence of large spherical cells, with eosinophilic granules, on the basal part of the gill lamellae was first described by Keys and Willmer (1932). Because of their morphological similarity to the hydrochloric acid-secreting cells of the gastric epithelium they were called "chloride secreting" cells, and suggested to be the site of the extra-renal salt excretion which characterizes salt balance in seawater fish. Pettengill and Copeland (1948) and Copeland (1948, 1950) have proposed that the chloride excretory cells of *Fundulus heteroclitus* may also be responsible for the active uptake of ions in freshwater. A similar view was expressed by Getman (1950) for the chloride cells of *Anguilla rostrata*. The first electron microscopic studies were made by Doyle and Gorecki (1961) who expressed doubts about the function of the "so-called chloride cells", despite their numerous mitochondria and branching tubular endoplasmic reticula. Comparative studies of the chloride cell in seawater and in freshwater species suggest that mitochondria are fewer in number and the tubular system is less developed in freshwater species. Differences in the apical region, which is generally enlarged in surface in freshwater living fish, are also of importance (Kessel and Beams, 1962; Philpott and Copeland, 1963; Threadgold and Houston, 1964; Petrik and Bucher, 1969; Shirai and Utida, 1970; Bierther, 1970; Doyle and Epstein, 1972). Philpott and Copeland (1963) were the first to demonstrate that the internal tubular system is continuous with the lateral and basal membranes. It thus represents a tremendous increase of the cell surface in close contact with the mitochondria. Continuity also between this system and the apical cell surface has been claimed by Shirai and Utida (1970). The apical cell region is generally filled with numerous vesicules which are in continuity with clefts originating from the apical membrane and with the tubular system (Philpott and Copeland, 1963). The amorphous glycocalyx which lines the outside of the apical membrane has been shown to contain acid or neutral polysaccharides or glycoproteins (Vickers, 1961; Philpott, 1967; Petrik and Bucher, 1969). This material is able to adsorb and accumulate Na^+ and Cl^- according to Bierther (1970) who suggests that the mucoid layer may play a role by decreasing the ionic gradient between the external and internal media. Histochemical demonstration of Na^+ in the mucous layer was confirmed recently by Shirai (1972) who found little Na^+ in the tubular system of the freshwater eel as compared with the seawater specimens.

The problem of the dynamics of cell proliferation in relation to seawater adaptation will be discussed in the second part of this review. In two euryhaline teleosts, *Anguilla anguilla* and *Cichlosoma biocellatum*, adaptation to distilled or deionized water is accompanied by an increase in the number of chloride cells. Olivereau (1971) suggests that these cells are derived by new cell formation from a stock of unspecialized "ionoblasts" while Mattheij and Stroband (1971) consider them to be transformed respiratory cells.

In the primitive cyclostomes, chloride cells are absent from the isoosmotic *Myxine glutinosa* (Morris, 1965). Such cells are however, present in the gills of freshwater *Lampetra fluviatilis*, which are able to hypoosmoregulate in 50% seawater, and in the gills of the freshwater-adapted sexually maturing animals. The seawater and freshwater cell types are different in size and it is doubtful whether the same cells serve as excretory and uptake cells by a reversal of their functional polarity (Morris, 1957).

Epithelial structures, presumably specialized in salt absorption, have been pin-pointed in the gills of various crustaceans with the help of the silver staining technique. Examples are in *Daphnia* (Gicklhorn and Keller, 1925), *Carcinus maenas* (Koch, 1934) and *Potamon* (Ewer and Hattingh, 1952), the latter being a species for which a differentiation of the staining properties was observed, suggesting a specialization of function in the gills. Krogh (1939) disputes the validity of this technique on the grounds that silver-absorbing cells are conspicuous in species of phyllopods (*Branchipus*) in which he did not find evidence of salt absorption. More recently sodium uptake has been demonstrated to occur, with the help of isotopic tracers, in a related phyllopod, *Chirocephalopsis* (Horne, 1968). Furthermore, Koch *et al.* (1954) showed that silver ions are inhibitors of NaCl uptake in the isolated gill of *Eriocheir*. It may be noted that staining with silver is also observed in the gills of *Artemia salina*, a species in which there is every reason to believe that excretion of salt occurs (Croghan, 1958b).

Electron microscopic investigations have been made on the gills of three decapods, *Callinectes sapidus* (Copeland and Fitzjarrell, 1968), *Gecarcinus lateralis* (Copeland, 1968) and *Astacus leptodactylus* (Bielawski, 1971), and an amphipod *Gammarus duebeni* (Lockwood *et al.*, 1973). *Gecarcinus* is unusual in that it is able to live on land and obtain its water and salts from damp sand. Bliss (1968) has shown that setae on the postero-lateral ventral surface of the crab absorb water from the moist environment. The water travels by capillary attraction into pericardial sacs that serve as stores from which the gills presumably absorb salt and water. In the two euryhaline crabs the "chloride cells" are found in specialized regions of the respiratory leaflets, varying in number along the branchial series. The cells are characterized in all three decapods by a considerable increase in the apical surface achieved by numerous folds immediately adjacent to the cuticula. A considerable increase also occurs in the basal surface by cellular interdigitations that contain numerous mitochondria in close contact with the basement membrane and haemolymph. Pinocytotic vesicles are numerous in the apical region. In *Callinectes* hypertrophy of the mitochondria is considerable. In *Gammarus duebeni* differences in the development of the apical lamellar folds, terminating in finger-like processes in contact with the cuticle, have been noted in relation to the salinity of the external medium. In seawater-adapted animals, the inter-lamellar spaces are reduced to narrow channels, while in 2% seawater-adapted individuals the spaces are large. The functional significance of these variations is not yet clear (Lockwood *et al.*, 1973). The enormous development of mitochondria with associated plasma membranes has also been observed in the salt-excreting *Artemia salina* (Copeland, 1967). In this hypo-osmotic regulator, the "chloride cells", or mitochondria-rich cells, are associated with mitochondria-poor "light cells" by means of numerous inter-digitations. Comparison of *Artemia* with the species mentioned above does not indicate any characteristic feature revealing the polarity of the cell function with respect to whether salts are absorbed or excreted.

In the air breathing aquatic Diptera larvae, the ultrastructure of the anal papillae of the mosquitoes, *Culex quinquefasciatus* and *Aëdes aegypti*, was studied by Copeland (1964) and Sohal and Copeland (1966). "Chloride cells" with numerous apical infoldings of the membrane on the cuticular side, manifesting a complicated pattern of internal canaliculi in relation to the distal (hemolymph) side are observed, together with numerous mitochondria clamped about the tubular system and oriented in precise relation to the basal fold. More recent investigations by Phillips and Meredith (1969) showed similar epithelial specialization in the anal papillae of *Aëdes campestris* larvae, which are able to osmoregulate in both freshwater and seawater and which retain functional anal papillae in salt water. In this species a reversal of functional polarity may, therefore, occur in the "chloride cell" in a manner similar to that of the euryhaline teleosts.

Anal papillae with conspicuous "chloride cells" also occur in various aquatic-breathing *Trichoptera* (caddisfly) larvae of the families *Philopota-midae* and *Glossosomatidae* (Nüske and Wichard, 1971, 1972). Those of the *Glossosomatidae* contain numerous tracheoles within the epithelium which suggests that these organs combine osmoregulatory and respiratory functions. In the *Limnophilidae*, however, the thread-like tracheal gills possess an epithelium which is purely respiratory, with tracheoles enclosed in extracellular slits which form deep invaginations of the basal membrane (Wichard and Komnick, 1971). The chloride cells are found in ten specialized epithelial fields on the dorsal and ventral sides of the 3rd to 7th abdominal segments. Histochemical and electron diffraction techniques have shown that the cuticle above the "chloride cells" accumulates chloride (Wichard and Komnick, 1973). Chloride and Na^+ ions have also been demonstrated histochemically within the apical and basal infoldings of the epithelial chloride cells of *Philopotamus* and *Anabolia* (Nüske and Wichard, 1971; Wichard and Komnick, 1973).

Finally, in another group of aquatic larvae, the *Ephemeroptera*, which like the *Trichoptera* are also aquatic-breathing, the tracheal gills, which undoubtedly serve a respiratory function, have chloride cell complexes irregularly distributed within the epithelium. Histochemical tests show a high concentration of Na^+ and Cl^- in the porous cuticle, as well as in the cell apex or apical cavity of the "chloride cells" where it is probably present in association with acid mucopolysaccharides or mucoproteins which can bind and accumulate these electrolytes from a hypotonic solution (Wichard and Komnick, 1971; Komnick *et al.*, 1972).

To my knowledge there have been no ultrastructural studies of the body surfaces of annelids and molluscs in relation to their osmoregulatory function. A recent ultrastructural study of the clam mantle by Istin and Masoni (1973) notes the presence in the palleal epithelium of cells with numerous microvilli and apical pinocytotic vesicles. However, mitochondria are not particularly abundant. The gill ciliated epithelium has been studied by Satir and Gilula (1970). The presence of "intermediate cells" with numerous microvilli was noted. Sodium ions were demonstrated histochemically in the septate junctions of the epithelial cells and in the microvillar infolding. Mitochondria are not particularly abundant, but a smooth tubular system is conspicuous in the apical side of the cell.

7. BIOCHEMICAL ASPECTS OF ACTIVE TRANSPORT

Three enzymes have been linked with the mechanism of transepithelial electrolyte transport: $Na^+ + K^+$ dependent ATPase, carbonic anhydrase and an anion-stimulated ATPase.

a. *$Na^+ + K^+$ dependent ATPase and amphibian osmoregulation*

Bonting (1971) has recently reviewed most of the studies on $Na^+ + K^+$ dependent ATPase activity in frog skin and the possible role of this enzyme in Na^+ transport across this epithelium. Only additional references will be discussed here.

Correlation between enzyme activity and transport activity in frog skin has been further confirmed by Kawada *et al.* (1969) and by Boonkoom and Alvarado (1971), who observed a much higher activity in adult than in larval skins of *Rana pipiens* and *R. catesbiana*. This difference is ascribed to the appearance of Na^+ transport during metamorphosis. Boonkoom and Alvarado (1971) showed in addition that salt depletion, which enhances Na^+ transport in adult but not in larval skins, produces a parallel increase in enzyme activity exclusively in adult skins. The gill of *R. catesbiana* tadpoles, which is the main site of active Na^+ transport, contains far greater $Na^+ + K^+$ dependent ATPase activity than the skin. Salt depletion is followed by an augmentation of enzymic activity which parallels transport activity.

Concerning the role of this enzyme in transepithelial sodium transport, the classical electron microscope study of Farquhar and Palade (1964, 1966), who tentatively localized "transport ATPase" in the intercellular membrane, is of importance. Histochemical techniques have only recently been developed which preserve $Na^+ + K^+$ dependent ATPase activity and allow for electron microscopic localization of the enzyme (Ernst and Philpott, 1970; Ernst, 1972*a*, *b*). The possible role of an enzyme-mediated $Na^+ + K^+$ exchange in transepithelial Na^+ transport, which has been discussed in a preceeding paragraph, is further illustrated by the effects of ouabain added on the serosal side. A simultaneous inhibition of Na^+ transport and K^+ influx from the extracellular fluids into the epithelial cells has recently been reported by Biber *et al.* (1972).

Ouabain has also been observed to inhibit the uptake of Na^+ at the mucosal border (Biber, 1971). Erlij and Smith (1973), reinvestigating this problem, suggest that this inhibition is a consequence of Na^+ accumulation resulting from the block of transepithelial Na^+ transport by a kind of cellular feed-back mechanism, and is not a result of a direct effect on Na^+ transport at the outer barrier. Preincubation of skins with ouabain in Na^+ free solutions is not followed by a block of mucosal Na^+ uptake when Na^+ is reintroduced in the bathing solutions. A similar block of the mucosal Na^+ uptake following accumulation of Na^+ in the epithelial cells has been observed by U. Katz (personal communication) in salt loaded *Bufo viridis* skins.

b. *$Na^+ + K^+$ dependent ATPase and teleostean osmoregulation*

$Na^+ + K^+$ dependent ATPase activity has been demonstrated in fish gills. The specific activity of this enzyme is greater in the gills of seawater-adapted euryhaline teleosts than in those of their freshwater-adapted relatives, and this difference has been associated with the increased salt load pumped out across the gills from the body fluids in marine forms. Euryhaline species which exhibit higher activities in the seawater forms include *Fundulus heteroclitus* (Epstein *et al.*, 1967), *Anguilla japonica* (Kamiya and Utida, 1968), *A. rostrata* (Jampol and Epstein, 1970; Butler and Carmichael, 1972), *A. anguilla* (Motais 1970*a, b*; Milne *et al.*, 1971), *Salmo gairdneri* (Kamiya and Utida, 1969; Pfeiler and Kirschner, 1972) and various salmonids (Zaugg and McLain, 1970, 1972; Zaugg *et al.*, 1972). Only in the labrid fishes *Crenimugil labrosus* and *Dicentrarchus labrax* (Lasserre, 1971) and in *Fundulus kansae* (W. R. Fleming, personal communication) has the enzyme activity been found to be higher in freshwater adapted individuals. When freshwater stenohaline species are compared with marine species, the branchial activity of the former is four to ten times lower (Kamiya and Utida, 1969; Jampol and Epstein, 1970). The marine elasmobranch, *Squalus acanthias*, resembles freshwater teleosts with respect to the low specific $Na^+ + K^+$ dependent activity of the gills (Jampol and Epstein, 1970).

A very extensive study of the branchial $Na^+ + K^+$ activated ATPase in *S. gairdneri* suggests that the enzyme extracted and purified from the freshwater adapted trout has distinct characteristics when compared with the enzyme extracted from the seawater-adapted fish. While the "freshwater enzyme" behaves at 37°C as a classical "transport ATPase" sensitive to ouabain, at 13°C, activation by K^+ and the response to ouabain is lost. The "seawater enzyme" retains both these characteristics at 13°C (Pfeiler and Kirschner, 1972). The affinities for Na^+ and K^+ of the "freshwater enzyme" have been found by Motais (1970*a, b*) to be significantly different from those of the "seawater enzyme" in the European eel.

Localization of the enzyme in the teleostean gill has recently progressed. A technique for separation of the chloride cells from the respiratory epithelial cells has been developed in two laboratories. Kamiya (1972), using gill filaments dissociated by elastase treatment followed by dextran density gradient centrifugation, showed that most of the $Na^+ + K^+$ dependent ATPase activity was confined to the chloride cells of the seawater Japanese adapted eel. No enzyme was found in the chloride cells isolated from freshwater adapted eels. M. Bornancin and J. Sargent (personal communication) using a different technique confirmed that in seawater European eels the activity is about 30 times higher in the chloride cells than in the other epithelial cells. In the freshwater eel some activity, about 7 times less than in seawater, is still to be found in the chloride cells. The enzyme activity of the other epithelial cells is 2.5 times lower.

Localization of the enzyme has also been attempted by histochemical techniques. With the limitations inherent to these techniques some caution is necessary when interpreting the results. Enrst and Philpott (1970) using formaldehyde-fixed gill tissues of seawater-adapted *Fundulus grandis* showed localization of the enzyme in the mitochondria of the chloride cells. Shirai (1972) using short term glutaraldehyde fixation, which inhibits only about 40% of the $Na^+ + K^+$ dependent ATPase activity of the gills of *Anguilla anguilla*, also localized large quantities of the enzymic reaction product in the chloride cells, but only small quantities were present in the respiratory cells of the seawater-adapted eel. In the freshwater eel, enzyme activity was also detected in the chloride cells but in smaller amounts. According to Shirai, most of the reaction product is located in the tubular system in both seawater and freshwater eels. The apical region of the cells in the freshwater eel is free of the reaction product and heavily stained in the seawater eel. The significance of the localization of the enzyme in the chloride cell of the seawater eel will be discussed in Section III.

In freshwater fish most of the enzyme is found in the infoldings of the lateral and basal membranes, a localization which is analogous to that suggested by Farquhar and Palade (1966) for frog skin. It is suggested that the function of the enzyme in transepithelial Na^+ transport is similarly related to a coupling between Na^+ efflux and K^+ influx on the serosal border. The observations of Pfeiler and Kirschner (1972) reported above do not seem, however, to fit this hypothesis. Direct evidence for the implication of Na^+-K^+ dependent ATPase in Na^+ uptake is reported by Richards and Fromm (1970). Addition of ouabain to the perfusion medium of an isolated trout gill preparation distinctly blocks net Na^+ uptake. Indirect evidence for a relationship between the enzyme and Na^+ uptake is afforded by the recent report of Butler and Carmichael (1972). These workers found that adaptation of the eel to deionized water, a procedure which is known to enhance the Na^+ pumping efficiency of the gill (Henderson and Chester Jones, 1967), induces a significant increase of the Na^+-K^+ dependent activity of the gill.

c. *$Na^+ + K^+$ dependent ATPase and crustacean osmoregulation*

Quinn and Lane (1966) were the first to report the occurence of this enzyme in the gill of the euryhaline land crab *Cardisoma guanhumi*. The effects of adapting the crab to media of varying salinities on the specific enzyme activity were unfortunately not tested.

Augenfeld (1969) discovered the enzyme in the brine shrimp *Artemia salina*. Microsomal preparations from whole animal homogenates were used. Enzymic activity is higher in animals reared at higher salinities, up to 400% sea water, than at lower salinities (50% sea water). This probably indicates that the increased enzymic activity is correlated with the increased salt load and the salt extrusion activity of the gill. In this respect *Artemia* resembles the euryhaline teleosts.

$Na^+ + K^+$ dependent ATPase activity has also been demonstrated in the gills of another euryhaline crustacean *Sphaeroma serratum* (Thuet *et al.*, 1969; Philippot *et al.*, 1972). The enzyme is located in high quantities in the endopodites of the fourth and fifth pleopods and in much smaller quantities in the corresponding exopodites or in the three first pleopods. Moreover, in endopodites 4 and 5 the enzymic activity is 2·5 times higher in animals reared in low external salinities (100mM) in which hyper-osmotic regulation is observed, than at higher salinities (seawater) when the body fluids are iso-osmotic with the external medium.

Kamemoto and Tullis (1972) have recently reported the presence of $Na^+ + K^+$ dependent ATPase in the kidney of *Procambarus clarkii* and in the gill of *Metopograpsus messor*. In both species, enzyme activity is influenced by environmental salinity, increasing in media of lower salinity and decreasing in higher salinities. These adaptive variations are probably under endocrine control as will be discussed below.

Investigations concerning the involvement of this enzyme in crustacean osmoregulation are practically non-existent. Smith and Linton (1971) report that the addition of ouabain to the perfusion medium of excised gills of *Callinectes sapidus* adapted to freshwater results in a slight increase in the bioelectric potential (negative inside). This change is ascribed by the authors to the inhibition of a "metabolically linked positive component, possibly an electrogenic inward movement of Na^+". Unfortunately no ionic flux measurements were performed at the same time.

There have as yet been no reports of $Na^+ + K^+$ dependent ATPase activity in the anal papillae or similar sites of active electrolyte transport in aquatic insect larvae.

d. *Possible importance of an anion-stimulated ATPase in Cl⁻ transport*

Wiebelhaus *et al.* (1971) discovered a considerable anion-ATPase activity in a purified microsomal fraction of the gills of *Necturus*. This activity was twenty times higher than that measured in the fundus or in the pancreas of the same animals. Unfortunately no information is available concerning gill function in *Necturus*. It is probable, however, that in *Necturus*, as in the tadpoles of *Rana catesbiana*, the gill is the site of active Na^+ and Cl^- transport.

Unpublished investigations by F. Garcia-Romeu and J. Ehrenfeld suggest that this enzyme also occurs in frog skin. Kristensen (1972) showed that SCN^- inhibits active Cl^- transport across frog skin without interfering with Na^+ transport; Br^- is even more effective, which suggests that both anions interfere with a chloride binding site.

Unpublished investigations of G. de Renzis and M. Bornancin showed that HCO_3^- stimulation of the ATPase activity of microsomal branchial preparations of the goldfish is ten times higher than $Na^+ + K^+$ stimulation.

The former enzymic activity was sensitive to SCN^- while Na^+-K^+ ATPase was insensitive. Addition of small concentrations(0.15mM) of SCN^- into the external medium produces a 60% inhibition of the Cl^- influx of the goldfish (Epstein *et al.*, 1973), while Na^+ uptake remains unchanged. Whether SCN^- interferes with the Cl^-/HCO_3^- exchange mechanism or substitutes for Cl^- in the exchange remains to be shown.

e. *Carbonic anhydrase and the NaCl uptake mechanisms*

Carbonic anhydrase activity has been detected in teleostean gills (Leiner, 1938; Maetz, 1956a) and crustacean gills (Maetz, 1946) but not in frog skin (Maren, 1967a, b).

Injection of acetazolamide is followed in the goldfish by inhibition of Na^+ uptake (Maetz, 1956b) and Cl^- uptake (Maetz and Garcia-Romeu, 1964). This suggests that hydration of respiratory CO_2 to H_2CO_3 which then dissociates to give $H^+ + HCO_3^-$ available for Na^+/H^+ and $Cl^-/$ HCO_3^- exchanges, is rate limiting in the uptake mechanisms. In the trout Kerstetter *et al.* (1970) confirmed that injection of acetazolamide inhibits Na^+/H^+ exchange, while ammonia excretion remains unchanged. Kerstetter and Kirschner (1972) could not, however, confirm inhibition of Cl^- uptake in the trout, suggesting that HCO_3^- diffusing from the blood into the epithelial cells, is readily available and is not rate limiting for Cl^-/HCO_3^- exchange at the apical border.

According to recent investigations (J. Ehrenfeld, in preparation) the crayfish resembles the trout in that acetazolamide injection blocks Na^+/H^+ exchanges, while Cl^-/HCO_3^- remains unperturbed. In the crayfish however, ammonia excretion is partially inhibited by the drug.

The specificity of the effects of acetazolamide is open to question because of the recent reports of Erlij (1971) and Kristensen (1972) showing that acetazolamide blocks Cl^- transport by frog skin *in vitro* without affecting Na^+ transport. In *Rana esculenta, in vivo*, however, both Na^+ and Cl^- uptake from low NaCl concentrations are inhibited by injection of acetazolamide (Garcia-Romeu and Ehrenfeld, 1972). In Na_2SO_4 solutions both Na^+ uptake and H^+ excretion by the skin are simultaneously inhibited, and in choline chloride solutions both Cl^- uptake and HCO_3^- excretion are blocked. Careful re-examination of carbonic anhydrase activity in isolated frog skin epithelia failed to detect significant amounts of enzyme (M. Istin, F. Garcia-Romeu and J. Ehrenfeld, personal communication).

B. Water permeability of the body surfaces

In freshwater animals, the body fluids are hyper-osmotic with respect to the external medium so water enters through the permeable body surfaces.

Water balance is maintained by the elimination of water through the excretory organs.

Comparisons of the Na^+ or Cl^- effluxes across the body surfaces of euryhaline aquatic animals have shown that acclimatization to freshwater involves a reduction of the passive flux of both ions. This condition holds whether euryhalinity involves iso-osmotic, hyper-osmotic or hypo-osmotic regulation in seawater. It is of interest to investigate whether changes in electrolyte permeabilities are paralleled by changes in water permeability. In order to compare permeabilities the various routes of water entry and exit must be identified and their relative importance assessed.

1. SITES OF WATER TRANSFER IN FRESHWATER ANIMALS

In aquatic animals with water as the respiratory medium, one would suspect that the region of the body surface specialized for respiratory gas exchange would also be the major site of water exchange. In air-breathing aquatic animals, such as the aquatic larvae of *Diptera* or adult amphibians, it appears that the region which is specialized in ion uptake is also the main site of water exchange.

A few examples will be given to illustrate these points. Wigglesworth (1933*a*, *c*) in a series of classical experiments demonstrated that the fairly large anal papillae, which were up to then taken to be gills, were in fact the main site of water entry in *Aëdes* mosquito larvae. He placed ligatures in different positions round the body and studied the rate of swelling or shrinking of the larvae in hyper- or hypo-osmotic solutions. For example, larvae were ligatured behind the neck to prevent drinking, and between segment nos 5 and 6 of the abdomen to prevent discharge of urine from the Malpighian tubules into the hindgut. When such larvae were placed in freshwater only the hindmost part of the body swelled. The Malpighian tubules and pyloric chamber became extremely distended, but the other parts did not swell. If the papillae were destroyed, no swelling and no distension of the Malpighian tubes occured. These experiments indicated that the general body surface is rather impermeable, that the main site of water entry is the anal papillae and that the excretory organs play an important part in water elimination.

More recent experiments making use of various techniques have shown that the gills are the main site of water exchange in wholly aquatic vertebrates and invertebrates.

One study concerns the freshwater eel. Motais *et al.* (1969) applied Fick's principle to the problem of HTO clearance by the gills of the eel after an intravenous injection of the tracer. They measured the HTO diffusional efflux from the whole animal and, by catheterizing the afferent and efferent branchial arteries, obtained the difference in HTO concentration due to gill clearance. If the gills are the only site of HTO efflux, the ratio of the HTO efflux to the difference in HTO concentration should be equal to the gill blood flow, *i.e.*, the cardiac output as measured independently by an electromagnetic flowmeter. Motais *et al.* (1969) showed that the contribution of the gills accounted for at least 90% of the diffusional water efflux.

Similar conclusions concerning the importance of gills in the euryhaline crab *Rhithropanopeus harrisi* were drawn by Capen (1972). Gill chamber perfusion experiments with D_2O showed that 90% of the total water influx is via the gill chamber. In the freshwater *Astacus leptodactylus*, Bermiller and Bielawski (1970) demonstrated that the gill is also the major site of water entry. The amount of urine produced was calculated from the rate of weight increase when the nephridial outlets were plugged, and the amount of water taken up by the isolated gills was measured by a colorimetric method. The two values agree to within 10%, which suggests that a certain amount of water is taken up elsewhere, presumably by the gut.

In recent years, with greatly improved techniques for measuring the drinking rate, it has become increasingly evident that drinking must occur in freshwater animals, disproving the original statement of Smith (1930) that drinking does not occur in freshwater fish. Among the new methods for evaluating drinking rate, those using radioactive γ emitter substances are to be recommended. Such substances include colloidial silver, [110]Ag (Dall, 1967; Dall and Milward, 1969), colloidal gold, [198]Au (Motais *et al.*, 1969), ^{125}I polyvinylpyrrolidone (Evans, 1968) and ^{131}I or ^{125}I phenol red (Maetz and Skadhauge, 1968). A few examples will illustrate the relative importance of the gut and the gills as sites of water entry.

In the freshwater eel, a drinking rate of $130\mu l/100g/h$ was recorded by Motais *et al.* (1969). As the urine flow amounts to $540\mu l$, as much as 25% of the total osmotic water uptake must occur by the gut, assuming that all the water drunk is absorbed. Recently Kirsch (1972*b*) re-evaluated the relative contribution of the gut and the gills in the silver eel. He claims that as much as 75% of the water inflow is by the gut.

Table IV summarises most of the data concerning drinking rates obtained in aquatic animals whether hypo-, iso- or hyper-osmotic regulators. It may be seen that the drinking habit has also been confirmed in crustaceans kept in waters of low salinity (Hannan and Evans, 1973) and in worms (Dietz, and Alvarado, 1970). In general, the drinking rate increases with the osmotic load, whether the animals' body fluids become iso-osmotic or remain hypo-osmotic to the external medium. The importance of the intestinal route of water entry in hypo-osmotic regulators will be discussed in the second part of this review. It is doubtful whether the increased

TABLE IV: Drinking rates (in $\mu l/100g/h$)

	Body Wt.	Temp. (°C)	Medium	Drinking Rate	References
ANNELIDA					
Lumbricus terrestris	3g	15	FW	40	Dietz and Alvarado (1970)
INSECTA					
Limnephilus affinis	20–50mg	14–17	33–66%SW	125–300	Sutcliffe (1961a)
Limnephilus stigma	12–26mg	14–17	33–66%SW	2000	Sutcliffe (1961b)
Philanisus plebeius	5–8mg	20	SW	1000	Leader (1972)
CRUSTACEAE					
Artemia salina	6mg	22·5	SW	2000	Smith (1969b)
			SW	3000	Thuet et al. (1968)
Uca pugilator	3g	24	5%SW	250	
			SW	600	Hannan and Evans (1973)
Penaeus duodarum	5·8g	24	SW	1730	
Metapenaeus bennettae	5g	20	66%SW	150	
			SW	700	
Metapograpsus gracilipes	5g	20	SW	600–700	Dall (1967)
Macrophthalmus crassipes	—	20	SW	600–700	
CHELICERATA					
Limulus polyphemus	5–9g	24	SW	730	Hannan and Evans (1973)
CYCLOSTOMES					
Lampetra fluviatilis	35–45g		FW	42·5	Pickering and Dockray (1972)
			SW	500	
TELEOSTS					
Aphanius dispar	0·5–3g	20	FW	760	Lotan (1969)
			SW	2000	
			200%SW	2700	
Cottus scorpius	150g	10	SW	495	Foster (1969)
Cottus bubalis	7g	10	33%SW	77	Foster (1969)
	7g	10	SW	188	

Species	Body weight	Temp (°C)	Medium	Value	Reference
Cottus morio	3–5g	10	FW	182	Foster (1969)
Paralichthys lethostigma	750–2100g	17–21	75%SW	718	Hickman (1968)
Salmo salar	85g	10	SW	460	Potts et al. (1970)
Salmo gairdneri	150–250g	20	SW	380	Oide and Utida (1968)
		17	SW	340	
		20	FW	0	Shehadeh and Gordon (1969)
			35%SW	175	
			50%SW	396	
			SW	538	
Anguilla japonica	185g	20	FW	0	Oide and Utida (1968)
			SW	370	
	80–225g		FW	135	Maetz and Skadhauge (1968)
			100%SW	125	
			200%SW	802	
	50–110g	15	35%SW	61	Maetz (1970)
			100%SW	167	
			FW	57	
Anguilla anguilla (yellow)	—	8	FW	230	Gaitskell and Chester Jones (1971)
	60g	12	FW	385	Motais and Isaia (1972a)
		23	FW	1	
		5	FW	52	
		15	FW	79	
	60g	25	SW	24	Motais and Isaia (1972a)
		5	SW	105	
		15	SW	542	
	—	25	SW	149	Gaitskell and Chester Jones (1971)
(silver)	390g	12	FW	215	Kirsch and Mayer-Gostan (1973)
		13	SW	140	
Fundulus heteroclitus	2–5g	20	FW	140–830	Potts and Evans (1967)
			SW	1540–2300	
Tilapia mossambica	0.5–3g	21–24	40%SW	260	Potts et al. (1967), Evans (1968)
			100%SW	440	
			200%SW	975–1110	
				1590	

TABLE IV: (continued)

	Body Wt.	Temp. (°C)	Medium	Drinking Rate	References
Xiphister atropurpureus	7–40g	13	10%SW	7	Evans (1967a)
			100%SW	34	
Pelates quadrilineatus	1–10·5 g	—	40%SW	576	Dall and Milward (1969)
			100%SW	957	
Periophthalmus vulgaris	0·1–3·4g	—	15%SW	1083	Dall and Milward (1969)
			40%SW	527	
			100%SW	355	
Monacanthus sp.	4·6–4·8g	—	SW	1330	Dall and Milward (1969)
Carassiops sp.	1–3·7g	—	FW	646	Dall and Milward (1969)
	15–80g	20	FW	80	
Carassius auratus	150g	5	40%SW	883	Lahlou et al. (1969)
		15	FW	8	
		25	FW	27	
			FW	79	
Serranus scriba and S. cabrilla	40g	10	SW	259	Isaia (1972)
		15	SW	461	
			SW	563	
Stephanolepsis cirrhifer	100g	20	SW	475	Oide and Utida (1968)
Goniistius zonatus	200g	20	SW	315	Oide and Utida (1968)

drinking rate observed when stenohaline freshwater species are transferred to salt water serves a useful purpose in terms of water balance (Lahlou *et al.*, 1969; Sutcliffe, 1962).

2. DIFFUSIONAL AND OSMOTIC WATER PERMEABILITIES OF THE BODY SURFACE
   IN RELATION TO ENVIRONMENTAL SALINITY

Table V summarises data concerning the rate of water exchange of various aquatic animals. Values obtained from seawater and freshwater animals, whether hypo-, iso- or hyper-osmotic regulators, have been included for comparison. It is difficult to compare these data without taking into account the body weight, the experimental temperature and the osmotic gradient across the body surface. Accordingly, such information is also given in Table V.

The relationship between body size and unidirectional water flux has been studied by plotting these two parameters on a log/log plot. Evans (1969b) demonstrated that the data from various teleosts fitted the equation $M = aW^x$ where M is the flux in ml/animal.hr, W is the body-weight in grams and a, x are constants. If the flux is related to the body surface, x will be equal to $2/3$; if it is directly related to body weight, $x = 1$. Evans found no evidence for either relationship, x being 0.88. Similar values were found for *Limulus*, the horseshoe crab, where $x = 0.89$ (Hannan and Evans, 1973). Hannan and Evans also calculated the x values from the data of Smith (1967, 1970d) for two decapods, *Rhithropanopeus harrisi* and *Carcinus maenus*, expressed by plotting the turnover rate as a function of the body weight. Values of 0.81 and 0.73 respectively were found. According to Evans (1969b) and to Hannan and Evans (1973), the x values are correlated to the change of gill surface as a function of body weight. Thus, recalculation of the data of Price (1931), Gray (1954) and Hughes (1966) for the gill surface area of fish, and of Gray (1957) for the gill surface area of crabs, both as a function of a body size, yield x values varying between 0.85 and 0.72. It seems reasonable to assume, therefore, that the water flux is mainly across the gill, a conclusion which is further supported by the wide variation in body shapes of the fish studied by Evans. Trout, eel and flounder all yield similar x values.

Studies of temperature effects on diffusional water permeability have already been discussed in the introductory remarks. *In vivo* studies yield relatively high Q_{10} values, up to 5.1 for the seawater eel (Motais and Isaia 1972), 3.8 for the alderfly larvae *Sialis lutaria* (Shaw, 1955b), 2.7 for the goldfish (Isaia, 1972) and 2.6 for *Limulus* (Hannan and Evans, 1973). Most of the values for teleosts and crabs are in the range 1.7–2.1 (Evans, 1969b; Hannan and Evans, 1973).

TABLE V: Diffusional water turnover rates (in % body water h⁻¹)

	Body Wt.	Temp. (°C)	Ext. medium	Turnover %h⁻¹	Osmotic gradient (mOsmoles/l)	References
ANNELIDS						
Nereis diversicolor	250mg	19	FW	600	250	Smith (1970b)
Nereis limnicola	100mg	19	50%SW	1200	0	Smith (1964)
Nereis succinea	100mg	19	50%SW	1200	0	Smith (1964)
INSECTS						
Sialis lutaria	30–100mg	20	FW	11	340	Shaw (1955a)
Philanisus plebeius	5–8gm	20	SW	41	−600	Leader (1972)
CHELICERATES						
Limulus polyphemus	5–9g	24	20%SW	159·6	230	Hannan and Evans (1973)
			50%SW	139·6	85	
			100%SW	202·5	0	
CRUSTACEANS						
Artemia salina	6mg	22·5	SW	15	−625	Smith (1969b)
Daphnia magna	25mg	?	FW	4900	130	H. Ussing (in Krogh, 1939)
Gammarus duebeni	60–90mg	18	2%SW	250	535	Lockwood et al. (1973)
			70%SW	420	0	
			SW	585	0	
			140%SW	650	0	
Astacus fluviatilis	30g	10	FW	20	400	Rudy (1967)
Metapenaeus bennettae	5g	20	SW	57–80	−220	Dall (1967)
	?	10	10%SW	55	410	
Palaemonetes varians			70%SW	64	0	Rudy (1967)
			120%SW	64	−375	
			40%SW	72	280	
Carcinus maenas	40g	10	40%SW	72	280	Rudy (1967)
	40g		SW	79	0	

						Reference
Carcinus maenas	1–25g	18·5	30%SW	176	320	Smith (1970*d*)
	1–25g		45%SW	207	265	Smith (1970*d*)
Macropipus depurator	30g	10	SW	236	0	Smith (1970*d*)
Rhithropanopeus harrisi	0·2–4g	18–20	SW	239	0	Smith (1967)
			1%SW	60	425	
			45%SW	70	250	
			SW	99	−50	
Rhithropanopeus harrisi	0·2–4g	20	10%SW	83	430	Capen (1972)
			75%SW	145	0	
Hemigrapsus nudus	2–50g	20	60%SW	129	75	Smith and Rudy (1972)
			95%SW	141	0	
Penaeus duorarum	5·8g	24	SW	76·5	−150	Hannan and Evans (1973)
	3g	24	3%SW	34·4	830?	
Uca pugilator			50%SW	36·9	350?	Hannan and Evans (1973)
			SW	33·1	−120	
Uca minax	7g	24	3%SW	30·0	830?	Hannan and Evans (1973)
			50%SW	33·8	350?	
			SW	33·1	−120	
Uca rapax	2·6g	24	3%SW	21·0	830?	Hannan and Evans (1973)
			50%SW	21·5	450?	
			SW	21·3	−120	

CYCLOSTOMES

Polistotrema stouti	54g	12	SW	325	0	Rudy and Wagner (1970)

ELASMOBRANCHS

Scyliorhinus canicula	200–400g	16	SW	157	0	Payan and Maetz (1970, 1971)
Ginglymostoma cirratum	100–1000g	24	SW	81	0	Carrier and Evans (1972)
Torpedo marmorata	200–400g	16	SW	97	0	Payan and Maetz (1971)
Raja montagu	200–800g	16	SW	167	0	Payan and Maetz (1971)
R. erinacea	400–1300g	13	SW	64	0	Payan et al. (1973)
R. radiata		13	SW	64	0	
Potamotrygon sp.	?	24	FW	96	310	Carrier and Evans (1973)

TABLE V: (continued)

	Body Wt.	Temp. (°C)	Ext. medium	Turnover %h⁻¹	Osmotic gradient (mOsmoles/l)	References
TELEOSTS						
Salmo salar	20–45g	10	FW	47·5	300	Potts *et al.* (1970)
			SW	32·5	−700	
Salmo trutta	3–200g	10	FW	36	300	
Salmo gairdneri	20–30g	10	FW	33	300	
Gasterosteus aculeatus	0·8–2g	10	FW	14·5	300	Evans (1969b)
			SW	18·0	−700	
Gobius niger	10–15g	10	SW	5	−700	
Ctenolabrus rupestris	2–20g	10	SW	10	−700	
Tilapia mossambica	0·5–3g	21–24	FW	189	300	Potts *et al.* (1967)
			SW	102	−700	
			200%SW	59	−1500	
Pholis gunnellus	1–10g	10	20%SW	11	100	Evans (1969a)
			SW	13	−700	
Xiphister atropurpureus	7–40g	13	10%SW	14	200	Evans (1967b)
			SW	14	−700	
Aphanius dispar	0·5–3g	20	FW	68	360	Lotan (1969, 1971)
			SW	92	−500	
			200%SW	69	−1400	
Fundulus kansae	0·5–2·5g	20	FW	138	360	Potts and Fleming (1970)
			33%SW	130	0	
			67%SW	108	−350	
			SW	87	−700	
			200%SW	74	−1500	
Cottus morio	3·5g	10	FW	41·5	360	Foster (1969)
			33%SW	24·9	0	
			75%SW	24·5	−400	

	Weight		Medium			Reference
Cottus scorpius	150g	10	33%SW	13·5	0	Foster (1969)
Cottus bubalis	7g	10	SW	16·8	200	
			10%SW	17·7	0	
			33%SW	16·8	−700	
Leptocottus armatus	141g	12	SW	16·0	−700	Rudy and Wagner (1970)
Platichthys flesus	220–245g	16	SW	31	−700	Motais *et al.* (1969)
			FW	31·1	265	
			SW	19·8	−825	
			SW	10·1	−800	
Platichthys platessa	1–70g	10	SW	19·4	+250	Evans (1969b)
		20	FW	78·5	+250	
Phoxinus phoxinus	1–3g	10	FW	139·0	+300	
		20	FW	42·9	+300	
	1–3g	10	FW	91·6	+300	
		20	FW	25·2	+300	
Carassius auratus	150g	5	FW	51·1	+300	Isaia (1972)
		15	FW	132·0	+300	
		25	SW	18·8	−820	
Serranus cabrilla and *S. scriba*	40g	10	SW	23·8	−820	
		15	SW	29·4	−820	
		20	FW	13	255	
Anguilla anguilla (yellow eel)	60g	5	FW	27	255	Motais and Isaia (1972a)
		15	FW	41	255	
		25	SW	14	−815	
		5	SW	35	−815	
		15	SW	56	−815	
		25	FW	8·6	250	
(silver eel)	60–120g	10	SW	5·7	−815	
	60–120g	10	FW	8·0	250	Evans (1969b)
		11	SW	8·0	−815	
Zoarces viviparus	300g	10	FW	9·5	250	Kirsch (1972b)
	30–100g	10	SW	10	−800	Evans (1969b)
Lumpenus lampretaeformis	3–9g	10	SW	1·8	−800	

In Table VI are summarized data on the osmotic water permeabilities of the body surface in freshwater and saltwater animals. The water flow is expressed in terms of % body weight per hour and per osmole of osmotic gradient. These values have mostly been calculated from the rate of urine flow in freshwater animals, and from the rate of drinking and urine flow in seawater animals, as will be discussed in the third part of this review (Section C). In some animals including worms, molluscs, small crustaceans and insect larvae, the rate of osmotic swelling observed when the animal is prevented from urinating and placed in a hypotonic medium, or the rate of shrinking when the animal is kept in a hyperosmotic medium, has been used for the calculation of the osmotic permeability.

A comparison of the water permeabilities in the various groups yields some interesting conclusions, in spite of the limitations imposed by the disparities in techniques, experimental temperatures and body sizes.

TABLE VI: Osmotic permeability of the body surfaces (in % body weight/hour osmole gradient).

	Ext. medium	O.P.	References
ANNELIDS			
Nereis diversicolor*	FW	14·5	Jørgensen and Dales
	50%SW	37	(1957)
Nereis limnicola*	50%SW	10	
Nereis succinea*	50%SW	50	Smith (1964)
Nereis virens*	50%SW	85	Jørgensen and Dales
Nereis pelagica*	50%SW	165	(1957)
Lumbricus terrestris*	FW	6·3	Dietz and Alvarado
			(1970)
MOLLUSCS			
Margaritana margaritifera	FW	100	Chaisemartin
			(1968a, 1969)
Viviparus viviparus	FW	20	Little (1965)
Doris sp.*	SW	40	Bethe (1934)
Aplysia juliana*	SW	45	van Weel (1957)
INSECTS			
Sialis lutaria*	FW	0·5	Shaw (1955a)
Limnephilus affinis*	BW	1·15	Sutcliffe (1961a)
Philanisus plebeius*	SW	1·1–1·5	Leader (1972)
Aëdes aegypti*	FW	4·5	Wigglesworth
			(1933b, c)
CRUSTACEANS			
Gammarus pulex	FW	2·8	Sutcliffe (1968)
Astacus pallipes	FW	0·8	Bryan (1960a)
Potamon niloticus	FW	0·004	Shaw (1959b)
Eriocheir sinensis	BW	0·6	Shaw (1961b)
Carcinus maenas	BW	6·2	Shaw (1961a)

TABLE VI: (continued)

	Ext medium	O.P.	References
Gammarus duebeni*	2%SW 40%SW SW	3·9 6·2 14	Lockwood and Inman (1973)
Maia squinado*	SW	10·0	Schwabe (1933)
Marinogammarus finmarchicus*	SW	6·5	Sutcliffe (1968)
Marinogammarus obtusatus*	SW	9·2	
Gammarus oceanicus*	SW	7·0	
Metapenaeus bennettae*	SW	34·7	Dall (1967)
Metaporgapsus gracilipes*	SW	5·0	
Artemia salina	SW	3·3	Smith (1969b)

CYCLOSTOMES

	Ext medium	O.P.	References
Lampetra fluviatilis	FW SW	3·0 0·85	Morris (1956)
Petromyzon fluviatilis	FW	4·00	Wikgren (1953)

TELEOSTS

		Ext medium	O.P.	References
Salmo gairdneri		FW	1·6	Homes and Stainer (1966)
Carassius auratus	25°C 15 5	FW — —	6·0 3·3 1·2	Isaia (1972)
Serranus cabrilla Serranus scriba	20 15 10	SW SW SW	0·30 0·28 0·20	
Anguilla anguilla (yellow)	25 15 5 25 15 5	FW FW FW SW SW SW	2·0 0·73 0·30 0·60 0·12 0·03	Motais and Isaia (1972a)
Platichthys flesus	16	FW SW	0·94 0·17	Motais et al. (1969)
Xiphister atropurpureus		10%SW SW	0·167 0·028	Evans (1967a)

AMPHIBIANS

	Ext medium	O.P.	References
Ambystoma tigrinum (larval)	FW	3·5	Kirschner et al. (1971)
(adult)	FW	8·0	Alvarado and Johnson (1965)
Triturus alpestris	FW	30·0	Bentley and Heller (1964)
Necturus sp.	FW	5·2	
Rana esculenta	FW	12·5	Mayer (1969)
Xenopus laevis	FW	4·9	Balinski and Baldwin (1961)

O.P., osmotic permeabilities were calculated from osmotic swelling experiments in all animals indicated by asterisk and from urine flow rates and drinking rates in the remaining species.

a. *Invertebrates*

Annelids are characterized by extremely high osmotic and diffusional water permeabilities of the body surfaces. Euryhaline polychaetes (*Nereis diversicolor*, *N. limnicola*) exhibit a smaller osmotic permeability than stenohaline species, such as *N. succinea*, *N. virens* or *N. pelagica*. According to Oglesby (1969), who recently reviewed the problem of the regulation of water content in Annelida, some caution is necessary in interpreting data obtained from swelling experiments. This is because the water osmotic influx is compensated for more or less rapidly by an increase of water elimination via the nephridia, which may intervene more rapidly in euryhaline species. According to Smith (1964), *N. limnicola* and *N. succinea* which differ widely in their rate of swelling, exhibit similar HTO turnover rates.

Jørgensen and Dales (1957) suggest that the osmotic permeability of the body surface of *N. diversicolor* is lower in specimens adapted to freshwater than to 50% seawater. Their experiments were criticized by Potts and Parry (1964a). Recent experiments by Smith (1970b), however, confirmed this difference in permeability. The rate of D_2O penetration is twice as fast in 50% seawater-adapted worms as in freshwater-reared animals.

Molluscs exhibit osmotic water permeabilities as high as annelids. The values given in Table V are calculated from the urine flow, measured in the freshwater species and in *Doris* from weight change values recorded under osmotic stress. Unfortunately no HTO or D_2O water turnover studies are available in molluscs.

The problem of the water permeability of the body surfaces of aquatic insects has been reviewed by Shaw and Stobbart (1963). Measurements of the rate of water loss in dry air indicate that the cuticle of aquatic larvae is more permeable than that of the terrestrial forms, the latter being characterized by highly orientated lipid water-proofing mechanisms. Weighing experiments on larvae with excretory apertures blocked clearly indicate that the air breathing Diptera larvae (*Aëdes*), in view of the highly permeable anal papillae, are far more permeable than Trichoptera larvae (*Limnephilus*) and Neuroptera larvae (*Sialis*), both of which have tracheal gills. As will be discussed in the third part of this review (section C), the salt water living Diptera larvae are far less permeable than their freshwater relatives, since in most species the anal papillae are far less developed. In the caddis larvae, osmotic permeability is similar in both salt water and freshwater forms (references in Tables V and VI).

In the chelicerate, *Limulus*, the HTO turnover values are in the same range as those found in Crustaceae of the same size that behave as iso-osmotic regulators in seawater and hyper-osmotic regulators in dilute seawater (e.g. *Carcinus maenas*). Hannan and Evans (1973) observe that, in the absence of an osmotic gradient, the diffusional water permeability is significantly higher.

In Crustaceae, an extremely wide range of diffusional water permeabilities is observed, the turnover rates ranging from 4900% in *Daphnia magna* to 15% in *Artemia salina*.

It can be seen in Table V that salt water Crustaceae which are hypo-osmic regulators (*Palaemonetes, Penaeus, Uca*) display a lower HTO turnover rate than the iso-osmotic regulators (*Gammarus duebeni, Carcinus, Rhithropanopeus, Macropipus*). The former, for example *Uca pugilator*, *U. rapax* and *U. minax* (Hannan and Evans, 1973), as well as *Palaemonetes varians* (Rudy, 1967), appear to maintain unchanged HTO turnover rates irrespective of the osmotic gradient to which they are submitted. Eury-haline species, which are iso-osmotic regulators in seawater and hyper-osmotic regulators in lower salinities, display a higher diffusional perme-ability in iso-osmotic conditions. This is clearly observed in *Rhithro-panopeus harrisi* (Smith, 1967; Capen, 1972) and in *Gammarus duebeni* (Lockwood *et al.*, 1973), but not always in *Carcinus maenas* (compare Smith, 1970*d* and Rudy, 1967). When osmotic permeabilities are con-sidered, differences between marine, brackish and freshwater species have been noted by Sutcliffe (1968) in his comparative study of gammarids (see Table VI). Marine species exhibit the highest permeability, while freshwater species show low permeabilities. When all the available data for freshwater Crustaceae are compared, huge differences may be noted. Thus *Potamon niloticus* has an osmotic water permeability that is 200-650 times lower than *Astacus* or *Gammarus pulex*. Similarly, when the water turn-over rates are compared, *Daphnia* exhibits a 200 times faster turnover than *Astacus*. Even when differences in weight (25mg compared with 30g) are taken into account, a 100-fold difference is still observed.

Thus, when all the data from invertebrates are considered, there seems to be no clearcut trend, indicating an adaptive variation of the water permeability of the body surfaces and internal media. It is noteworthy that freshwater species with very high water permeabilities, such as *Daphnia* and *Margaritana*, also exhibit relatively low body fluid osmotic concentrations. Conversely, species such as *Potamon* and *Astacus* with low osmotic permeabilities, maintain high blood concentrations.

Nevertheless in a few species, mostly crustaceans, the occurence of feed-back regulatory mechanisms, allowing for readjustments of the water permeability of the body surfaces in response to external salinity changes, has been firmly established. Two reports suggest that such regulatory responses are very rapid and are probably not under endocrine control. Bielawski (1964) observed in the crayfish gill, *in vitro*, a decrease in osmotic permeability in parallel to a decrease in the osmotic gradient. This readjustment probably results from an automatic response of the epithelial membrane, the mechanism of which remains to be studied. Lockwood *et al.* (1973) showed in *Gammarus duebeni* that, the permeability change measured by the HTO turnover, occurs within five minutes when animals acclimatized to seawater are transferred to dilute media. Furthermore, transfer to mannitol solutions isotonic to seawater induced permeability changes to a level close to that found in animals placed in dilute saline. Thus, it seem that the stimulus initiating the change is not an osmotic one. The authors suggest that there is either a nervous control of permeability, or a direct effect of ions on the permeability of the membrane.

A third report by Capen (1972) concerns the water permeability changes incurred by the crab *Rhithropanopeus harrisi* in response to a change in external salinity. Acclimatization to 75% seawater from 10% seawater and vice versa was recorded against time, water uptake being measured with HTO. Measurable changes in water uptake occurred within 30 min. In two hours half the permeability change from high to low salinity was completed, whereas for the reverse transfer seven hours were necessary for the same degree of adaptation. Several days more were needed for full acclimatization. In this species, therefore, an endocrine control of the permeability changes seems probable.

Capen (1972) also investigated the localization of the receptors which perceive the salinity change from 10% to 25% seawater. Gill chamber perfusion experiments with external media of various salinities showed that no change in permeability occurs. Thus, the receptors are not located in the gill chamber, and the regulation system responsible for the HTO flux readjustments must be composed of a receptor with some means of communication (nervous or hormonal) with the effectors (presumably the gill). Various observations suggest that the anterior region, which is known to bear structures with a possible chemoreceptor function, is responsible for the stimulus influencing HTO turnover. Experiments after removal of antennules or chelae show that these sensory organs are not concerned. It is probable that the mouth parts are the sites of the receptors.

Further experiments demonstrated that salinity rather than osmotic pressure change is the stimulus which intervenes at the receptor site. Transfer to 10% seawater made iso-osmotic to 75% seawater by the addition of sucrose is followed by a significant decrease of the HTO turnover rate. Thus, in this respect, *Rhithropanopeus* reacts in the same way as *Gammarus duebeni*.

b. *Vertebrates*

In lower vertebrates, the highest HTO turnover rates are observed in fish which are iso-osmotic to the environment in seawater. The best examples are the myxinoid *Polistotrema stouti* (Rudy and Wagner, 1970) and various elasmobranchs including *Scyliorhinus canicula* and *Raja montagu* (Payan and Maetz, 1971), as well as *Ginglymostoma cirratum* (Carrier and Evans, 1972). The more euryhaline *Torpedo marmorata*, *Raja erinacea* and *R. radiata* exhibit lower HTO permeability according to Payan *et al.* (1973), who suggest that this difference may have a significant bearing on the ecology of these fishes. The freshwater *Potamotrygon* however, exhibits a relatively high water turnover (Carrier and Evans, 1973).

In the teleosts, Evans (1969*b*) and Motais *et al.* (1969) note that freshwater species with a 300 milliosmolar gradient between external and

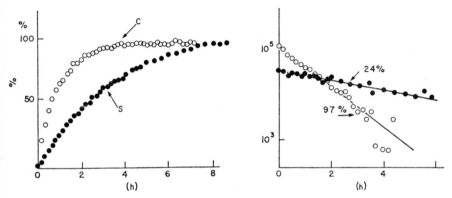

FIG. 12. Comparison of external HTO appearance curves of a freshwater fish *Carassius* (C) and a marine fish *Serranus* (S). The analysis of the curves is given in the right hand side of the figure (according to Motais *et al.*, 1969).
Ordinate (left): external radioactivity in % of equilibrium value.
Ordinate (right): log $(Q_{eq}-Q_t)$, Q_{eq} and Q_t being external radio-activities at equilibrium and at time t.
Abscissa (right and left): time in hours.
The calculated turnover rates in % h^{-1} are given.

internal media generally show a faster HTO turnover rate than marine species with a gradient of 600-800 milliosmoles. This difference is illustrated in Fig. 12 which compares HTO turnovers in *Carassius* and in *Serranus*. This adaptive change of the branchial diffusional water permeability is particularly well documented in euryhaline species compared after adaptation to high or low external salinities. Potts *et al.* (1967) were the first to note a three-fold increase in the HTO turnover rate in *Tilapia mossambica* transferred from 200% seawater to freshwater. A two-fold increase is observed in *Fundulus kansae* for the same range (Potts and Fleming, 1970). Less striking increases—by about 50%—are noted for

Salmo salar (Potts *et al.*, 1970) and for *Platichthys flesus* and the yellow eel, *Anguilla anguilla* (Motais *et al.*, 1969) transferred from seawater to freshwater.

These observed differences strongly suggest that the diffusional water permeability of the gill is highest in environments corresponding to the smallest osmotic gradient. Some data are, however, in disagreement with this suggestion. For instance in her study of various members of the family Cottidae, Foster (1969) observes that animals kept in iso-osmotic media (33% seawater) do not exhibit higher turnover rates than animals kept in higher or lower salinities. No differences in HTO turnover rates are observed in the silver form of *Anguilla anguilla* when compared in freshwater or seawater (Evans, 1969b); or in the yellow form in a more recent study by Motais and Isaia (1972a); or in the cypronodont *Aphanius dispar* compared in freshwater and 200% seawater (Lotan, 1969); or in either *Xiphister atropurpureus* or *Cottus bubalis* compared in 10% seawater and 100% seawater (Evans, 1967a; Foster, 1969). When some species such as *Gasterosteus aculeatus* or *Aphanius dispar* are compared in freshwater and seawater, the freshwater adapted individuals exhibit the lowest HTO turnover (Evans, 1969b; Lotan, 1969). If, however, the osmotic permeabilities are compared in freshwater -and seawater-adapted teleosts by taking into account the drinking rate, urine flow and the difference in the osmotic concentration between the external and internal media, the P_{os} value is considerably higher in the freshwater fish (see Table VII). This striking difference is also observed in fishes such as the yellow eel (Motais and Isaia, 1972) and *Xiphister* (Evans, 1967a, b), which are not characterized by an increased diffusional water permeability in relation to adaptation to media of lower salinities. Moreover, the comparison between the osmotic and diffusional permeabilities indicate that the ratio P_{os}/P_d is always higher in freshwater than in seawater species. Table VII collects the P_{os} and P_d values and their ratio for the various species for which the surface available for water movement is known. It must be emphasized, however, that the calculation of the P_{os} value depends on the assumption that the membrane which is the site of water transfer is semi-permeable. This important point will be discussed in the third part of this review.

For amphibians, most of the data related to the diffusional and osmotic permeabilities, obtained on isolated ventral skins, are given in Table VII. The values given may be overestimates because the ventral skin is generally more permeable than the skin from other areas (Bentley and Main, 1972). Values of the osmotic permeability obtained on whole animals are also given in Table VI. From both Tables VI and VII it is apparent that amphibian skin is more permeable than the teleostean gill. The difference is more

TABLE VII: Permeability coefficients, P_d and P_{os}, in various aquatic animals.

	Medium	P_{os}	P_d (in $\mu.sec^{-1}$)	P_{os}/P_d	References
ANNELIDS					
Nereis diversicolor	50%SW	—	—	4	Jørgensen and Dales (1957) Smith (1970b)
Nereis limnicola	50%SW	—	—	1	
Nereis succinea	50%SW	—	—	5	Smith (1964)
Lumbricus terrestris	FW	2	—	—	Dietz and Alvarado (1970)
INSECTS					
Opifex fuscus	SW	—	0·015		Nicholson (in Leader, 1972)
Sialis lutaria	FW	0·05	0·05	1	Shaw (1955a)
CRUSTACEANS					
Artemia salina	SW	0·96	0·069	13·2	Smith (1969b)
Gammarus duebeni	2%SW	—	—	1·16	Lockwood and Inman (1973)
	40%SW	—	—	1·16	
	SW	—	—	1·97	
Uca pugilator	SW	—	0·09	—	Hann and Evans (1973)
Astacus fluviatilis	FW	—	0·09	—	Rudy (1967)
Astacus leptodactylus	FW	0·12	—	—	Bergmiller and Bielawski (1970)
ELASMOBRANCHES					
Scyliorhinus canicula	SW	8·1	0·56	14·5	Payan and Maetz (1971)
Raja erinacea *Raja eglanteria*	SW	0·84	0·21	4·0	Payan et al. (1973)
TELEOSTS					
Opsanus tau	FW	—	0·14	—	Lahlou and Sawyer (1969c) Lahlou (1970)
	SW	—	0·07	—	
Xiphister atropurpureus	10%SW	—	—	1·6	Evans (1967a, b)
	SW	—	—	0·15	
Salmo salar	SW	—	—	1·5	Potts et al. (1079)
Carassius auratus (15°C)	FW	1·29	0·25	5·2	Isaia (1972)
Serranus cabrilla	SW	0·16	0·12	1·3	Isaia (1972)
S. scriba (15°C)	SW	0·16	0·12	1·3	
Platichthys flesus (16°C)	FW	0·70	0·27	2·59	Motais et al. (1969)
	SW	0·14	0·17	0·83	
Anguilla anguilla (15°C) (yellow)	FW	0·38	0·18	2·1	Motais and Isaia (1972a)
	SW	0·064	0·23	0·28	
AMPHIBIANS (*skin*)					
Xenopus laevis	FW	2·8	0·3	9·3	Maetz (1968)
Rana esculenta	FW	5·6	0·55	10·1	
Rana temporia	FW	3·3	0·65	5·1	Dainty and House (1966a, b)
Bufo regularis	FW	15·5	0·57	27·2	Maetz (1968)
Bufo bufo	FW	23·6	1·48	16·0	Koefoed-Johnsen and Ussing (1953)
Rana cancrivora	FW	4·85	—	—	Dicker and Elliott (1970)

striking when osmotic permeabilities are considered, because the ratio P_{os}/P_d is generally higher in amphibians than in freshwater teleosts. Comparison of the various amphibians reveals that aquatic amphibians (*Xenopus*, *Necturus* or larval *Ambystoma*) exhibit the lowest permeabilities, and terrestrial species (*Bufo*) the highest. Adaptation to dry environments is known to produce a striking increase in the osmotic permeability of amphibian skins, especially in the semi-terrestrial species, and far smaller or no changes at all in aquatic species (see review by Bentley, 1971*a*). Studies on adaptation to a saline environment have given contradictory results with respect to osmotic permeability changes. A decrease has been observed in *Bufo bufo* (Ferreira and Jesus, 1973), whereas there is no change in *Rana esculenta* (Jard, 1958). *Rana cancrivora*, which survives in seawater, exhibits a skin permeability similar to that observed in other *Ranidae* (Dicker and Elliott, 1970).

C. Endocrine control of salt and water metabolism

The problem of whether the adaptive changes in electrolyte transfer and water permeability of the body surface in relation to salt loading and salt depletion are under endocrine control will now be considered briefly.

1. LOWER VERTEBRATES

The problem of the endocrine control of salt and water metabolism has been reviewed recently (Maetz, 1968; Hoar and Randall, 1969; Mayer, 1970; Bentley, 1971*a*; Henderson and Chester Jones, 1972). Therefore only the more recent literature concerning the subject will be reported.

a. *Amphibia*

Recent evidence suggests that the decrease in Na^+ transport which accompanies prolonged exposure to saline media may be associated with changes in the molting cycle of the amphibian skin. This decrease results from a diminution of the Na^+ conductance of the outer barrier. (Hornby and Thomas, 1969) or, more specifically, of the amiloride-sensitive component of the Na^+ conductance (Katz and Lindemann, 1972). The latter authors observe that the reduced sensitivity to amiloride is correlated with a slowing down of the rate of shedding. These observations will be considered in the light of our knowledge of the control of molting in Amphibia. Jørgensen (1949) observed that the ability of the skin to absorb Na^+ actively is temporarily lost during molting. Moreover, Jørgensen and Larsen (1964) showed that the adenohypophysis-adrenal system participates in the control of molting in *Bufo bufo*. Extirpation of the adenohypophysis blocks molting. Injection of ACTH, corticosterone or aldosterone initiates shedding a few hours after injection. Myers *et al.*

(1961) found that removal of the anterior pituitary from *Rana pipiens* is followed by a decrease in active Na^+ transport which is rectified by treatment with ACTH or aldosterone. Maetz *et al.* (1958) showed that aldosterone injections fully restore Na^+ transport in *R. esculenta* skin depressed by saline exposure. This observation was confirmed by Crabbé (1964) and Crabbé and de Weer (1964). Enhancement of Na^+ transport by aldosterone was observed on isolated skins within 30–60 min. More recently Nielsen (1969) showed that small doses of aldosterone induce molt in isolated skins of *R. temporaria*, and that molt is accompanied by characteristic changes in active Na^+ transport although this occurs after a delay of several hours. Both molt and activation of Na^+ transport were abolished by actinomycin D treatment. Hormonal stimulation of Na^+ transport in the isolated skin of *Bufo bufo* was studied in more detail by Hviid Larsen (1971). He showed that slough formation depends on pre-existing conditions. In some skins, stimulation of Na^+ transport is observed within 30–60 min of aldosterone addition, while no shedding occurs. In other skins, stimulation is delayed until several hours after shedding. Stimulation is observed even if the aldosterone is removed 15 min after application. Slough formation interferes with sites responsible for the regulation of Na^+ entrance to the transporting cells, because both responses to oxytocin and to amiloride are temporarily blocked as shedding of the *stratum granulosum* commences (Nielsen and Tomilson, 1970; Hviid Larsen, 1971). Electron microscopic studies by Voûte *et al.* (1972) suggest that the mitochondria-rich cells of the *stratum granulosum*, which respond to variations in the salt-load, also respond to aldosterone treatment by augmentation in the number of mitochondria, ribosomes, vesicles and microvilli. Voûte suggests that these cells must be involved in the desquamation process. The frog bladder, however, although not subjected to cyclic changes, responds in much the same way as the skin.

As to the mechanism of action of aldosterone at the cellular level there seems to be general agreement that aldosterone induces DNA-dependent RNA synthesis (Sharp and Leaf, 1966; Edelman, 1968). This hypothesis is supported by numerous observations on the effects of inhibitors, but so far there has been no convincing direct evidence for the synthesis of new RNA or protein (Vancura *et al.*, 1971). If the hypothesis is correct, then the role of such new "induced" protein may be that of a "permease", enhancing the entry of Na^+ across the apical barrier into the transporting cell (Crabbé and de Weer, 1969; Leaf and Macknight, 1972; Handler *et al.*, 1972). The problem is still unresolved as to whether aldosterone also affects the activity of the Na^+ pump located at the intercellular or basal border of the epithelial cells (Edelman, 1968; Edelman and Fanestil, 1970; Kirsten *et al.*, 1968). There is now overwhelming evidence that the short-term stimulation of Na^+ transport does not result from an increased specific $Na^+ + K^+$ dependent ATPase activity, or from a change in the affinity constant of the enzyme (Hill et al., 1973). Long-term effects may, however, involve such changes. This is suggested by variations in enzyme activities observed in the gills of R. catesbiana tadpoles, and in the skin of the adult frogs in response to salt depletion and salt loading (Boonkoom and Alvarado, 1971).

b. *Teleosts*

Concerning the freshwater teleosts, recent work shows that corticoids and a prolactin-like adenohypophysial factor are implicated in the control of salt and water metabolism.

Mineralocorticoids may intervene in the modulation of the branchial Na^+ uptake in response to variation in the salt load. Prolonged treatment with aldosterone restores Na^+ influx in salt-loaded goldfish (Maetz, 1964). Aldosterone is, however, not a natural corticoid in this fish. Adrenalectomy in freshwater eels is followed by a net loss of Na^+ resulting from a decreased Na^+ uptake by the gill and an impairment of kidney function. Cortisol enhances Na^+ uptake by the gill and reabsorption by the kidney, as well as renal free-water clearance. In high doses, it induces salt loss by the gill (Chester Jones et al., 1969; Maetz et al., 1969b). Hypophysectomy produces similar effects, but in addition to the depression in corticoid function, the removal of the prolactin-like hormone secreted by the adeno-hypophysis results in additional defects, as will be discussed below. Hypophysectomy is followed by a depression of the branchial Na^+ uptake in *Fundulus heteroclitus* (Maetz et al., 1967a) and in *Anguilla anguilla* studied a few days after ablation (Langford 1971). A few weeks after hypophysectomy the Na^+ influx is similar to that of control animals in *A. anguilla* (Maetz et al., 1967b) and in *Carassius auratus* (Lahlou and Sawyer, 1969a). Injection of cortisol enhances Na^+ influx in hypophysectomized eels studied a few days after ablation. Table VIII illustrates the effects of hypophysectomy and cortisol treatment in such animals. Butler and Carmichael (1972) recently showed that, in the American eel, hypophysectomy is followed within two days by a significant decrease of the branchial $Na^+ + K^+$ dependent ATPase activity. Injection of cortisol prevents this decrease. More work is, however, necessary before the hypothesis of a pituitary-interrenal control of branchial Na^+ uptake via $Na^+ + K^+$ dependent ATPase activity can be accepted. This is especially so since Milne et al. (1971) failed to observe any change in the specific activity of this enzyme in the gill of the European eel following hypophysectomy, and Butler and Carmichael (1972) could not confirm a decrease in enzyme activity of the gill of *A. anguilla* after interrenalectomy.

TABLE VIII: Sodium ion exchanges in *Anguilla anguilla* adapted to freshwater. Effects of hypophysectomy and cortisol treatment (Langford, 1971).

Group	n	J_{in} (m ± s.e.) (μequiv/100g/h)	J_{net}	J_{out} (total)	$J_{in}/(Na)_{ext}$
Control					
$(Na^+)_{ext}$ ~0·1mM	7	3·1 ± 0·70	−0·4 ± 0·55	3·5 ± 1·05	28·0 ± 3·55[b]
~1mM	7	15·6 ± 1·75[c]	+8·5 ± 1·69[a]	7·1 ± 1·10	20·3 ± 3·43[c]
Hypophysectomized (4 days)					
~0·1mM	8	2·2 ± 0·44	−2·1 ± 0·91	4·3 ± 1·17	14·7 ± 2·79
~1mM	8	9·7 ± 1·60	+0·7 ± 1·06	8·0 ± 1·13	9·4 ± 1·38
Hypophysectomized (6 days + 100μg cortisol per 100g body wt. daily last 2 days)					
~0·1mM	4	5·0 ± 0·80[b]	−3·4 ± 1·05	8·4 ± 0·72	36·8 ± 4·79[a]
~1mM	4	18·8 ± 3·02[c]	+7·4 ± 2·31[c]	11·3 ± 2·26	19·8 ± 3·75[b]

Statistical comparison with hypophysectomized group
[a] P <0·001
[b] P <0·01
[c] P <0·025

Note: 1) 4 days after hypophysectomy, the effects of prolactin deficiency are not apparent.
2) The urinary papilla was not catheterized.

The possible role of the prolactin-like hormone in the control of salt and water balance of freshwater fish has recently been reviewed (Lam, 1972; Ensor and Ball, 1972). Only a few additional points will be raised.

Hypophysectomy results in decreased HTO diffusional permeability and osmotic inflow of water in the goldfish. Urine flow is simultaneously reduced, an effect which may be independent of the diminution of the osmotic water entry across the gill. Both HTO turnover and urine flow are restored to control values by prolactin treatment (Lahlou and Sawyer, 1969a, b; Lahlou and Giordan, 1970). Similar observations were obtained on the euryhaline *Fundulus kansae*, adapted to freshwater (Stanley and Fleming, 1966a, b, 1967; Potts and Fleming, 1970). According to Potts and Fleming (1970) two factors may intervene in the observed increase in osmotic and diffusional permeabilities of the gill in relation to freshwater life: an increased rate of endogenous prolactin secretion and a decreased external Ca^{2+} concentration. Concerning the role of prolactin, however, some recent observations do not fit with this generalization. For instance, Macfarlane and Maetz (1974) report that, in the flounder *Platichthys flesus*, a species which is characterized by an increased HTO turnover in relation to freshwater adaptation, hypophysectomy remains without effect on HTO turnover or urine flow. Furthermore Lam (1969), working on the isolated gill of *Gasterosteus aculeatus*, and Ogawa *et al.* (1973), investigating the isolated gill of *Carassius auratus*, claim that prolactin *decreases* the osmotic permeability in these preparations. In the teleostean urinary bladder, which is a terminal expansion of the fused archinephric duct, and which may provide information that can be extrapolated to renal tubule function in fish, prolactin also decreases the osmotic permeability of the epithelium (Hirano *et al.*, 1971; Johnson *et al.*, 1972; Utida *et al.*, 1972). Thus, the mode of action of prolactin on the water permeability of epithelial membranes remains to be investigated further.

The main role of prolactin is the control of the branchial Na^+ efflux in freshwater, a function which is of the utmost importance in the survival of some euryhaline teleosts after ablation of the hypophysis. This Na^+-sparing effect, first demonstrated in *Fundulus heteroclitus* (Potts and Evans, 1966; Maetz *et al.*, 1967a) has been confirmed in *Poecilia latipinna* (Ensor and Ball, 1972), in *Tilapia mossambica* (Dharmamba and Maetz, 1972) and in *Platichthys flesus* (Macfarlane and Maetz, 1974). In all these species, although prolactin is of critical importance for the survival of the hypophysectomized fish, no effect on the active component of the Na^+ balance, the Na^+ influx, is observed. Prolactin only depresses the Na^+ efflux. A similar effect is also observed in the European eel, a species which survives in freshwater after hypophysectomy (Maetz *et al.*, 1967b). In

F. kansae, prolactin treatment also reduces Na^+ loss in intact fish transferred to deionized water, while in hypophysectomized fish kept in Na^+-enriched tap water, it increases Na^+ influx and remains without effect on Na^+ efflux (Fleming and Ball, 1972).

Stimulation of active Na^+ transport by prolactin has also been reported in the teleostean bladder by Johnson *et al.* (1972) and Utida *et al.* (1972). Such an effect may well be the result of a stimulation of the $Na^+ + K^+$ dependent ATPase activity, which seems associated with Na^+ influx, as shown by the inhibitory effect of ouabain on this process (Johnson *et al.*, 1972). Prolactin treatment was observed to increase by 60% the $Na^+ + K^+$ dependent ATPase activity of the kidney of hypophysectomized *Fundulus heteroclitus* after three days in freshwater (Pickford *et al.*, 1970a). Simultaneously, a significant decrease (65%) of the branchial $Na^+ + K^+$ dependent ATPase activity was demonstrated. This suggests that prolactin may participate in the regulation of the activity of the Na^+ excreting pump by hastening the "shut-down" of this pump during freshwater adaptation.

In amphibians, the role of the neuroendocrine hormones appears to be mainly related to adaptation to a terrestrial environment. In fishes, the involvement of neuroendocrine factors secreted by the neurohypophysis or by the urophysis in the osmoregulatory processes is still in doubt. Maetz and Lahlou (1974) and Lacanilao (1972a, b) have recently reviewed the literature on this subject.

2. INVERTEBRATES

Neuroendocrine control of osmoregulation in various invertebrates has recently been the subject of an increasing number of investigations. The most clearcut results have been obtained in Crustaceae.

a. *Crustaceae*

Scudamore (1947) was the first to suggest the possibility of a neuroendocrine control of water balance in Crustacea. He noted that removal of the eyestalk, which contains the sinus gland, results in an increased water content of the crayfish. This change is prevented by re-implantation of the gland. Similar results were reported for *Carcinus maenas* by Carlisle (1956), and for *Eriocheir sinensis* by de Leersnyder (1967) who observed a 200% increase of the urine flow after eyestalk removal. A 60–100% increase in urine flow was reported in *Procambarus clarkii* by Kamemoto and Ono (1969). A change in the water permeability of the body surface is also suggested to occur by Thompson (1967) who observed an increased HTO exchange flux in *Pseudotelphusa jouyi*. Water balance in the land crab *Gecarcinus lateralis* also seems to be regulated by neuroendocrine factors (Bliss *et al.*, 1966).

Kamemoto *et al.* (1966) and Kamemoto and Ono (1969) confirmed the earlier observations by Scudamore (1947) on the crayfish, using animals with obstructed nephropores. Removal of eyestalks resulted in an enhancement of the weight increase and body fluid dilution. Injection of eyestalk extracts partially prevented the weight increase, and the Na^+ and Cl^- concentration of the body fluids increased. Operated animals transferred to hyperosmotic media increased their Na^+ and Cl^- concentrations more rapidly than controls. Kamemoto postulated that a factor controlling the water permeability of the body surfaces is secreted by the brain and transported to the eyestalks for release. More recently, however, Kamemoto and Tullis (1972) observed that purified fractions of the brain of *Procambarus* produce an *increase* rather than a decrease of the water permeability as measured by HTO influx. They propose a new scheme suggesting that the brain secretes a substance stimulating water entry, and that the eyestalks produce an inhibitor of the brain factor. A similar scheme was proposed for *Metapograpsus messor* by Kato and Kamemoto (1969).

The experimental changes of electrolyte concentrations of body fluids after eyestalk removal as observed by Kamemoto and Ono (1969) can only partially be the result of a perturbed water balance. Neuroendocrine factors may also intervene in the control of the electrolyte exchange across the body surfaces. The experimental evidence is, however, contradictory. On the one hand Bryan (1960*b*) reports that eyestalk removal does not affect the ability of the crayfish to alter the rate of Na^+ uptake in response to salt loading or salt depletion. On the other hand Ramamurthi and Scheer (1967) report the existence of a substance extracted from the cephalothorax of *Pandulus jordani*, which inhibits the Na^+ efflux of *Hemigrapsus nudus*. More recently, Kamemoto and Tullis (1972) prepared extracts from the brain and from the thoracic cord of *Procambanus*, which stimulate the Na^+ influx in this animal. Moreover, a substance which inhibits Na^+ influx in the crayfish may be extracted from the brain of various marine species, e.g. *Metapograpsus messor* and *Thalamita crenata*.

Neuroendocrine factors may increase diffusional water and electrolyte permeabilities in isolated crustacean effector-organs. Such a stimulating effect was demonstrated by Mantel (1968) on the foregut of *Gecarcinus*, using extracts from the ventral ganglion. Mantel suggests that the net movement of water and salt is towards the lumen of the gut for about two days after ecdysis, and in the reverse direction several days later. Bliss (1968) considered that this organ may serve as a temporary water store during ecdysis. The mechanism whereby the fluid movement occurs, in all probability "local osmosis", remains to be investigated.

Very recent research centres on the mode of action of the neuro-endocrine factors on effector tissues. Kamemoto and Tullis (1972) report that in *Metapograpsus messor*, eyestalk ligation results in a decrease in the $Na^+ + K^+$ dependent ATPase activity of the gill. Injection of brain homogenates into animals with their eyestalks removed restores this activity. Similar observations were obtained for the $Na^+ + K^+$ dependent ATPase activity of the antennal glands of the crayfish.

b. *Molluscs*

In freshwater molluscs the best evidence for a neuroendocrine involvement in osmoregulation has been obtained in the gastropod *Lymnaea*. This is one of the few molluscs in which both the physiology of osmoregulation and the structure of the neuroendocrine system have been studied. Weendelaar Bonga (1972) produced a quantitative analysis at the ultrastructural level, indicating that two types of neurosecretory cells are involved. In snails exposed to deionized water (a condition which according to Greenaway (1970) stimulates water elimination and uptake of Na^+), a release of neurosecretory material from the so-called dark green cells in the pleural ganglia and from the yellow cells in the parietal ganglia of the central nervous system was observed. In animals exposed to hypertonic saline the activity released declined and secretory granules accumulated, while water elimination and ion uptake were reduced. Some secretory axons, similar in size and appearance to the yellow cells, form a network around the ureter. These axons reacted in a way similar to those observed in the axon endings of yellow cells in the central neurohaemal areas. Lever *et al.* (1961) report that removal of pleural ganglia (with the dark green cells) leads to a considerable swelling of the snails. Injection of homogenates of these ganglia into intact animals reduced body weight. The authors suggested, therefore, the presence of a diuretic factor in the ganglia. These observations have been confirmed by Chaisemartin (1968*b*) in *Lymnaea limosa*. Cauterization of the pleural ganglia resulted in an increased body weight and a decreased rate of inulin clearance, which is indicative of a reduced urine flow. Very little effect on Na^+ fluxes was noted. Chaisemartin (1968*b*) suggests that these ganglia control water metabolism and urine flow. The diuretic effect may be related to an increased blood pressure and filtration rate, since changes in the heart rate after destruction of the pleural ganglia are observed.

Chaisemartin (1968*b*) also reports that cauterization of the parietal ganglia resulted in a decreased Na^+ turnover rate as measured with $^{22}Na^+$. He suggests that the ganglia produce an endocrine factor regulating ion transport mechanisms. The Na^+ efflux decreased by about 70%. Unfortunately the Na^+ influx, the active component of the exchanges, was not studied.

In other Mollusca most of the data concern marine species, such as the opisthobranch, *Aplysia rosea* (Vincente, 1969) and the lamellibranch, *Mytilus galloprovincialis* (Lubet and Pujol, 1963, 1965). These species are both osmoconformers when transferred to dilute media.

c. *Annelids*

In annelids, the best evidence for neuroendocrine control of osmotic and ionic regulation has been obtained in earthworms by Kamemoto (1964). Removal of the brain in *Lumbricus terrestris* caused an increased body weight and decreased Na^+ concentrations in the blood and coelomic fluid. The latter decreases were counteracted by either transplantation of the brain or by injection of brain homogenates. It is suggested that there is a chemical factor in the brain, possibly of neurosecretory origin, which is involved in osmotic and ionic regulation in earthworms. Obviously further studies are necessary to determine the function of the brain factor in osmotic and ionic regulation.

No similar experiments are available for polychaetes although considerable work has been done on the osmotic relations of these animals. Kamemoto *et al.* (1966) report that in certain nuclei situated in the anterior part of the brain of *Nereis virens*, the number of neurosecretory cells increases with decreasing external salinity. *Nereis virens* is known to be an excellent volume regulator under such conditions. Further studies are necessary to elucidate the possible relationship of neurosecretion and osmoregulation in polychaetes.

d. *Aquatic insect larvae*

The control of Na^+ uptake by the anal papillae of the mosquito larvae, *Aëdes aegypti*, in relation to salt depletion has been investigated by Stobbart (1971*b*). The time course of the exchange fluxes and the net flux readjustment has been studied in control and Na^+ deficient larvae upon readaptation in two molar NaCl solution. Influx, efflux and uptake rates declined exponentially with time, which is suggestive of a negative feedback mechanism controlling the transport system. Uptake was impaired when certain nervous tissues were destroyed, ligatured or clamped out of circulation. Thus a ligature between the thorax and abdomen, or pinching or freezing of the thoracic ganglia all blocked net uptake, the influx being less affected than the efflux. These effects are explicable in terms of impaired circulation of blood through the papillae. They suggest that a hormone, apparently produced by the thoracic ganglia and the retrocerebral complex (*corpora allata* and homologues of the prothoracic glands) is involved in the control of Na^+ uptake. The effects after interruption of the nerve cord suggest that an abdominal centre, sensitive to internal stimuli, conveys information forwards along the nerve cord to a control centre situated in the thoracic ganglia which inhibits or increases hormone production. Evidence for a second hormone produced in the head and also involved in the control of Na^+ transport is not conclusive, since a ligature at the neck produced only a slight effect.

Stobbart (1971a, c) suggests that control of both salt absorption by the anal papillae and water associated with salt resorption from the rectal fluid is effected by the same hormone, issued presumably from the thoracic ganglia. He notes that the rectum and anal papillae have a common embryological origin and that in both organs salt is transported in the same direction, *i.e.* from the apical to the serosal side. For maximal salt conservation or uptake in salt-depleted larvae both these structures require to be activated at the same time.

III. Adaptation to Hyper-osmotic environments

Animals which are able to maintain the osmolarity of their body fluids below that of the environment are found in both the sea and in inland saline water. The inland waters may originate from various sources and differ widely in salinity and ionic composition, although Na^+ and Cl^- are usually the most abundant electrolytes.

The general concensus of opinion is that hypo-osmotic regulators descended from freshwater or brackish water ancestors, because all these animals are characterized by blood Na^+ and Cl^- concentrations lower than those of seawater and in the same range as those of their freshwater relatives.

I shall not discuss marine or inland water animals, which are more or less iso-osmotic to the environment, although the Na^+ and Cl^- concentration of the blood is less than that of seawater. These animals have achieved iso-osmolarity by the addition of non-electrolytes. Urea retention has evolved independently in three groups: elasmobranchs, coelacanths and a few species of amphibians. Elasmobranchs have evolved very different ways to achieve their salt and water balance as compared with teleosts (see Payan and Maetz, 1970, 1971 and 1973). Therefore, of the primarily aquatic vertebrates I shall deal only with lampreys and teleosts which have independently involved identical osmoregulatory mechanisms.

Most of the species of invertebrates living in salt water are iso-osmotic regulators. Hypo-osmotic regulation has evolved only in two groups: Insecta and Crustacea. Insect larvae, although secondarily aquatic animals will nevertheless be considered to allow for comparison with Crustaceae. The contrasting adaptive features of the salt water and freshwater forms are also of special interest in this group.

The best studied insects in this respect are larval forms of two families which differ widely in their respiration physiology: the dipterous mosquito larvae, which have spiracles and make contact with atmospheric air for their supply of oxygen, and the caddis-fly larvae (Trichoptera), which depend upon diffusion of dissolved oxygen through the cuticle. In the second group one type of larva, such as *Limnephilus affinis*, adapts to salt water (up to 75% seawater) for a short period by adding a non-electrolyte component to the body fluids to give iso-osmolarity (Sutcliffe, 1961a). The other type, *Philanisus plebius*, is able to live constantly in seawater with body fluids very much hypotonic to the external medium (Leader, 1972). The osmoregulatory ability of the latter larva matches that observed in most of the dipterous larvae (*Aëdes detritus, Cricotopus vitripennis, Ephedra riparia, Tanypus nubifer*), or that of the nymph of the dragonfly *Enallagma clausum* (Ramsay, 1950; Sutcliffe, 1960; Lauer, 1969). The osmoregulatory performance of *Ephydra cinerea* is as exceptional as that of the brine shrimp (Nemenz, 1960), but the mechanisms remain to be studied in this species. I shall refer to the two best studied species, *Aëdes* and *Philanisus*, in comparison with the best investigated Crustacea. In the latter group a wide variety of species has been recognised as hypo-osmotic regulators, including *Artemia salina* (Croghan, 1958a–d; Smith, 1969a, b), which is the best investigated example, various copepods (Brand and Bayly, 1971), the isopod *Sphaeroma* (Harris, 1970, 1972; Charmentier, 1971) and decapods ranging from prawns, *Palaemonetes varians* (Potts and Parry, 1964b) and *Macrobrachium australiense* (Denne, 1968), to shrimps *Penaeus duorarum* (Bursey and Lane, 1971) and crabs, *Metapograpsus gracilipes* (Dall, 1967) or *M. messor* (Kato and Kamemoto, 1969).

Osmoregulatory balance in hypo-osmotic regulators is dominated by the osmotic water loss encountered by these animals. This loss is most severe in animals which have to rely on gills or expansions of the body surfaces to extract dissolved oxygen. Water loss has to be made good by compensatory water uptake, and this is obtained by drinking and "uphill" water transport through the gut. This water uptake involves salt uptake which taxes the mineral balance of the animals, adding an extra load to the salt entering along the chemical gradient across the body surfaces. To compensate for these various salt inflows, mechanisms of active salt excretion have evolved. In some cases the kidney has developed the ability to cope with the surplus salt, although this situation is encountered only in a few insect larvae. In general, extrarenal mechanisms of salt excretion occur, the teleostean gill being the best known example of such an effector organ. Unfortunately, the routes of extrarenal salt excretion are still a matter of speculation in most hypo-osmotic regulating invertebrates.

To analyse the mechanisms of salt and water balance in hypo-osmotic regulators, I shall deal in detail with the roles of the gut and gills in fish, but will not discuss the role of the excretory organs. Comparative aspects concerning the invertebrates will be discussed for each effector organ.

A. Electrolyte and water transfer across the gut

Homer Smith (1930) was the first to propose that marine teleosts drink seawater to replace water lost osmotically across the body surface. The evidence for this compensatory mechanism was as follows:

(1) The dye phenol red dissolved in the ambient water accumulated in the intestine and became three to five times more concentrated in the rectum than in the external medium;

(2) ligaturing the oesophagus prevented phenol red accumulation in the intestines;

(3) even in fasting animals the intestine contained a colorless fluid more or less isosmotic to the body fluids and which was discharged from the anus;

(4) chemical analyses of this fluid revealed progressive absorption of Na^+, K^+ and Cl^- and concentration of Ca^{2+}, Mg^{2+} and SO_4^{2-};

(5) the urine contained high concentrations of Mg^{2+} and SO_4^{2-} which were shown to be derived from the inevitable absorption of small amounts of these ions from the intestinal content;

(6) very little of the Na^+, Cl^- and K^+ absorbed in the intestine was found in the urine; Smith (1930) argued that these ions must be excreted extra-renally, probably by way of the gills, in order to maintain salt-balance.

This fundamental contribution by Homer Smith is, according to Hickman (1968), "now a central and accepted principal of hypo-osmotic regulation". Recent qualitative and quantitative aspects of seawater drinking, and subsequent water and ion transfer in the gut in different species of teleosts will be discussed and compared with other marine vertebrates and invertebrates.

1. DRINKING

Table IV (pp. 66–68) presents some of the most significant recent results on drinking rates obtained for fish and invertebrates adapted to various media ranging from freshwater to hypersaline seawater.

It can be seen that the rates are extremely variable. This is due to differences in the size of the animals and experimental temperatures. Shock effects due to handling of the fish are certainly responsible for variations in drinking rates. Recent evidence concerning the effects of shock on the osmolality of the body fluids as a result of an increased permeability of the body surfaces, may throw some light on the variations in water ingestion due to such phenomena (Pickford *et al.*, 1971*a-c*; Stevens, 1972; Pic, 1972; Pic *et al.*, 1973). The most important factor inducing changes in the rate of water swallowing is the external salinity, as can be seen by comparing freshwater and marine fish or from the rates recorded in euryhaline fish as a function of increased osmotic stress. Increased drinking rates are to be expected as they have to compensate for increased passive water losses. In the stenohaline *Carassius auratus*, an important augmentation of the drinking rate is also observed with increased osmotic stress (Lahlou *et al.*, 1969). There are, however, notable exceptions to this generalization in the mudskipper, *Periophthalmus vulgaris* (Dall and Milward, 1969) and the silver eel (Kirsch and Mayer-Gostan, 1973).

Not much is known about the stimulus inducing fish to swallow increased amounts of water. The observation that augmented drinking is induced by increased external salinity leaves undecided whether external factors such as salt content, or osmolality, or internal factors such as increased internal osmolality intervene. Sharratt *et al.* (1964) suggest that it is the *osmolality* rather than the *salinity* of the external medium which is important. Eels are unable to adapt to hypertonic sucrose solutions. Increased drinking occurs, however, since the gut is found to be distended by external fluid while the animal loses weight. Similar observations were reported by Motais (1967) for flounders kept in hypertonic mannitol solutions. Such experiments thus demonstrate that in fish, as in mammals, increased osmolality of the body fluids induces drinking, even though the fluid absorbed is strongly hyperosmotic to the body fluids.

Recently Hirano (1974) was able to record the rate of water ingestion in the eel by cannulating the oesophagus and collecting the ingested water in a drop-counter. He showed that changes in external salinity and, in particular, transfer to seawater caused an immediate increase of the drinking rate. Moreover it is the increase in Cl^- concentration, irrespective of the accompanying cation, which is the specific inductor of the reflex, the minimum effective concentration being about 10 mM. Drinking in this case, therefore, is induced by media of concentrations much lower than those of the body fluids, an observation which excludes osmotic stimuli and suggests specific chemical stimulation. Hirano also showed that transfer from freshwater to sucrose or mannitol solutions iso-osmotic to seawater induce drinking only after 8–10 h exposure. Thus both external sensory stimuli, by way of undetermined chemo-receptors, and internal stimuli, by way of unknown osmo-receptors, are concerned with the regulation of water ingestion.

Very recently Kirsch and Mayer-Gostan (1973), studying the European silver eel, found a biphasic response of the drinking rate following transfer from freshwater to seawater. There is an immediate increase lasting only a few hours, so marked that it causes a transitory augmentation of the body weight. Almost certainly this early response is triggered off by a sensory stimulus of the type analysed by Hirano (*loc.cit*). After this initial phase the eel almost stops drinking and loses weight. Within two days there is a second drinking phase, lasting up to one week after transfer, during which the fish recovers its original weight or even overcompensates the weight loss. Previous investigations by Keys (1933) and Oide and Utida (1968) on the eel have demonstrated a drinking phase, occuring from the second to the seventh day following transfer into seawater, which coincides with a recovery from the maximal weight loss registered on the second day. Oide and Utida (1967), as will be discussed below, showed that this phase corresponds to an increase in water absorption by the gut. Kirsch and Mayer-Gostan (1973) suggest that this second drinking phase is triggered off by the augmented internal osmolality resulting from dehydration and salt influx.

Pic *et al.* (1973) showed that adrenalin injections into seawater-adapted mullet produce a considerable decrease in the water ingestion rate, probably owing to contraction of the smooth musculature of the gut. This suggests that the initial drinking response after transfer to salt water is not mediated by a non-specific shock effect, causing adrenalin release.

2. INTESTINAL HANDLING OF SALT AND WATER

Various publications have recently appeared concerning the fate of the water and electrolytes entering the body as swallowed sea water.

a. *Balance sheet studies*

Not all the water swallowed is absorbed by the gut. Two techniques have been used to measure the quantity of water actually absorbed. Smith (1930) ligatured the anal opening to avoid fluid loss and compared the amount of fluid swallowed, using phenol red as an indicator, with the amount of fluid remaining in the gut. The calculated quantity absorbed may be slightly overestimated, as normally some fluid is lost through the anus. Oide and Utida (1968) and Pickering and Morris (1970) used the same technique. Hickman (1968) and Shehadeh and Gordon (1969), however, collected the rectal fluid by cannulating the anal opening. The volume collected for a given period was compared with the amount ingested as measured independently. The calculated quantity absorbed may thus be underestimated. Nevertheless, it can be seen from Table IX

TABLE IX Intestinal absorption of water in per cent of the amount swallowed

Species	milieu	Water	Na⁺	Cl⁻	K⁺	Ca²⁺	Mg²⁺	SO₄²⁻	Method	References
Paralichthys lethostigma	SW	75·8	98·8	93·9	98	68·5	+15·5	+11·3	C	Hickman (1968)
Anguilla rostrata	SW	78							A	Smith (1930)
Myoxocephalus octodecimspinosus	SW	63							A	
Anguilla japonica	SW	74							A	
Stephanolepsis cirrhifer	SW	62							A	Oide and Utida (1968)
Goniistius zonatus	SW	68							A	
Salmo gairdneri	SW	76							A	
Salmo gairdneri	SW	80	99	98	52	64	−5	−7	C	
	50%SW	78	91	94	86	86	+9	+3	C	Shehadeh and Gordon (1969)
	33%SW	66	66	85	98	95	+50	+23	C	
Lampetra fluviatilis	50%SW	75	94	86	—	—	−15	−29	A	Pickering and Morris (1970)

C: cannulation of anal opening.
A: anal opening blocked.
The negative values indicate the possibility of excretion of the electrolyte.

that both methods yield similar values. Between 60 and 80% of the water ingested is absorbed. In dilute sea water, the fraction reabsorbed decreases with increasing dilution (Shehadeh and Gordon, 1969).

Analyses of the rectal fluid or intestinal contents show that Na^+ and Cl^- are absorbed more rapidly than water. The concentration of Na^+ attains 20 mM in the flounder (Hickman, 1968) and the trout (Shehadeh and Gordon, 1969), which is more than 20 times less than that of seawater and at least 7 times less than that of the body fluids. About 99% of the Na^+ swallowed is absorbed and 94–98% of the Cl^-. The fraction absorbed is significantly less in animals kept in dilute seawater. While the Na^+, Cl^- and K^+ and, to a lesser extent, the Ca^{2+} concentrations decrease along the intestine, the Mg^{2+} and SO_4^{2-} concentrations increase. When compared to seawater the concentration ratios for Mg^{2+} and SO_4^{2-} attain 2·4 and 3·7 respectively in the trout and 3·6 and 3·9 in the seawater adapted flounder. Nevertheless in these two species Mg^{2+} and SO_4^{2-} are absorbed, but to a far smaller extent than Na^+ or Cl^- or even Ca^{2+}.

In some species such as *Lampetra fluviatilis* (Pickering and Morris, 1970) and *Pelates quadrilineatus* (Dall and Milward, 1969), the marked increase in the concentration of divalent ions in the gut is higher than can be accounted for by the rate of water absorption. Mg^{2+} and SO_4^{2-} seem to be excreted by the gut of lampreys, while Ca^{2+} is excreted by the intestine of *Pelates*. Several years ago Mashiko and Jozuka (1964) had suggested such an excretory function for the gut of the marine teleost, *Duymeria flagellifera*. Ca^{2+} excretion may also occur in other teleosts such as the flounder and the trout. In the flounder Hickman (1968) reports that the rectal fluid contains a crystalline precipitate tentatively identified as $CaCO_3$. In the trout, Shehadeh and Gordon (1969) record that a "mucus tube" is excreted by the anus, consisting of a meshwork of fibres with an embedded precipitate of calcium carbonate. It is not clear whether this excretory product has been accounted for in the balance sheet given by these authors.

Both Shehadeh and Gordon (1969) and Hickman (1968) noted that the gut fluid is at least as alkaline as seawater, which accounts for the calcium carbonate precipitate. Shehadeh and Gordon (1969) suggest that chloride absorption occurs by way of a Cl^-/HCO_3^- exchange, a point which will be discussed below.

b. *Mechanisms of Na^+ and Cl^- absorption*

Smith (1964, 1966) measured a potential difference (P.D.) of 5–6 mV across a sac made from freshwater goldfish intestine, the serosal side being positive to the mucosal side. The addition of glucose to the bathing fluid augmented the P.D., while ouabain inhibited it. A net transfer of Na^+ and Cl^- occured from the mucosal to the serosal side, and the absorption rate increased after the addition of glucose. Thus it was concluded that the fish intestine behaves in a manner similar to the rabbit ileum (Schultz and Zalusky, 1964a, b). The positive electrical potential probably corresponds to active Na^+ movement from the mucosal to the serosal sides.

The intestine of the marine *Cottus scorpius*, however, exhibits no P.D. difference. Isotopic studies revealed a flux ratio $\dfrac{mucosa \rightarrow serosa}{serosa \rightarrow mucosa}$ of nearly 3 for Na^+ and 2 for Cl^-, which demonstrates active transport of both ionic species. Further investigations with Na^+ accompanied by the impermeant anion $SO_4{}^{2-}$, and with choline chloride showed that the transport mechanisms are linked since the net Na^+ and Cl^- absorption rates from these solutions were considerably reduced or were nil. The results were interpreted as showing that the net transfer of Na^+ and Cl^- ions between identical Ringer solutions is achieved by some process whereby these ions are pumped in the form of electrically neutral sodium chloride (House and Green, 1965).

More recently Huang and Chen (1971) studied ion transport across the intestinal mucosa of the flounder, *Pseudopleuronectes americanus*. Isolated small intestine, bathed on both sides by a Ringer solution and mounted in an Ussing chamber, exhibit a P.D. ranging from 1 to 5 mV with the serosa being negative to the mucosa. The addition of glucose to the bathing solution increased the negativity. After replacement of the CO_2-O_2 gas mixture with nitrogen, the P.D. dropped to zero, this effect being reversible. When the bathing fluid is replaced by Ringer containing Na_2SO_4 the P.D. reverses from negative to positive. Ion flux measurements with isotopes revealed a net mucosal to serosal flux of Na^+ in Ringer containing Na_2SO_4, the flux ratio being 1·25, while the transfer of $^{35}SO_4{}^{2-}$ was negligible. In choline chloride there is a net Cl^- flux with a flux ratio of 1·45. The net transport rate is higher for Cl^- than for Na^+. In NaCl-Ringer the net Cl^- uptake was doubled. Both Na^+ and Cl^- ions are actively transported across the intestine. Studies on ion movement across intestinal sacs between identical Ringer solutions also revealed a significant uptake of Na^+ and Cl^- accompanied by fluid movement. An increase in the $HCO_3{}^-$ concentration in the lumen is observed suggesting a $Cl^-/HCO_3{}^-$ exchange. The presence of $HCO_3{}^-$ in the serosal bathing solution is essential for optimal transmucosal transport of the Cl^- ion. Acetazolamide inhibits the absorption of fluid, ion transport and $HCO_3{}^-$ secretion into the lumen, suggesting that carbonic anhydrase plays a role in the transport of anions across the fish intestine, as it does in mammalian intestine. The problem of the interrelationships of Cl^-, $HCO_3{}^-$, Na^+ and H^+ transport in the human ileum has been discussed recently in the light of a simultaneous double exchange—$Cl^-/HCO_3{}^-$ and Na^+/H^+—model, Turnberg *et al.*, 1970). If Cl⁻ movement exceeds Na⁺ movement, HCO_3^- accumulates in the lumen and the electric potential is serosal negative. Acetazolamide simultaneously blocks Na⁺, Cl⁻ and fluid transport. Such a model may also apply to the flounder intestine.

More recently Ando and Utida (in Utida *et al.*, 1972) made a preliminary survey of the potential differences in various marine and freshwater fish intestines. In four out of five marine species, a serosal negative potential was found. In the fifth the potential was practically nil, the same as in *Cottus scorpius* studied by House and Green (1965). In four out of five freshwater species a serosal positive potential was observed, as in *Carassius auratus* studied by Smith (1964). In *Anguilla japonica* adapted to fresh water, the potential, as well as the short circuit current, was slightly negative They remained negative but increased markedly in the posterior intestine upon adaptation to sea water, while the electrical resistance decreased significantly. The presence of a Cl⁻ pump in the seawater eel is further confirmed by ion-flux studies with isotopes on the short-circuited preparation. In the gut of the freshwater eel Na⁺ and Cl⁻ ions were both actively transported and had identical net fluxes, though the influx (mucosa→serosa) and efflux were greater for Na⁺ than for Cl⁻. In the gut of the seawater eels, unidirectional fluxes of both ions increased markedly but only the net absorption flux of Cl⁻ increased in comparison to the freshwater eels. These recent results seem to be in contradiction to earlier reports from Utida's school (Oide, 1967) showing that net Na⁺ absorption increases by 100% when isolated intestinal sacs from freshwater eels are compared with those from seawater eels (see Table X). "Open circuit" intestines were used in the earlier studies. It is probable that the increased negative potential which accompanies seawater adaptation favours Na⁺ absorption. Thus the discrepancy between earlier and recent results may be explained. "Closed circuit" experiments clearly showed that both Na⁺ and Cl⁻ are actively transported across the eel intestine, thus confirming the reports on the flounder and *Cottus scorpius*.

The results obtained with the eel, a euryhaline fish, strongly suggest that salt absorption is much higher in seawater adapted fish than in freshwater adapted ones. That the gut displays adaptive functional features in relation to external salinity changes is well illustrated by the results obtained *in vivo* by Skadhauge (1969) with the European eel. This author used an ingenious intraluminal perfusion technique, and compared the salt adaptation rates of freshwater, seawater and double-strength seawater-adapted eels. When dilute seawater is used as the perfusion fluid, the net absorption rate of Na⁺ and Cl⁻ ions was more or less independent of the NaCl concentration, over the range 90–300 mM. When eels were trans-

ferred from freshwater to seawater this rate doubled. A further augmentation was observed after transfer to double strength seawater. The net
Cl^- absorption rate was only slightly higher than that of Na^+. The ratio
of influx to efflux remained constant irrespective of the adaptation medium.
Table X gives the values obtained, showing conclusively that the salt
pumping efficiency increases in relation to osmotic stress. This table also
illustrates the fact that the rates of absorption are 4–5 times faster *in vivo*
than *in vitro*, although the experimental temperatures were identical in
both types of experiments.

These results, from experiments on a euryhaline fish, contrast with
those obtained with the stenohaline goldfish. Ellory *et al.* (1972) showed
that adaptation to a saline environment reduced the transflux of Na^+
to one quarter the value found for freshwater fish. The authors demonstrated that adaptation to saline involves a decrease in the Na^+ permeability
of the microvillar membrane of the mucosal cell, rather than a change in
the pump activity located on the basal or intercellular membrane. Lahlou
et al. (1969) observed that adaptation to a saline environment is accompanied by an increased drinking rate as discussed above. The adaptive
regulation of Na^+ movement described by Ellory *et al.* (*loc. cit.*) partially
offsets the increased salt load incurred by the fish.

During migration from seawater to freshwater, the river lamprey
Lampetra fluviatilis has been shown to lose the drinking reflex and the
intestinal salt absorption mechanism which characterizes life in seawater
(Morris, 1958; Pickering and Morris, 1970; Pickering and Dockray,
1972). "Fresh-run" river lampreys may be readapted to 50% seawater
and retain salt balance for a certain time. Pickering and Morris (1973) using
isolated intestinal preparations from such animals observed a large active
Na^+ ion and water absorption flux in the anterior intestine. Its magnitude
decreased towards the posterior end of the intestine. Dinitrophenol and
ouabain blocked this uptake mechanism. In the posterior region a net
excretion flux of Ca^{2+} and SO_4^{2-} into the lumen was recorded, the rate of
which was at least 200 times smaller than that of the monovalent ion flux
in the anterior region. No comparison was made with isolated intestines
from fully freshwater adapted animals. Pickering and Dockray (1972)
calculated and compared the amount of Na^+ and Cl^- swallowed and
absorbed by fresh-run and long term freshwater-adapted lampreys
replaced in 50% seawater for 24 h. The rate of Na^+ and Cl^- absorption
was 28–30 $\mu M/100g/day$ in the fresh-run animals, and 2 μM in the
typical freshwater animals.

During the transition from freshwater to seawater, the kinetics of adaptive readjustments of the salt uptake were studied in the isolated intestines
of the Japanese eel by Oide and Utida (1967). Figure 13 shows that there

TABLE X: Intestinal absorption mechanisms of the eel in relation to external salinity

Medium	Ionic fluxes					Water transport				
	Na		Cl		T.P.	P_{os}	Solute linked water flow		Mole ratio H_2O/Na	
	F_{net}	F_{in}/F_{out}	F_{net}	F_{in}/F_{out}			At T.P.*	At iso	At T.P.*	At iso
In vivo studies (Skadhauge, 1969)										
FW	16·6±1·7	2·60±0·36	20·5±2·4	2·74±0·20	73±3	0·37±0·05	26	65	39	99
SW	36·3±3·3	2·70±0·16	42·3±3·7	2·95±0·24	126±5	0·72±0·10	90·5	162	64	114
200%SW	64·0±11·0	2·95±0·23	67·6±14·9	—	244±32	—	—	—	—	—
In vitro studies (Oide, 1967)										
FW	3·6±0·49	—	—	—	—	—	17·8±2·9	—	—	270
SW	7·0±0·52	—	—	—	—	—	41·6±3·3	—	—	331

Fluxes in μ-equiv h^{-1}cm^{-2} (at 20°C). F_{in}, F_{out}, F_{net}: influx (mucosa serosa), outflux and net flux.
Osmotic permeability (P_{os}) in μl h^{-1} m-osmoles^{-1} cm^{-2}.
Mole ratio in moles of water/mole Na$^+$ absorbed.
Water flow in μl. h^{-1}cm^{-2} at T.P. or at isotonicity (iso).
T.P.: turning-point osmotic gradient (in m-osmoles) corresponding to zero-net water flow.
FW, Freshwater; SW, salt water; 200%SW, double strength seawater.
* (according to the double flow hypothesis of Skadhauge, 1969).

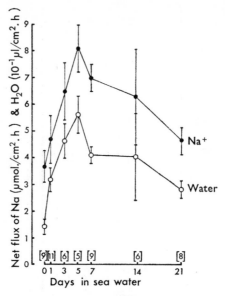

FIG. 13. Changes in water and Na$^+$ net flux (mucosa → serosa) in isolated intestines
of cultured eels after transfer to seawater.
Ordinates: flux of Na$^+$ (μmoles/cm^{2}h) or water ($10^{-1}\mu$l/cm^{2}h).
Abscissa: days in seawater.
In parentheses: number of animals tested.
(according to Oide and Utida, 1967).

was an increase in the salt absorption capacity with a maximum occurring
on the 5th day after transfer. During the course of seawater adaptation
the water absorption capacity parallels the sodium movement.

c. *Water movements across the intestine*

Balance sheet studies suggest that marine teleosts absorb water from
swallowed seawater, after dilution of this medium in the stomach and later
in the intestine. The common view is that an osmotic influx of water from
blood to lumen rapidly brings the mucosal solution to isotonicity with the
blood, and then water is absorbed by some mechanism dependent on the
uptake of salt into the blood.

Various authors have tested this hypothesis on *in vitro* preparations of
teleost gut. Sharratt *et al.* (1964) were the first to observe that intestinal sacs
from European eels, filled with seawater and placed in Ringer solutions,
rapidly gained weight. House and Green (1965) studying the weight change
in intestinal sacs from seawater-adapted *Cottus*, filled with seawater and
placed in Ringer solutions, found that in five out of six preparations a

weight gain was followed by a weight loss within 1–3 h. Similar experiments were performed by Utida et al. (1967a) who suspended intestinal sacs from the Japanese eel in Ringer solution. When the lumen was filled with hypertonic saline (2/3 seawater and seawater), water first entered passively but a "turning point" corresponding to a reversal of the water flow was observed. This reversal was more rapid in 2/3 seawater than in seawater, while absorption was observed immediately in isotonic 1/3 seawater. These authors also compared intestines taken from freshwater and seawater adapted eels. The "turning point" occurred earlier in seawater preparations in which the movements of water were also swifter in both directions. Thus the increased ability to absorb salt from the intestines taken from seawater eels, is paralleled by an increased ability to absorb water. This is illustrated in Table X and in Fig. 13. These observations strongly suggest that water movement is secondary to ion absorption, a hypothesis confirmed by the following experiments. Addition of a metabolic poison such as cyanide halts the solute and solvent movements between identical Ringer solutions across intestinal sacs of Cottus (House and Green, 1965). Acetazolamide, which partially inhibits Na^+ and Cl^- uptake across the intestine of the flounder, reduces accordingly the water absorption rate across sacs of flounder intestine (Huang and Chen, 1971). Cyanide, iodoacetate, dinitrophenol and ouabain also partially block solvent and solute movements across the intestines of seawater and freshwater eels (Oide, 1967).

Several observations suggest that, although water movement is secondary to ion absorption, the molar ratio H_2O/Na^+ of the absorbate may change in relation to the physiological needs of the animal. Thus Oide (1967) comparing intestinal sacs from freshwater and seawater eels noted that, while the rate of Na^+ absorption from a Ringer solution doubled after salt water adaptation, the water absorption rate increased 2.5 to 3 fold. Hence the ratio H_2O/Na^+ increased by 25 to 50% on seawater adaptation, a phenomenon which is of physiological significance. Oide and Utida (1967) investigated the kinetics of the functional readjustment of the gut after transfer of the eel from freshwater to seawater. They noted that both Na^+ and water absorption rates pass through a maximum at the fifth day. The H_2O/Na^+ ratio is also maximal, about 80% higher than in freshwater.

This change in the relative amounts of water and salt absorbed simultaneously, indicates that probably the intestinal osmotic permeability varies in relation to the external salinity. Sharratt et al. (1964) already suggested such variations, because the osmotic inflow of water into intestinal eel sacs filled with seawater and placed in Ringer solutions was faster in seawater than in freshwater eels.

The most complete information on the water absorption mechanism in the eel intestine was obtained by Skadhauge (1969) with the help of his *in vivo* intraluminal perfusion technique, which allowed for better controlled thermodynamic conditions. The rate of water movement was measured by the help of a "water marker" (phenol red) which is practically impermeant. When 1/2 or 2/3 seawater was used as the perfusing medium, water first entered the gut along the osmotic gradient. A reversal of flow was soon observed, however, being seen as a decrease, followed by an increase in the dye concentration. At the "turning-point" (corresponding to zero net water flow) of the phenol red concentration curve, the "turning point osmolality" of the perfusion fluid was higher than the plasma osmolality by 73 mOsmoles in freshwater animals, 126 mOsmoles in seawater fishes and 244 mOsmoles in double-strength seawater eels. Thus the osmotic gradient against which water absorption becomes possible, increases with adaptation to higher salinities, a situation of obvious physiological significance. A remarkably constant ratio between the NaCl pumping rate and the osmotic gradient corresponding to zero net flow of water was found, suggesting that the two phenomena are closely related. Perfusion experiments with impermeant solutes show that, in the absence of NaCl, water movement is essentially a passive process. From these experiments the passive intestinal osmotic permeability was estimated. The permeability was higher in seawater-adapted eels than in freshwater animals. "Rectification" of flow is suggested by the fact that the serosal to mucosa permeability was less than that in the opposite direction. These observations support the view that "solute-linked water flow" occurs in the gut of fish, as in other epithelia such as the gall-bladder. This flow is secondary to the salt movement and results from the osmotic pressure produced by local accumulation of salt within the membrane (see introductory remarks). The model of 'local osmosis' seems adequate to explain water transport in the absence of, or even against osmotic gradients, and also clarifies the mechanism of functional adaptation displayed by the gut of the eel. If the solute-linked water flow varies proportionally with the salt transport rate and the osmotic permeability, then the increased molar ratio H_2O/Na^+ of the transported fluid can be explained on a quantitative basis. In any case the fluid reabsorbed is hyperosmotic to the body fluids although less so in seawater-adapted animals. It must be concluded that fluid absorption from the gut is effected at the expense of the mineral balance of the body as a whole and excess salt has to be excreted elsewhere.

3. CELLULAR AND BIOCHEMICAL ASPECTS OF INTESTINAL FUNCTION

Surprisingly, there is almost no information concerning the morphological aspects of the function of the intestines in fish in relation to salinity adaptation. Hirano (1967) observed that the intestinal wall seems thinner and the surface area smaller in the seawater-adapted eels. Hence it does not seem possible that the increased rate of water and salt absorption could be the result of an augmentation of the available surface. More recently, Pickering and Morris (1973) studied the anterior intestine of fresh-run lampreys with light and electron microscopic techniques. They suggest that of the three cell-types occuring in this region, the columnar cell is most probably the site of salt and water transfer. This cell is characterized by a brush border and by a tubular, branching, smooth-membraned endoplasmic reticulum that is well developed in the basal part of the cell. Large elongated mitochondria occur in this region. The columnar cell rests on a convoluted basement membrane and the lateral cell membranes are often folded in a complex manner. Occasional large intercellular spaces are found. Such a cell corresponds to the 'forward-type' membrane of Diamond (1971) with the intercellular spaces or intracellular canaliculi serving as the sites of the standing gradient of fluid transport.

Differences in the enzymic content of the intestinal epithelium in relation to the adaptive behaviour of the gut have been recorded. Two enzymes have been found to have higher activities in seawater: $Na^+ + K^+$ dependent ATPase (Oide, 1967; Jampol and Epstein, 1970) and alkaline phosphatase (Utida, et al., 1968). The role of the former enzyme in salt transport has been established by experiments demonstrating the inhibitory effects of ouabain as discussed in the preceding section. As to the role of alkaline phosphatase, it is of interest that NaCl activation, which characterizes this enzyme, is far more important in extracts from seawater-adapted fish than from the freshwater living species (Utida et al., loc.cit.). Oide (1970) prepared highly purified enzyme extracts from the intestinal mucosa in order to verify whether seawater adaptation induces the synthesis of a new type of alkaline phosphatase. She found that there was little difference between the properties of freshwater and seawater enzyme preparations. Upon adaptation to seawater, the specific activity of this enzyme is increased three-fold without any change in its affinity for the substrate. Recent experiments by Mrs. Oide (S. Utida, personal communication.) indicate that, in intestinal sacs from both freshwater and seawater eels, water absorption from mucosa to serosa is increased five fold when the luminal pH is increased from 7.2 to 9, presumably as a result of increased phosphatase activity. Furthermore, inhibitors of this enzyme, such as EDTA, borate or cysteine, introduced in the lumen depress water transport by 70 to 80%. It is clear that salt and water movement across the intestine are mediated by both $Na^+ + K^+$ dependent ATPase and alkaline phosphatase. The fact that ouabain is more active when added to the serosal side and EDTA is more potent when added to the luminal side suggest different cellular localization of the enzymes.

Finally a third enzyme, carbonic anhydrase, may be of functional significance, as suggested by the results of Huang and Chen (1971) reported above.

4. CONTROL OF INTESTINAL FUNCTION: ROLE OF ENDOCRINES

Utida and his colleagues have demonstrated that the hypophysial-interrenal axis is responsible for the adaptive changes of the intestinal tract induced by salt water absorption. Hirano *et al.* (1967), Hirano (1967) and Hirano and Utida (1968) showed that, after removal of the hypophysis, there is no longer the enhancement of water absorption after transfer into seawater. Among all the hormones tested for replacement therapy only ACTH and cortisol proved capable of restoring normal water movement when administered at low doses. Cortisol also proved effective in augmenting the rate of Na^+ absorption. More recent investigations by Utida *et al.* (1972) suggest that, independently of the salt pumping efficiency, water permeability is increased by cortisol, since the molar ratio H_2O/Na^+ of the fluid passing across the intestine nearly doubles after cortisol injection into a freshwater eel, the ratio becoming almost identical to that recorded in seawater-adapted fish. Cortisol also increases the serosal negative P.D. and the short circuit current of the isolated gut preparation.

Although mammalian prolactin has no effect on intestinal water absorption when injected into freshwater eels (Hirano and Utida, 1968), injection into seawater eels depresses the rate of salt and water absorption measured in the isolated intestine. The molar ratio H_2O/Na^+ or Cl^- is reduced which suggests that prolactin decreases both ion pumping and water permeability (Utida *et al.*, 1972). Thus prolactin acts as an antagonist to cortisol on the intestine.

Recent investigations on the level of circulating cortisol in the eel demonstrate that salt water adaptation is accompanied by a transitory increase in the hormonal concentration, suggesting a temporary activation of the adreno-corticotrophic function of the hypophysis. Hirano and Utida (1971) report a short-term increase 2–4 h after transfer to salt water in the Japanese eels. Ball *et al.* (1971) studying the European eel and Forrest *et al.* (1973*b*) studying the American eel observed a later response (1–5 days), with a maximal level 2 days after transfer. Thus, the observed rise of the cortisol level may well trigger off the adaptive changes exhibited by the intestinal tract.

There are, however, additional unsolved problems. For example, all three groups of investigators agree that the plasma cortisol level is the same in freshwater- and in seawater-adapted animals. Since the gut nevertheless maintains a much higher rate of water movement and salt absorption in long term adapted seawater animals than in freshwater animals, the factors involved in long-term adaptation remain to be defined. Non-correlated changes in plasma cortisol levels and intestinal absorption rates are also reported by Hirano and Utida (1971). Thus after treatment of freshwater eels with small doses of ACTH and cortisol, the plasma cortisol concentration increased immediately and returned to its initial level within 24 h. An increase in intestinal water movement was observed after a delay of 10 h and lasted for at least a week. Such a latent period in the response to cortisol suggests induction of some metabolic process underlying the increased rates of salt and water movement.

Pickford *et al.* (1970*b*) advance the possibility of an increase in the specific activity of the $Na^+ + K^+$ dependent ATPase observed in the intestinal mucosa. The augmentation found after cortisol treatment in hypophysectomized, seawater adapted *Fundulus heteroclitus* is, however, modest. Nor do the authors give any indication of the effects of hypophysectomy on the enzyme level. Further investigations are warranted concerning the role of the hypophysis in the modulation of the $Na^+ + K^+$ dependent ATPase activity since, in a later report, Pickford *et al.* (1970*a*) show that, within 3 days of transfer into freshwater, the enzyme activity decreases by 50% in hypophysectomized *Fundulus.*

A few studies have been made on endocrine control of the intestinal function in freshwater-adapted fish. They are interesting to compare with the above observations on seawater adaptation. Gaitskell and Chester Jones (1970) have reported that cortisol is required for maintaining water absorption from the gut of eels in freshwater. Adrenalectomy produced a significant drop in the rate of absorption which was restored to normal values by injection of cortisol. That the hypophysial-interrenal axis may also regulate salt uptake in the gut of stenohaline freshwater teleosts is suggested by a recent report by Ellory *et al.* (1972). In the goldfish, hypophysectomy reduced the Na^+ transepithelial flux, leaving the Na^+ influx across the mucosal barrier unchanged. Injection of cortisol restored the Na^+ transflux to the control level without interfering with the permeability of the mucosal barrier to Na^+. The authors suggest that cortisol is concerned with regulation of the Na^+ pump located at the serosal or intercellular barrier of the epithelial cell. It is not clear, however, how cortisol exerts its effect. A metabolic control by way of the delivery of energy to the pump or a functional control by way of altering the number of pumping sites via enzyme synthesis may be envisaged. In any case, adaptation to saline, which involves regulation of the Na^+ permeability of the mucosal barrier, is certainly effected by a feed back mechanism which does not include the hypophysial-interrenal system in this stenohaline species, unless a second unknown mineralocorticoid hormone intervenes in the process. Thus the functional adaptation of the intestinal tract in response to a salt load is quite different in the goldfish in comparison to that of the euryhaline eel.

A recent report suggests the possibility that other corticoid hormones intervene in the regulation of gut function. Pickering and Dockray (1972) recently showed that gonadectomy inhibits the breakdown of marine osmoregulation which accompanies anadromic migration in *Lampetra fluviatilis*. In intact and sham-gonadectomized lampreys maintained for 2 months in freshwater and returned to 50% seawater for 24 h, the drinking and intestinal salt and water absorption rates are 10 to 20 times lower than those observed in "fresh-run" lampreys returned to 50% seawater for 24 h. Gonadectomized animals, although adapted for 2 months to freshwater, behave as "fresh-run" lampreys. Thus, maturation of the gonads reduces the capacity to transport the monovalent ions in the intestine, but the nature of the factors involved remain to be elucidated. It would be worthwhile to investigate whether, during anadromic migration of the salmon, maturation of the gonads also interferes with gut function. In the lamprey the intestine degenerates during upstream migration and gonadectomy prevents this atrophy (Dockray and Pickering, unpublished).

When catadromic migration occurs in relation to the reproductive cycle, an atrophy of the gut also occurs. Such a situation holds in the silver eel (*in* Kirsch, 1971). Most of the above reports on the regulation of gut function concern cultured eels not ready for migration. The observations by Kirsch (1971) and Kirsch and Mayer Gostan (1973), that the drinking rate in long term adapted silver eels becomes very small, poses intriguing problems which suggest that the model given by Smith (1930) for teleostean osmoregulation may have to be revised (see Motais and Garcia-Romeu, 1972).

5. COMPARATIVE ASPECTS: THE ROLE OF THE GUT IN INVERTEBRATES

In contrast to the wealth of information concerning fish, the facts on the role of the gut are very scanty and rather controversial for invertebrates.

Table IV (pp 66–68) shows that the drinking rates are of similar magnitude in invertebrates and vertebrates. Dall (1967) argues, however, that identical drinking rates are found in the penaeid shrimp, *Metapenaeus bennettae*, with a serum osmolality about 60% of that of seawater, and in various crabs which are only slightly hypo-osmotic and have little need to gain water to offset osmotic dehydration. The rate of dehydration is, however, not only a function of the gradient, but also of the permeability coefficient of the body surfaces which is often higher in iso-osmotic animals (section IIB). It may, nevertheless, be seen that the iso-osmotic *Limulus polyphemus* exhibits a rate of water ingestion in the same range as that of the above-mentioned species (Hannan and Evans, 1973). In addition, *Metapenaeus* kept in iso-osmotic, 66% seawater still drinks $150\mu l/100g/h$. These considerations led Dall (1967) to conclude that at least part, if not all of the water ingested is serving a function other than osmotic regulation. Dall suggests that it may serve to remove salt excreted into the gut, or else divalent ions such as Ca^{2+} (Dall, 1965) or even monovalent ions (Dall, 1967). One set of data in favour of this hypothesis is the analysis of the concentration of Na^+ or Cl^- in various gut sections compared with that of blood. The Na^+ concentration was higher in the anterior diverticulum and the Cl^- level higher in the proventriculus, the anterior diverticulum and the mid-gut. Unfortunately comparison of the specific radioactivities of Na^+ or Cl^- in these various sections and in the blood, after injection of tracer, did not yield convincing results. A second set of data concerning the Na^+ and Cl^- turnover fluxes with the anterior (branchial) and abdominal parts of the body separated by a partition were, however, more significant. In prawns, only one tenth of the Na^+ efflux and one-third of the Cl^- efflux occur in the abdominal region presumably *via* intestinal excretion. The main efflux occurs in the cephalothoracic region, presumably across the gills. After injection of a salt load, however, a 100% increase of the Cl^- efflux and a 30% increase of the Na^+ efflux occurs in the abdominal region, while Cl^- and Na^+ effluxes are depressed by 14 and 50% respectively in the anterior region. Thus one may suspect that, in *Metapenaeus*, the gut participates in the salt balance by excreting surplus salt.

The results obtained by Dall (1967) do not agree with the observations made by Croghan (1958*a-c*) on *Artemia salina*, the brine shrimp. Croghan compared the osmolality of the Na^+ and Cl^- concentrations of the gut contents, the haemolymph and the external medium, in animals adapted to a range from 45mM to 5400mM external NaCl concentration. In the lower range, the blood is hyperosmotic to the external medium by as much as 90 mOsmoles, while in the higher range the blood is hypo-osmotic by almost 10 000 mOsmoles. Gut fluid is hyperosmotic to blood at all the ranges of external salinities by about 400–500 mOsmoles. It is hyperosmotic to the external medium at external NaCl concentrations up to 390 mM, and hypo-osmotic above this value. The Na^+ and Cl^- concentrations were measured in the gut and found to be much lower than those present in the blood. Since phenol red added to the external bath is ingested and strongly concentrated in the gut fluid, Croghan concludes that, in the brine shrimp, water is absorbed along with Na^+ and Cl^-. The high osmotic pressure is probably caused by the divalent ions, which are absorbed at much slower rate and left behind. A fine precipitate, tentatively identified as $CaSO_4$, was observed in the lumen.

When compared with teleosts, there are obvious analogies in *Artemia* but also differences which require further investigation. For instance, absorption of water must occur against an osmotic gradient (400 mOsmoles) far steeper than that encountered by Skadhauge (1969) in the eel adapted to 200% seawater (245 mOsmoles). Studies on fish withstanding higher external salinities such as *Aphanius dispar*, which thrives in 400% seawater would be of interest. Unfortunately the recent report by Lotan and Skadhauge (1972) was not concerned with the 'turning point osmolality' in this species. Furthermore, the gut content is hyperosmotic to the external medium when *Artemia* is hyperosmoregulating. The Na^+ concentration of the gut fluid is comparable to that of the haemolymph, whereas the Cl^- concentration is about half. Obviously in such conditions the gut serves another function, which Dall (1967) suggested was probably excretory. Drinking rate measurements and gut fluid analysis for bivalent ions over the whole range of external salinities are clearly needed. Comparative studies of gut function in crustacean iso-osmotic regulators would also be of interest.

In the salt-water species of insect larvae "there can be no doubt that the gut plays an important part in the uptake of ions" according to Shaw and Stobbart (1963). "Drinking of the saline medium normally occurs and the gut is the major route of salt entry". The best studied group is the caddis larvae. In the near iso-osmotic *Limnephilus affinis* a small quantity of water is drunk (125–300µl/100g/h) (Sutcliffe, 1962), whereas in the hypo-osmotic regulating *Philanisus plebeius* the drinking rate is higher (1000µl) (Leader, 1972). The fate of the salt water in the gut has only been studied in *Philanisus*. Here the mid-gut fluid is hyperosmotic to the haemolymph by about 50–75 mOsmoles for external concentrations ranging from 60 to 600 mM NaCl. At 900 mM NaCl external salinity, the difference between the mid-gut fluid and the haemolymph attains 275 mOsmoles. This is a much less steep concentration gradient than that found in *Artemia salina* by Croghan. Clearly, water absorption along with electrolytes takes place in such conditions. The fact that such an uphill water transport really occurs is shown by the use of amaranth as a water marker. Leader (*loc.cit.*) observed that this dye concentrates in the gut. In addition, larvae prevented from drinking by ligaturing the mouth lose weight, and the osmotic pressure and Cl^- concentration of the body fluid rise. Unfortunately the Na^+ and Cl^- concentrations of the mid-gut contents were not analysed, but it is highly probable that these electrolytes are absorbed as in other insect larvae. The fate of the bivalent ions swallowed is also not known.

The dipterous larvae, for example *Aëdes detritus*, kept in hypersaline solutions also drink water, and according to Beadle (1939) the major if not the only uptake of salts occurs via the mouth and gut wall. Unfortunately he did not measure the drinking rate. Ramsay (1950) showed that the mid-

gut fluid is only slightly hyperosmotic to the hemolymph: osmotic equilibrium appears, therefore, to take place in the gut. In the mid-gut caeca,
where presumably water absorption takes place, the fluid is somewhat more
concentrated. Thus, in the mosquito, as in the caddis larvae, absorption of
water must occur against a chemical gradient for water. Again, no information is available concerning the monovalent electrolyte concentration in the
gut fluid, but it is highly probable that water uptake is secondary to Na^+
and Cl^- absorption. Stobbart (1971c) compared body volume control in the
freshwater *Aëdes aegypti* and saltwater *A. detritus*. The latter differs
markedly from the former in that the rate of urine production is not
dictated by the rate of osmotic inflow of water. On the contrary, the rate
of drinking and water absorption by the gut is governed by the rate of
urine production. Larvae with their mouths sealed continue to produce
urine and, consequently, shrink. Under these conditions they continue to
excrete urine as concentrated as possible, and maintain their osmotic
pressure below that of the medium at the expense of a continually reduced
haemolymph volume. This process occurs in nature when the larvae are
in brine at a high concentration beyond the presumptive absorption
capacity of the mid-gut, in which case they cease to drink until the medium
becomes diluted again.

B. Salt secretion across the gills or other extra-renal effector organs

In teleosts, whether adapted to freshwater or to seawater, the gill is the
most important organ for the maintenance of the salt balance. In a hyper-
osmotic environment, its function is concerned with the excretion of Na^+
and Cl^- in order to compensate for the intestinal absorption of these
electrolytes which is linked with the "uphill" transport of water discussed
in the previous section.

1. EVIDENCE FOR SALT EXCRETION

Salt secretion was found to be localized in the head region of the eel by
Keys (1931a, b) with the help of his perfused heart-gill preparation. The
demonstration of a net excretion of Cl^- through the gills into the buccal
cavity depended entirely upon the great precision of the measurement of
Cl^- concentrations. The rate of Cl^- extrusion was governed by its
concentration in the perfusion medium, the maximal rate observed being
about 100 $\mu Eq/100g/h$. Keys thought that the excretion of Cl^- was
to some degree associated with water movement. As the external medium
was sea water, water moved along the osmotic gradient and the volume of
fluid bathing the gills increased during the experiment. Schlieper (1933a, b)

confirmed a net Cl⁻ excretion rate between identical Ringer solutions and showed that no movement of water occured under these conditions.

For almost 40 years there has been no other attempt to demonstrate directly the salt excreting activity of the gills of seawater fish. Kirschner (1969) made a survey of the difficulties encountered in developing a heart-gill preparation. The latter deteriorates rapidly, especially if the heart is replaced by an artificial cardiac pump.

Only two techniques have been described using isolated gills. The incubated gill preparation proposed by Bellamy (1961) and used by Kamiya (1967) and Utida *et al.* (1967*b*) suffers the disadvantage of not being perfused. When kept in seawater exosmosis of water must progressively increase electrolyte concentrations in the internal medium, which may interfere with epithelial function. This may explain why a zero net flux of Na⁺ and Cl⁻ is obtained in such conditions. The perfused gill preparations are of two types, the first of which is the constant pressure type used by Richards and Fromm (1970) on freshwater rainbow trout, and by Rankin and Maetz (1971) for the study of the vascular action of hormones. This preparation was used also in seawater and freshwater eel gills. The eel preparation was found by us to be rather leaky and unsuitable for efflux and net excretion rate measurements. The second type of preparation involves the constant flow technique developed by Shuttelworth (1972) and results in perfused gills with practically no leaks. On such preparations a net excretion rate of Na⁺ or Cl⁻ amounting to 10 μEq/100g/h animal was obtained with seawater gills. This rate, which is 5 to 10 times less than that expected from *in vivo* flux measurements (Maetz, 1971; Kirsch and Mayer-Gostan, 1973) agrees, however, with that obtained by Keys (1931*b*) using his heart-gill preparation.

2. SODIUM AND CHLORIDE EXCHANGES AND THE ELECTROCHEMICAL GRADIENT

The most complete information has been obtained in the seawater-adapted eel and this will be discussed in detail first. Figure 14 illustrates schematically the comparative Na⁺ exchanges in the freshwater and the seawater-adapted eel, and the various routs of exchange. Table XI also records the values for the Cl⁻ and K⁺ exchanges and the flux ratios. All these values were obtained from small sized animals that had been studied. The outfluxes were measured directly after an intraperitoneal injection of tracer. The influxes were calculated from the outflux values minus the net excretion rates deduced from the independently measured drinking rates. The urinary excretion rates of Na⁺, Cl⁻ and K⁺ may be considered to be negligible. The fluxes across the gill given in Fig. 14 account for a turnover rate of about 30% of the exchangeable Na⁺ per hour found by Mayer and Maetz (1967) in handled animals. More recently

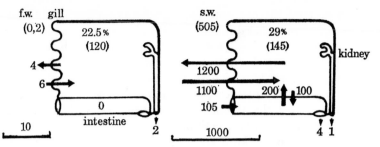

FIG. 14. Comparative Na$^+$ exchanges in fresh water (FW) and sea water (SW). External and internal Na$^+$ concentrations are given in brackets. Internal Na$^+$ space is given in % body weight. Fluxes are given in μEquiv/100g/h. Note different scales. Only net Na$^+$ loss by the kidney is given (according to Maetz, 1971).

Kirsch (1972a) and Kirsch and Mayer-Gostan (1973) have published values for Cl$^-$ effluxes of about 230–250 μEq/100g/h in unstressed eels, with a drinking rate corresponding to a net flux of about 75 μEq. The J_{in}/J_{out} value is about 0.7. More recent evaluations of the drinking rate yield an even smaller net flux. It is probable that for Na$^+$, as for Cl$^-$, the unidirectional fluxes are smaller in unstressed animals.

Gill potential measurements, obtained by inserting a microelectrode into the filamental afferent blood vessel in anaesthetized eels, yielded values of $+ 18 \pm 2.9$ mV (positive inside) according to Maetz and Campanini (1966) and of $+ 22.5 \pm 2.7$ according to House and Maetz (1974).

In Table XI the experimental flux ratios obtained on stressed fish are compared with the theoretical ratios calculated from Ussing's relation,

TABLE XI: Experimental and theoretical flux across the gill of the seawater adapted eel.

Ion species	C_{out}	C_{in}	J_{out}	J_{in}	J_{in}/J_{out} exp.	J_{in}/J_{out} theor.
	(μequiv/ml)		(μequiv/100g/h)			
Na$^+$	505	145	1310[a]	1200[b]	0·92	1·42
Cl$^-$	600	140	1150[c]	1030[b]	0·90	10·56
K$^+$	10·5	3·5	55[d]	53[b]	0·96	1·22

a. Maetz *et al.* (1967b).
b. J_{in} determined from ($J_{out}-J_{net}$), where the net efflux, J_{net}, is taken to be equal to the drinking rate (200μl/100g/h; Maetz, 1970 and Motais and Isaia, 1972a) multiplied by the concentration of Na$^+$, Cl$^-$ and K$^+$ in seawater.
c. Epstein *et al.* (1973).
d. Maetz, unpublished.
exp.: experimental; theor.: theoretical.

taking into account the concentrations gradients for Na$^+$, Cl$^-$ and K$^+$ and the recorded potential difference. All the theoretical values are larger than the experimental ratios. These disparities mean that all three electrolytes are actively pumped out across the gill epithelium. The evidence is particularly clear for Cl$^-$ which is pumped out against both a concentration and an electrical gradient, but is more debatable for Na$^+$ and for K$^+$ in view of the fact that the flux ratios and the potentials were obtained under different experimental conditions.

The fact that the body fluids are positively charged with respect to the external medium raises the possibility that the potential may result from an active electrogenic efflux of Cl$^-$ ions. Alternatively it may correspond to a Nernst potential, resulting from a high permeability to cations such as Na$^+$ or K$^+$ which would tend to diffuse down their concentration gradient.

Values of the Na$^+$ and Cl$^-$ exchange fluxes obtained in other species of teleosts have been collected in Table XII for comparison with the eel. The flux values are extremely variable as a result of differences in experimental procedure and the size of fish. The effects of temperature have been studied in detail by Maetz and Evans (1972) on the flounder. The effects of handling have been discussed by Mayer and Maetz (1967), Evans (1969b), Kirsch (1972a) and Dharmamba et al. (1973). Variations of exchange fluxes in relation to the size of fish may be appreciated from the data concerning *Mugil capito* (Pic et al., 1973) and *Tilapia mossambica* (compare Dharmamba et al., 1973 and Potts et al., 1967). As expected, the fluxes vary with external salinity, a point which will be discussed below. The Cl$^-$ turnover is faster than the Na$^+$ turnover in some species such as the salmon trout (Potts et al., 1970), and slower in other such as *Pholis gunnellus* (Evans, 1969a) or *Platichthys flesus* (Motais, 1967).

Potential measurements have only been made in two other species, *Blennius pholis* (House, 1963) and *Pholis gunnellus* (Evans, 1969a). The values given for these fish are practically identical to those obtained with the eel.

3. THE CHLORIDE PUMP AND CHLORIDE ION EXCHANGES

Until recently very little was known about the mechanism responsible for the extrusion of Cl$^-$ across the gill. Because SCN$^-$ inhibits active transport of halide in other tissues (thyroid, stomach and cornea), its effect on Cl$^-$ efflux was examined in *Anguilla* by Epstein et al. (1973). An intraperitoneal injection, sufficient to give a 7 mMolar concentration in the body fluids, produced a 65% inhibition of the Cl$^-$ efflux within 2 to 6 minutes as illustrated in Fig. 15. Simultaneously the net excretion of Cl$^-$ ($-120\,\mu$Eq/100g/h) is replaced by a net gain ($+300\,\mu$Eq.) as shown by the increase in plasma Cl$^-$.

TABLE XXX. Electrolyte exchanges across the body surfaces of hypo-osmotic regulators.

Species	Body weight	Temp (°C)	External medium	Electrolyte	J_{out} (μequiv/100g/h)	J_{in} (μequiv/100g/h)	References
CRUSTACEANS							
Artemia salina	5–6mg	25	SW	Na	4250	2650	Thuet et al. (1968)
			SW	Cl	7500	5650	
Artemia salina	5–6mg	22·5	SW	Na	6800	5800	Smith (1969a, b)
			SW	Cl	7000	5800	
Palaemonetes varians	?	15	SW	Na	15 000	—	Potts and Parry (1964b)
			65%SW	Na	7350	—	
			2%SW	Na	1800	—	
TELEOSTS							
Poecilia latipinna	?	?	SW	Na	3720	4760	Ensor and Ball (1972)
			33%SW	Na	810	720	
Pholis gunnellus	1–2g	10	SW	Na	5220	3640	Evans (1969a)
			20%SW	Na	755	560	
Salmo salar (smolts)	20–45g	10	SW	Cl	4180	420	Potts et al. (1970)
			20%SW	Cl	525	665	
Salmo gairdneri	150(?)g	13(?)	SW	Na	540		Greenwald and Kirschner (1971)
			SW	Cl	810		
			SW	Na	168		
Aphanius dispar	0·5–3g	20	SW	Na	6850	5900	Lotan (1969)
			200%SW	Na	13 930	11 230	
			200%SW	Cl	2530	1480	
Cottus bubalis	4–15g	10	200%SW	Na	5400	3300	Foster (1969) (recalculated)
			SW	Na	1315	1235	
			10%SW	Na	100	97	
	12–30g	25	SW	Na	2070		Dharmamba et al. (1973)
Tilapia mossambica	0·5–3g	25	200%SW	Na	7000	5660	Potts et al. (1967)
			SW	Na	3330	2860	
			40%SW	Na	500	425	
Mugil capito	18g	16	SW	Na	5120	—	
			SW	Cl	2425	—	
	8g	16	SW	Na	7770	—	Pic et al. (1973)
			SW	Cl	6075	—	

TABLE XII: (continued)

Species	Body weight	Temp (°C)	External medium	Electrolyte	J_{out} (μequiv/100g/h)	J_{in} (μequiv/100g/h)	References
Xiphister atropurpureus	7–40g	13	SW	Na	460	445	Evans (1967b)
			SW	Cl	300	280	
			10%SW	Na	40	40	
			10%SW	Cl	65	65	
Fundulus heteroclitus	2–5g	20	SW	Na	3640	2825	Potts and Evans (1967)
			SW	Cl	5340	4390	
			40%SW	Na	820	480	
			40%SW	Cl	740	345	
Fundulus kansae	9–20g	16	SW	Na	2020	1850	Maetz et al. (1967a)
	0·5–2·5g	20	SW	Na	12 400	8000	Potts and Fleming (1971)
			150%SW	Na	19 400	8600	
Oryzias latipes	0·3–0·6g	?	SW	Na	~3450	—	Kado and Momo (1971) (recalculated)
			50%SW	Na	~1000	—	
Platichthys flesus	60–220g	16	50%SW	Na	2600	2250	Motais (1967)
			25%SW	Na	970	690	recalculated from Motais (1967)
			SW	Cl	160	—	Motais (1967)
			SW	Na	1120	—	
Serranus cabrilla / Serranus scriba	10–80g	16	SW	Na	2650	—	Motais (1967)
			25%SW	Na	560	—	
			SW	Cl	2050	—	Maetz (unpublished)
Pelates quadrilineatus	?	?	33%SW	Na	49%	(turnover rates in %/h.)	Dall and Milward (1969)
			145%SW	Na	18%		
Periophthalmus vulgaris	?	?	10%SW	Na	61%		Maetz and Skadhauge (1968) and Maetz et al. (1967b)
			200%SW	Na	7%		
Anguilla anguilla	60–120g	16	SW	Na	3550	2880	Epstein et al. (1973)
			50%SW	Na	1320	1135	
			SW	Cl	260	—	
			50%SW	Na	1200	—	
Blennius pholis	?	17–20	SW	Na	3600	—	House (1963) (recalculated)
			10%SW	Na	720	—	
Opsanus tau	200–700g	18–20	SW	Na	805	—	Lahlou and Sawyer (1969d)
	200–700g	18–20	50%SW	Na	255	—	

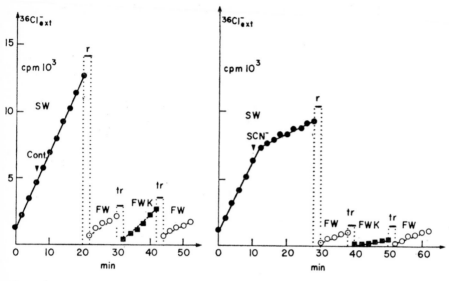

FIG. 15. The rate of appearance of $^{36}Cl^-$ in seawater adapted *Anguilla anguilla*.
On the left, control experiment with NaCl injection (arrow).
On the right, effect of NaSCN injection (arrow).
Efflux measured successively in seawater (SW), after 2 min rinse (r), in fresh-
water (FW), in K^+-enriched freshwater (FWK: 10 mequiv/l K as K_2SO_4) and in
FW. Tr signifies transfer.
Ordinate: external $^{36}Cl^-$ concentration in 10^3cpm.
Abscissa: time in minutes. The print-out of the external $^{36}Cl^-$ was every two minutes.
Note the reduction of efflux upon transfer into FW and the effect of K^+ addition
on $^{36}Cl^-$ efflux; also the depressing effect of SCN^- on Cl^- efflux in SW and the
absence of a K^+ effect in this fish. The flux in SW following SCN^- injection is
similar to the FW flux.
(According to Epstein *et al.*, 1973).

If the gill potential were due to an electrogenic Cl^- pump, injection of
SCN^- should reduce the potential within minutes. No such effect was
observed by House and Maetz (1974). Hence Cl^- transport is probably
not electrogenic.

One interesting aspect of the functioning of the Cl^- pump is the exchange diffusion component which characterizes Cl^- exchanges across the
gill. Motais *et al.* (1966) and Motais (1967) observed that in various fish,
both euryhaline and stenohaline, sudden transfer from seawater to freshwater was accompanied by an instantaneous decrease of the Na^+ and Cl^-
effluxes, which were reversible upon return to seawater providing that the
test-period in freshwater is of short duration, for example about 10 min.
Table XIII summarizes results of Na^+ and Cl^- efflux readjustments in

Table XIII: Relative Na^+ and Cl^- effluxes in various teleosts transferred from sea water to fresh water or other media (in % control values).

Species	Transfer media	Na^+	Cl^-	References
Platichthys flesus	FW	15·9±2·0	16·0±3·0	Motais et al. (1966)
	CaCl₂SW	11·7±2·5	72·6±7·5	
	Na₂SO₄SW	106·0±12·1	26·4±3·1	
Serranus scriba	FW	58 ±5·3	53	
Anguilla anguilla	FW	19·2±1·9	38·0±5·0	Motais et al. (1966), Epstein et al. (1973)
Xiphister atropurpureus	FW	53±5·7	122±17	Evans (1967b)
Pholis gunnellus	FW	39±1	93±13	Evans (1969a)

CaCl₂SW: calcium chloride solution with mannitol iso-osmotic to SW.
Na₂SO₄SW: sodium sulphate solution with mannitol iso-osmotic to SW.
Mean relative values ±s.e.
According to Motais et al. (1966).

various fish. It may be seen that, in some species (*Xiphister, Pholis*), the extent of the Na^+ and Cl^- flux reductions is very different. In the flounder, which is characterized by parallel readjustment of both effluxes, transfer into artificial media containing Na^+ or Cl^- at their respective concentrations in seawater but accompanied by the impermeant co-ions SO_4^{2-} or Ca^{2+}, causes independent "switching-off" of the Na^+ or Cl^- effluxes. These "Na^+-free" or "Cl^--free" effects were considered by Motais et al. (1966) to be the result of exchange-diffusion, that is to say coupling of the influx and efflux of the same ionic species *via* a common carrier, as discussed in the introductory remarks. There is, however, an important factor to be considered, the possibility of changes in the transepithelial potential, which may intervene and hence explain such instantaneous flux adjustments.

Maetz and Campanini (1966) observed in the seawater adapted eel transferred from sea water to fresh water a considerable potential change from about $+20$ mV (positive inside) to about -40 mV. Figure 16 illustrates a typical experiment of this kind. Thus, the reversal of potential may account, at least in part, for the sudden reduction of the Na^+ efflux, but not for the Cl^- efflux readjustment. If Cl^- movements were to be explained by simple diffusion, the Cl^- efflux would have increased. The fact that it decreased rather indicates the occurence of an exchange diffusion mechanism.

Some experiments performed by Epstein et al. (1973) suggest that the Cl^- pump and the exchange diffusion component are closely linked.

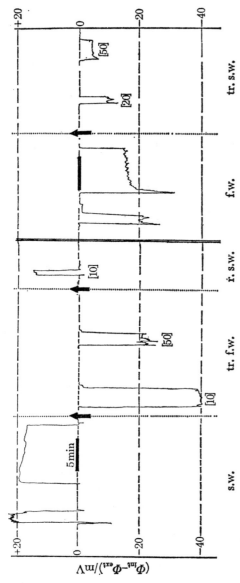

FIG. 16. Recordings of potential measurements across the branchial epithelium of the eel *in vivo*; successive impalements (according to Maetz, 1971 and Maetz and Campanini, 1966).
On the left: seawater adapted eel in seawater (SW) and after transfer into freshwater (tr. FW) for 10 min or 50 min, or 10 min after return to seawater (r. SW).
On the right: freshwater adapted eel in freshwater (FW) and following a 20 or 50 min transfer into seawater (tr. SW).
Ordinates: $\Phi_{int} - \Phi_{ext}$ in mV.
Abscissa: the time scale for 5 min is indicated.

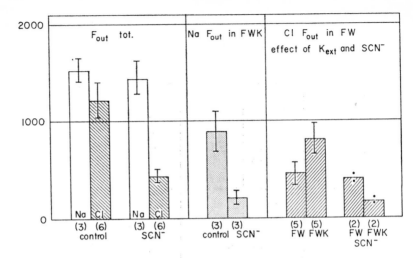

FIG. 17. Summary of experiments concerning the effects of SCN^- on Na^+ and Cl^- effluxes in the seawater adapted eel (according to Maetz, 1973b).
On the left: comparison of the effects of SCN^- on total Na^+ or Cl^- effluxes in sea water.
In the middle: effect of SCN^- on the Na^+ efflux observed in FWK (10 millimolar as K_2SO_4).
On the right: Cl^- efflux upon transfer from sea water into fresh water; effect of K^+ addition to the external medium; effect of SCN^- inhibition.
Ordinates: fluxes in μequiv/100g/h. Number of fish given in brackets.

Figure 17 shows that the Cl^- efflux observed in the seawater eel, after SCN^- treatment, is identical to that observed in intact eels immediately after transfer into freshwater. Moreover the "leak flux" in freshwater is insensitive to SCN^-. It may be concluded that SCN^- injection simultaneously blocks the exchange-diffusion component of the Cl^- efflux and the Cl^- pump.

One last point concerning the Cl^- pump needs to be emphasized. As will be discussed below, the presence of K^+ in the external medium is of importance in regulation of the Na^+ exchanges. Epstein et al. (1973) demonstrated that external K^+ is also essential for the functioning of the Cl^- pump. Figures 15 and 17 show that, after transfer of seawater adapted eels into fresh water, the addition of 10 millimolar K^+ to the external medium is followed by a 100% increase in the Cl^- efflux, even though K^+ is added with the impermeant SO_4^{2-} co-ion. This increase is not observed in thiocyanate-treated eels, rather the reverse. Moreover House and Maetz (1974) observed that addition of K^+ to the external medium, after transfer of seawater adapted eels into freshwater, is accompanied by a considerable

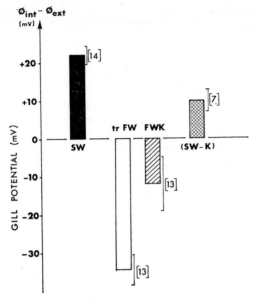

FIG. 18. Gill potentials ($\phi_{int} - \phi_{ext}$) in sea water and various transfer media in the seawater adapted eel.
The number of experiments and standard error of the means are given. (According to House and Maetz, 1974).

decrease in the transepithelial potential which cannot account for the increased Cl^- efflux. Figure 18 illustrates the effect of the addition of K^+ to freshwater on the branchial potential. It is thus probable that K^+ brings about an activation of the Cl^- pump, the mechanism of which remains to be investigated. The Cl^- extrusion mechanism needs further studies of this kind on other species of teleosts.

4. THE SODIUM PUMP AND SODIUM AND POTASSIUM EXCHANGES ACROSS THE GILL

In the preceding section the possibility that the branchial transepithelial potential is the result of an electrogenic Cl^- pump has been discussed and shown to be improbable. The alternative hypothesis, that it results from the diffusion of K^+ or Na^+ down the concentration gradient, will now be discussed. This possibility may be tested experimentally by observing the effects of changing the external ionic concentrations on the gill potential. Experiments of this kind performed by House and Maetz (1974) yielded results which are illustrated in Fig. 18 and which have already been discussed in the preceding section. In seawater eels transferred to freshwater, the gill potential switches from positive ($+ 22.5$ mV inside) to negative, attaining about -35 mV. This large difference (about 55–69 mV) and reversal could be ascribed to the reduction of the external Na^+ or K^+ concentration or both. When the K^+ concentration was raised to its value in seawater (10 mM), a considerable reduction in the negative potential by about 22.5 mV was observed. This decrease in gill potential after addition of K^+ suggests that the gill is relatively permeable to this ion. Moreover, an identical increase of the Na^+ concentration does not significantly alter the potential. Thus the gill appears to be far more permeable to K^+ than to Na^+. However, the K^+ gradient across the gill of fish in fresh water with added K^+ solution (FWK) is identical to that of fish in sea water. and yet the potential is negative in FWK and positive in sea water, This difference also indicates that the gill is permeable to Na^+ but, to a lesser extent than to K^+. The importance of K^+ in the generation of the gill potential is also illustrated by the decrease in potential from $+22.5$ mV to the $+10.5$ mV recorded when seawater eels are transferred into K^+-free sea water. The role of passive Cl^- movements was not examined by House and Maetz (1974), but clearly the branchial permeability to Cl^- must be relatively low as otherwise the gill potential would to be negative in sea water and positive in fresh water. House and Maetz (1974) proposed that the Goldman equation (eq. 21) serves as a quantitative guide to the relative ionic permeabilities of the eel gill. Thus

$$[\phi_{ext} - \phi_{int}] = \frac{RT}{F} \ln \frac{(Na)_{int} + \alpha(K_{int}) + \beta(Cl)_{ext}}{(Na)_{ext} + \alpha(K_{ext}) + \beta(Cl)_{int}}$$

with α, the P_K/P_{Na} ratio varying from 34 to 5.8, and β, the P_{Cl}/P_{Na} ratio varying from 0.16 to 0.30. The exact values are dependent on the experimental conditions. These permeability ratios are rough estimates which need to be substantiated by more thorough investigations.

If the branchial potential is in fact a diffusion potential resulting from the diffusion of Na^+ and K^+ and, to a lesser extent of Cl^- down their respective concentration gradients, the question arises as to how these chemical gradients are maintained across the gill. House and Maetz (1974) suggest that the steady state is obtained by active transports of Na^+ and K^+ as well as of Cl^-, but that the pumps are probably neutral pumps that involve one for one exchanges of the different ionic species.

Maetz (1969) proposed that Na^+ extrusion is effected by a Na^+/K^+ exchange pump. Flounders kept for 24 h in renewed, K^+-free sea water showed a progressive increase of the internal Na^+ content amounting to about 100μEq./100g/h, a value which would be expected if the Na^+ extrusion mechanism had failed. Simultaneously the Na^+ efflux was reduced by approximately this amount. After return to seawater for 48–72 h the internal sodium level declined significantly as if the pump had been "turned-on" upon addition of external K^+. Thus, this Na^+/K^+ exchange component would account for about 5% of the total Na^+ efflux observed in seawater adapted flounders. The remaining efflux would be mediated by a Na^+/Na^+ exchange which also operates in K^+-free seawater.

The existence of Na^+/K^+ exchanges across the gill is further shown by comparing the Na^+ effluxes of seawater adapted fishes transferred from sea water either to fresh water or to fresh water with added K^+ in concentrations ranging from 0 to 50 millimoles. Figure 19 illustrates such

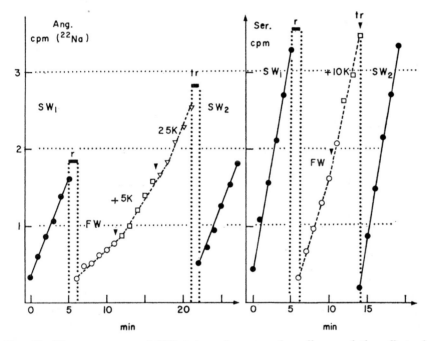

FIG. 19. The appearance of $^{22}Na^+$ into the external medium and the effect of external K^+.
On the left: an experiment with seawater adapted *Anguilla;* transfer from sea water (SW) to fresh water (FW). The effect of adding KCl into the external medium: 5 millimolar at first arrow and 25 millimolar final concentration at second arrow; return to sea water (SW₂). Note the important reduction of the efflux upon transfer into FW.
On the right: an experiment with stenohaline marine *Serranus;* the reduction of the efflux upon transfer to FW is only slight. At arrow, the addition of KCl to a final concentration of 10 millimolar; r: rinse in FW to remove external Na^+; tr: transfer without rinse.
Ordinate: external radioactivity in 10^3cpm.
Abscissa: time in minutes. Print-out of external ^{22}Na every minute. (According to Maetz, 1973*b*).

experiments performed on the euryhaline *Anguilla* and the stenohaline *Serranus*. Addition of K^+ to fresh water is followed by an instantaneous increase of the Na^+ efflux which attains values similar to those observed in sea water. These increases are clearly related to the external K^+ concentration. The relationship between the external K^+ concentration, the K^+ influx and the external K^+-dependent Na^+ efflux were recently investigated in the flounder by Maetz (1973*b*). Figure 20 illustrates the curves obtained. The relationship can be interpreted in terms of Michaelis-Menten kinetics

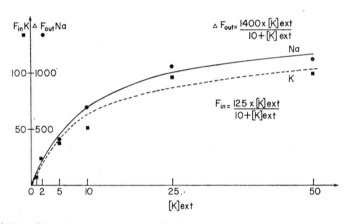

FIG. 20. Relationships between K^+ influx and K^- dependent Na^+ efflux, and external K^+ concentration in the seawater adapted flounder.
Ordinates: fluxes in μequiv/100g/h.
Abscissa: concentration of K^+ in millimoles litre^{-1}.
The equations of the two curves are given.
The F_{out} or K^+-dependent Na^+ efflux is the difference between the fluxes measured in freshwater and in the various KCl concentration in freshwater, immediately following transfer from sea water. (According to Maetz, 1973*b*).

suggesting the presence of a Na^+/K^+ exchange carrier with the same affinity constant for both fluxes.

Maetz (1969) noted that the Na^+/K^+ exchange mediates Na^+ effluxes which differ widely in importance when these fluxes are compared in sea water and K^+-free sea water, or in fresh water with added K^+. In the flounder they mediate 5% of the total efflux in Na^+-rich water and 50% in Na^+-free water. In the eel and the sea perch, as shown in Fig. 19, the Na^+/K^+ exchange may mediate as much as 100% of the flux normally observed in sea water. Maetz (1969, 1971, 1973*b*) ascribed this difference to a competition between external Na^+ and K^+ for a common carrier. The

K^+ influx is about 2 to 4 times higher in fresh water with 10 millimolar K^+ than in sea water which contains the same amount of K^+ but is also 500 millimolar with respect to Na^+.

Comparison of the Na^+ effluxes in sea water, fresh water and fresh water with added Na^+ in concentrations ranging from 0 to 800 millimoles,

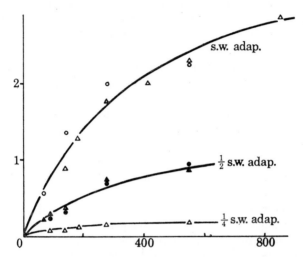

FIG. 21. Relationship between the Na^+ influx and the external Na^+- dependent component of the efflux, and the external Na^+ concentration in the seawater-adapted flounder (according to Motais et al., 1966).
Circles: influx.
Triangle: Na^+-dependent component of efflux.
Comparison of seawater adapted (SW adap), half strength and quarter strength seawater adapted (1/2 SW adap—1/4 SW adap) flounders.
During "instantaneous" readjustment upon transfer of fish from the adaptation medium to a new medium, the fluxes "move" along the three curves drawn.
During "delayed or secondary" regulation, the fish being kept in the new medium, the fluxes would slowly move from one curve to the other along an "isoconcentration" line, a process which needs several days. The rapid adjustment pattern was thought to be dependent on the characteristics of the carrier (the number or the turnover of sites and their affinity), while the delayed regulation process expresses changes in the number or the turnover of sites without a change in affinity.
Ordinates: fluxes in mequiv/100g/h.
Abscissa: concentration in mequiv/l.

also distinctly demonstrates the existence of a Na^+-dependent Na^+ efflux. Figure 21 illustrates the experiments performed by Motais et al. (1966) on the flounder showing that the Na^+ influx and the external Na^+-dependent Na^+ efflux are both related to the external Na^+ concentration. Again the relationship was interpreted in terms of saturation kinetics and taken as evidence for the existence of a Na^+/Na^+ exchange carrier, with the same affinity constant for both fluxes. It can be seen by comparing Figs 20 and 21 that the affinity of the carrier for K^+ is about 40 times higher than that for Na^+.

In the light of the observations obtained by House and Maetz (1974) concerning the effects of external Na^+ or K^+ on the gill potential in the eel (see Fig. 18), it is doubtful whether the hypothesis of Na^+ or K^+ carriers can be accepted without reservation. The gill potential changes explain, at least qualitatively, the above mentioned flux changes. For instance, the differences in the apparent affinity constants may simply reflect differences in the permeability coefficients of the gill for Na^+ and K^+. Whether gill potential changes and experimental concentration changes could explain the flux changes quantitatively, and whether in fact equations such as eq. 18 (p. 8), describing unidirectional flux as a function of potential and concentration, could yield curves exhibiting "saturation kinetics" remains to be seen. This question will be raised again below (p. 140), as such equations were indeed developed by Smith (1969b) for *Artemia salina*.

Several observations indicate that different mechanisms operate in Na^+/Na^+ exchanges and Na^+/K^+ exchanges across the gill, and that the Na^+/K^+ exchange may be an active process. Maetz and Evans (1972) observed in the flounder that the branchial Na^+/K^+ exchange is far more temperature sensitive than the Na^+/Na^+ exchange. Thus, transfer of flounders from 16°C (adaptation temperature) to 6°C produces a 50% reduction of the Na^+/Na^+ exchange, and a 6-fold reduction of the Na^+/K^+ component. Simultaneously, the internal Na^+ content of the fish increase as a result of the impairment of the Na^+ extrusion pump. Such differences in Q_{10} values argue in favour of different mechanisms, although they do not prove the active nature of the Na^+/K^+ exchange.

Epstein *et al.* (1973) observed in the seawater eel that injection of SCN^- is followed by a simultaneous increase of the Cl^- and Na^+ contents of the eels which suggests that inhibition of the Cl^- pump also perturbs the Na^+ balance of the fish. The linkage between Cl^- excretion and Na^+ excretion was investigated further. As shown in Fig. 17, in contrast to their pronounced action on the Cl^- efflux, injections of SCN^- produced a negligible effect on the total Na^+ efflux measured in seawater. SCN^-, however, greatly depressed the Na^+ efflux in eels transferred into fresh water with 10 millimolar K^+. This suggests that the Na^+/K^+ exchange, and not the Na^+/Na^+ exchange is inhibited by thiocyanate. Epstein *et al.* (1973) conclude that it is probable that the Cl^- pump is closely linked to the reciprocal transport of Na^+ and K^+ across the gill, a linkage which is also suggested by the stimulating effects of external K^+ on the Cl^- pump.

The most convincing evidence for the existence of a Na^+/K^+ exchange pump, possibly mediated by a $Na^+ + K^+$ dependent ATPase, will be discussed below.

5. CELLULAR AND BIOCHEMICAL ASPECTS OF SALT EXCRETION

Conte (1969) has reviewed the cellular aspects of salt excretion in the fish gill, and the structure of the chloride cells in gills in relation to their probable function in active transport has been briefly discussed in the preceding section. The presence of Cl^- has been confirmed by histochemical techniques in the "seawater cell", such Cl^- being located in the "pit" characterizing the mucosal border of the cell in some fish as well as inside the microtubuli (Philpott, 1965; Petrik, 1968). More recently, the presence of Na^+ also was confirmed by histochemical techniques, this ion being located in the tubular system, the nucleus and in the numerous vesicles which are observed in the neighbourhood of the mucosal border (Shirai, 1972). After injection into the blood, heavy metal complexes such as lanthanum chloride, or fairly big organic molecules such as horse radish peroxidase have been located inside the tubular apparatus (Philpott, 1967; Ritch and Philpott, 1969). Experimental changes in the external salinity are followed by conspicuous changes in the diameter and arrangement of the microtubules (Lasker and Threadgold, 1968; Newstead, 1971). It is interesting to note that the report by Lasker and Threadgold concerns the larval sardine which is remarkable in that chloride cells are absent from the gill epithelium but present in the skin. The presence of "chloride cells" in specialized areas other than in the branchial epithelium has been recently noted in various marine siluroid fish, which possess behind the urogenital opening a "dentric organ". Ultrastructural and histochemical studies suggest that this organ may function as a salt excreting gland (van Lennep and Lanzing, 1967; van Lennep, 1968). It would have been of interest to verify whether in such fish the gill is devoid of "chloride cells" as in the larval sardine.

According to Conte and Lin (1967) and Conte (1969), the chloride cells of the gill "wear out" and are replaced, faster in seawater than in freshwater. Autoradiographic observations with [³H] thymidine serving as a marker of cell division, show a faster cellular turnover in seawater- than in freshwater-adapted *Oncorhynchus* gills. Earlier observations of Conte (1965) on the same fish also suggest a more rapid cellular turnover in seawater as shown by far greater radiation sensitivity in this medium. Histopathological changes in the gill epithelium and osmotic imbalance induced by X-rays in seawater fish both strongly suggest a failure of the gill salt-excretion cells, while in freshwater fish the osmoregulatory imbalance produced by X-rays is slight. Parallel experiments have been made on the eel in our laboratory with the use of actinomycin D, an inhibitor of DNA-dependent RNA synthesis and cell division. Maetz et al. (1969) showed that, in sea water, this antibiotic produced a complete failure of the salt excreting mechanism associated with a progressive reduction in the Na^+ turnover rate. Death followed within a week. In fresh water on the other hand, the active component of the Na^+ uptake remained unchanged while the outflux was increased. The osmoregulatory unbalance is, however, slight and the freshwater eels survive much longer. Maetz and his colleagues suggested that actinomycin D intervenes specifically by inhibiting RNA polymerase, thus blocking the transcription mechanism in the nucleus and so protein synthesis, rather than by blocking cellular division. Motais (1970a,b) completed these experiments by studying the effects of actinomycin D in freshwater and in seawater eels on the $Na^+ + K^+$ dependent ATPase activity of the gill. He showed that actinomycin D remains without effect on the enzyme level in the freshwater eel, while it produces a significant reduction of the enzyme level in the seawater eel. A comparison with the absence of effects of actinomycin D on the Na^+ pump in fresh water, and the impairment of the Na^+ pump in sea water indicates that this enzyme plays an important role in the branchial ion excretion mechanism and that the half-life of this pump is relatively short in seawater.

    The distribution of the enzyme in the chloride cell has been discussed in a preceding section of this review (p. 60). Figure 22 illustrates the model proposed by Shirai (1972) for the functioning of the chloride cell in seawater fish, which takes into account the localization of both Na^+ and $Na^+ + K^+$ dependent ATPase by histochemical techniques. According to this model, the tubular "membrane ATPase" would serve to extrude intracellular sodium into the lumen of the tubules or into the intercellular spaces. The numerous mitochondria associated with this network would supply energy for the enzyme. The Na^+ originating from the blood vessels would thus accumulate in the tubular lumina and be transferred to the apical part of the cell. Since the small vesicles abundantly distributed in the apical cytoplasm also contain Na^+ as evidenced by a precipitate of Na^+ pyroantimonate formed during the histochemical reaction, these vesicles would appear to serve as final transporting organelles. Shirai assumes that the tubular system near the apex of the cells continually buds off vesicles which transport the Na^+ to the apical pit of the cell by a pinocytotic-type of mechanism. On the apical part of the cell the ATPase located in the plasma membrane would serve to eliminate the Na^+ which continually enters the cell along the concentration gradient from the external medium. In all these sites, whether tubular, vesicular or apical, Na^+ would be extruded from the cell by way of a Na^+/K^+ exchange. The K^+ would

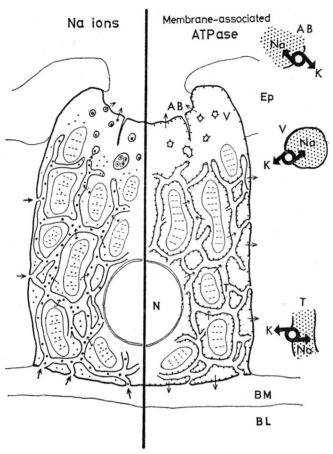

Na ions   Membrane-associated
                 ATPase   A B

Fɪɢ. 22. A model for the functioning of the teleostean chloride cell in seawater (modified from Shirai, 1972).
On the left, distribution and pathway of Na^+ ions.
On the right, tentative localization of membrane-associated ATPase and the pathway of active Na^+ transport.
Inserts: Na^+/K^+ exchange effected by the transport ATPase in the tubular system (T), in the apical vesicles (V) and in the apical border (AB).
BM: basement membrane. N: nucleus.
BL: blood lumen.

enter the cell *via* the pump to compensate for the K^+ continually lost by the cell along the concentration gradient.

In theory such a model proposes Na^+/K^+ exchange pumps operating at the apical and basal borders in opposite directions. In practice, because of the numerous infoldings of the basal border, the system functions with a

basal to apical polarity. Such a model explains the contradictory observations concerning the effects of ouabain, which interferes with Na^+
extrusion whether added to the outside medium or injected into the body
fluids.

Kamiya (1967) and Kamiya and Utida (1968), using the "incubated gill"
technique, which maintains a zero net flow of Na^+, showed that injection of
ouabain prior to gill isolation induces the preparation to gain Na^+. This
finding suggests that functioning of an internal $Na^+ + K^+$ dependent
ATPase is essential for Na^+ extrusion. Motais and Isaia (1972b) observed
that ouabain added to the external medium of the seawater-adapted eel
produces a progressive increase of the internal Na^+ level, suggesting an
impairment of the Na^+ extrusion pump. It produces simultaneously a
partial inhibition of the total Na^+ efflux and of the external K^+-dependent
Na^+ efflux. Additional investigations indicate that in sea water both K^+ and
Na^+ compete with ouabain for the active site of the enzyme. This may
explain why in some species such as the flounder ouabain is not effective.

Carbonic anhydrase, like $Na^+ + K^+$ dependent ATPase, is found in
higher specific activities in the gills of seawater than in freshwater fish
(Maetz, 1956a). Injection of acetazolamide in the sea perch, *Serranus*, was
observed to produce a transient increase of the internal Cl^- level. Maetz
(1956a, 1971) suggested that acetazolamide temporarily blocks the Cl^-
extrusion pump, and that this pump operates by way of a Cl^-/HCO_3^-
exchange mechanism located at the serosal border of the chloride cell and
its tubular infoldings. The HCO_3^- to be exchanged against the Cl^-
would be produced in the cytoplasm of the chloride cell from the enzyme-
catalysed hydration of respiratory CO_2. Recent observations, however,
cast some doubt on the validity of such a hypothesis. M. Bornancin and
M. Istin (unpublished data) comparing the carbonic anhydrase activity
of the isolated chloride cells and of the respiratory epithelial cells, observed
that they were similar. Investigations on the possible effects of carbonic
anhydrase inhibitors on branchial Na^+ and Cl^- effluxes yielded contradictory results. Forrest *et al.* (1971) reported that methazolamide decreases
the Na^+ efflux in the seawater eel by 50% within an hour of intraperitoneal
injection. However, no effect on serum Na^+ was observed. Epstein *et al.*
(1973) did not find an effect of acetazolamide on the total Cl^- efflux within
30 min of injection. New studies are clearly necessary to clarify the possible
role of carbonic anhydrase in the gill of seawater teleosts, and its implication in a Cl^-/HCO_3^- exchange mechanism. Such an exchange could
also hypothetically function in conjunction with an anion-sensitive
ATPase, as the experiments concerning SCN^- inhibition of the Cl^-
pump would suggest. This enzyme activity, however, has not been
reported in the seawater gill.

6. ADAPTIVE FEATURES OF GILL FUNCTION AFTER TRANSFER FROM FRESH WATER
TO SALTWATER

The morphological, biochemical and functional behaviour of the gill in
relation to salinity adaptation has been particularly well studied in the eel.
In the Japanese eel, Utida et al. (1971) observed that both the $Na^+ + K^+$
dependent ATPase activity and the number of chloride cells increase 5 to
6 fold during the first two weeks from fresh after transfer to seawater.
Histometric and electron microscopic studies by Shirai and Utida (1970)
indicated two types of chloride cell in the gill epithelium of the freshwater
animal, the B-type, which is weakly acidophilic and which is twice as
abundant as the A-type, which in turn is strongly acidophilic. Both types
increase in number after transfer but the B-type shows a maximal number
on the third day, followed by a decline and complete disappearance within
14 days. The A-type becomes more abundant than the B-type on the third
or fourth day and finally reaches a plateau within 14 days. The authors
suggest that the B-type is a transitional stage of the A-type which is
typical of seawater adaptation.

Olivereau (1970) working with the European eel confirmed the increase
in the number of chloride cells, but her technique did not allow for the
separation into two cell types. She noted that the cell number increases
within a few hours of transfer although signs of cell division are only
apparent within two days. During this initial phase chloride cells differentiate from pre-existing epithelial cells. A "wave" of cell divisions occurs
between 2 and 7 days with a maximum on the fourth day, after which the
rate of mitosis declines to a low level. During this second phase an induction
of cell division from a stock of unspecialized cells is evident. It would be of
interest to repeat these observations using an injection of [³H]thymidine
to tag the newly formed nuclei.

Maetz et al. (1969b) observed that actinomycin D injected into the eel
one day prior to seawater transfer upsets the pattern of readjustment of the
Na^+ efflux which normally accompanies adaptation to seawater. The Na^+
efflux remains very low and the internal Na^+ level becomes abnormally
high. The animals die within a week of transfer. The authors suggest that
molecular and probably cellular renewal of the eel gill epithelium are
necessary for salt water adaptation. Unfortunately the morphological
effects of the treatment were not checked. More recently M. Bornancin
(unpublished data) demonstrated that repeated injections of mitomycin, an
inhibitor of cell division, also interfered with the normal pattern of
readjustment of the Na^+ efflux. The perturbance only became evident 4 to
6 days after transfer and the eels died within two weeks due to a
complete breakdown of the Na^+/K^+ exchange mechanism.

Bornancin and de Renzis (1972) have investigated the biochemical and functional readjustment of the gill in the European eel following transfer from fresh to seawater. They found rapid increases in the plasma Na^+, the total Na^+ efflux and the passive Na^+ leak after transfer to seawater, these parameters attaining maxima at 2, 4 and 7 days after transfer respectively. The parameters declined thereafter and steady values were attained after 1 to 2 weeks' adaptation. The Na^+/K^+ exchange increased more slowly to be maximal 7 to 14 days after transfer. The $Na^+ + K^+$ dependent ATPase activity varied in parallel with the Na^+/K^+ exchange in a manner comparable to that found by Utida *et al.* (1971) in the Japanese eel. Thus, during freshwater to seawater transition, the rapid increase of the total Na^+ efflux initially reflects an increase in the diffusional Na^+/Na^+ exchanges. The Na^+/K^+ exchange, which presumably results from the $Na^+ + K^+$ dependent ATPase activity and represents the active Na^+ extrusion component, develops more slowly and independently of the Na^+/Na^+ exchange, presumably in parallel with the increase in the number of active chloride cells. Forrest *et al.* (1973a) made similar observations on the American eel. They found a close correlation between the increase in total Na^+ efflux and the $Na^+ + K^+$ dependent ATPase activity of the gill during fresh water to sea water adaptation

More recently Kirsch and Mayer Gostan (1973) studied the kinetics of the total Cl^- efflux during fresh water to sea water transition in the silver eel. The Cl^- efflux readjustment was similar to that of the Na^+ efflux described by Bornancin and de Renzis (1972) with a maximum (500 $\mu Eq./$ 100g/h) at 4 days, followed by a decline. The steady state value (230 $\mu Eq.$) was attained within two weeks. The absolute value of the flux was 2–4 times less for Cl^-, presumably as a result of differences in size and handling. Kirsch and Mayer-Gostan suggest that the Cl^- pump is already operative within 24 h of transfer, because they observed a transient decline of the exchangeable Cl^- between the first and fourth day of transfer. The internal Cl^- increased by about 80% during adaptation to sea water. The relative contribution of the intestinal and branchial routes of salt entry cannot be evaluated with precision from their data, as the quantity of Cl^- swallowed during the process of adaptation may not be immediately absorbed. It is probable that the increase in plasma electrolyte concentration is an essential factor for initiating active ionic excretion by the gills, as the previous studies by Mayer and Nibelle (1970) and Kirsch (1972a) have indicated.

Figure 16 illustrates the very slow readjustment of the branchial transepithelial potential which occurs in the eel during freshwater to seawater transition. The potential remains negative (inside) for 1 to 2 h, reverses and attains the positive value characteristic of seawater adaptation within 24 h. This change of polarity reflects a progressive change in the passive permeability to monovalent cations which accompanies adaptation.

Transfer from seawater to freshwater is accompanied by a slow degeneration of the strongly acidophilic cells which characterize seawater adaptated gills, while the $Na^+ + K^+$ dependent ATPase activity exhibits a parallel decline (Utida et al., 1971). Functional studies by Motais et al. (1966) and Motais (1967) showed that immediately after transfer to fresh water there was a decline in the Na^+ and Cl^- effluxes. Such declines were reversible, providing the "back-transfer" was within 20–30 min of the initial transfer. Figure 16 illustrates the reversal of potential which accompanied transfer into freshwater. If the fish was returned to its original medium the blood again became electropositive. In the preceding section, the significance of these instantaneous efflux and potential readjustments has been discussed. These changes reflect the diffusive nature of the potential with respect to Na^+ and K^+, while the Cl^- flux readjustments result from an exchange-diffusion effect. If the seawater adapted fish was kept for longer periods in fresh water, a progressive decrease of the Na^+ and Cl^- effluxes was observed within 20–30 min which was termed by Motais and his colleagues "delayed regulation". If, after several hours in fresh water, the fish was replaced in sea water, the total Na^+ and Cl^- effluxes were much smaller than those originally recorded, and several hours were necessary for the return to the steady state values characteristic of seawater adaptation. Maetz and Campanini (1966) observed that delayed regulation coincided with a slow readjustment of the potential. The potential attained the steady-state value characteristic of freshwater adaptation within about 24 h. This slow shift of the potential obviously corresponds to a slow decrease in the passive permeability of the gill to Na^+ ions which accompanies freshwater adaptation. On the other hand this slow potential shift explains, at least in part, the decrease in the Cl^- efflux. It is probable, however, that a decrease of the Cl^- permeability of the gill also occurs.

The salmon is another euryhaline species which has been relatively well studied. In both Pacific and Atlantic salmon, an increase in the number of chloride cells was observed in the smolt stage upon adaptation to seawater (Conte and Lin, 1967; Threadgold and Houston, 1964; Van Dyck, 1966). The $Na^+ + K^+$ dependent ATPase activity of the gill increased by about 100% in association with parr-smolt transformation (Zaugg and McLain, 1970, 1972). Such an increase was observed in early spring while the fish were still in fresh water. The preference for higher salinities which occurs during transformation in the salmon may be associated with an increase in "transport-ATPase" activity. When the fish enters salt water, this activity continues to increase for 30–35 days to reach a level 3–4 times that of freshwater animals. Adams et al. (1973) showed recently that the temperature level was critical for obtaining the early spring rise in ATPase activity in the smolt. If the temperature were too high (15°C or more) the two fold increase was not observed.

Studies on the Atlantic salmon were also made by Potts et al. (1970). In smolts, a rapid Na^+ turnover was found in seawater adapted animals, 11% h^{-1} in early smolts and 24% h^{-1} in late smolts. Only late parrs could be studied for a few days in sea water. They were characterized by a very low rate constant (3% h^{-1}). Thus the parr-smolt transformation coincides with an increase in the branchial turnover fluxes. When transferred from fresh water to sea water the rate of Na^+ efflux in the salmon smolt, as in the eel, remained constant for a few hours then rose slowly reaching 10% h^{-1} in about 24 h. Complete adaptation took several days. The Cl^- efflux readjustment occured more rapidly. A return to fresh water was accompanied by an instantaneous decline of both Na^+ and Cl^- effluxes by about 50%. A delayed regulation occured bringing the fluxes back to the freshwater 2% turnover level within a few hours.

The functional readjustments of the Na^+ exchange fluxes on transfer from high to low salinities have been studied in various other fishes. The flounder resembles the eel in its Na^+ exchange adjustments (see Fig. 21) but the Cl^- efflux changes have not been studied. The morphological, biochemical and bioelectrical aspects of branchial adaptation also remain to be investigated in this species. A recent study of the Na^+ effluxes in the Japanese medaka (*Oryzias latipes*) suggests that the pattern of adjustment is very similar to that observed in the flounder (compare Motais, 1961a,b and Kado and Momo, 1971).

The killifish, *Fundulus heteroclitus*, contrasts with the eel and flounder in that the instantaneous reduction for the Na^+ efflux upon transfer from sea water to fresh water is much smaller, amounting to about 30% of the total flux (Motais et al., 1966; Potts and Evans, 1967). The importance of external K^+ in the Na^+ efflux mechanism has not been tested. The importance of the delayed regulation process in the reduction of the Na^+ efflux, and limitation of Na^+ loss in fresh water has been noted by both groups of investigators. Only 2 days are necessary to attain the steady-state freshwater fluxes according to Maetz et al. (1967a). Differences in the "Na^+-free" effects which are observed when *Fundulus* and *Anguilla* are compared, may be explained by differences in the evolution of the gill potential in relation to seawater-freshwater transfer. However, these differences were ascribed by Motais et al. (1966) to different relative importances of the exchange diffusion components in the two species. The kinetics of the morphological and biochemical changes in the gill in relation to salt and freshwater adaptation have not been studied so far. The investigations of Pickford et al. (1970a) on the mechanism of action of prolactin

In hypophysectomized *Fundulus* transferred into fresh water, showed that the $Na^+ + K^+$ dependent ATPase activity decreased by 65% within 3 days. The level reached was similar to that observed in long term freshwater adapted animals, providing exogenous prolactin was injected. Thus adaptation may prove to be much faster in *Fundulus* than in *Anguilla*.

There have been relatively few investigations on the osmoregulatory behaviour of the gill in stenohaline species transferred to water of lower salinities. *Serranus cabrilla* and *S. scriba*, for instance, are able to survive in one quarter strength sea water. Motais *et al.* (1966) observed that, upon transfer from sea water to 1/4 sea water, there was an instantaneous regulation process followed by a delayed one, which together reduced the unidirectional Na^+ fluxes to about 25% of their original value. The kinetics of the readjustment were not studied. In the euryhaline *Platichthys* and *Anguilla*, the steady state flux observed in 1/4 sea water was 10% or less of that in sea water. The authors concluded that probably in euryhaline species the impermeabilization of the gill is much more extensive. Similar observations were subsequently made by Dall and Milward (1969) on *Pelates* and *Periophthalmus*, which behave like *Serranus*. They suggested that such an increased reduction observed in euryhaline species corresponds to a switch-over to essentially freshwater mechanisms. This interesting suggestion deserves further investigation.

7. CONTROL OF ELECTROLYTE TRANSFER AND THE ROLE OF ENDOCRINES

The existence of feed-back mechanisms maintaining the internal Na^+ level in the seawater-adapted eel has been clearly demonstrated by Mayer and Nibelle (1970). Intravenous infusions of hypersaline or hyposaline Ringer solutions, which produce an increase or a decrease of the plasma level of NaCl, were followed within 30 to 60 min by an increase (after hypersaline injection) or a decrease (after hyposaline) of the Na^+ efflux. The maximal response (a 75% increase or a 50% decrease) was attained within 2– 3h and the initial flux level was regained after 6 h or more. The Na^+ influx remained unchanged at the time when the maximal response of the Na^+ efflux was observed. Whether the relatively swift adjustments of the Na^+ efflux are the result of changes in the electrochemical gradient across the gill, or of variations in the activity of the Na^+ excreting pump remains to be investigated. Endocrine activity may be involved in these rapid responses. Neurohypophysial hormones, such as arginine-vasotocin for example, increase the Na^+ efflux within an hour in the flounder (Motais and Maetz, 1964, 1967). Adrenalin inhibits the Na^+ and Cl^- effluxes within a few minutes of injection in the mullet (Pic *et al.*, 1973). Such effects are mediated by the adenyl cyclase system which has recently been discovered in the gill epithelium (Cuthbert and Pic, 1973). Unfortunately the doses of neurohypophysial hormones and adrenalin used were high. Such investigations should be carried out on unstressed animals with cannulated blood vessels, allowing for blood sampling or intravenous infusion of hormones. It would also be of interest to measure, if possible, the circulating hormone level before and after experimental changes in the osmotic balance of the fish. It may be mentioned that Bentley (1971b) failed to detect levels of vasotocin similar to those found in Amphibia in the plasma of the eel, *Anguilla rostrata*, under osmotic stress. Whether the neurohypophysis in teleosts secretes hormones more peripherally than the adenohypophysis remains in doubt. This problem has been discussed by Maetz and Lahlou (1974).

Other endocrine principles have been implicated in long-term actions on ionic pumps, or on the passive permeability of the gill which accompany the "delayed" regulatory processes mentioned in the preceding section. Of particular interest are the corticosteroids secreted by the interrenals under ACTH stimulation, and the prolactin-like factor secreted by the adenohypophysis. These hormones have been shown to have long-lasting effects on ionic turnover in the gill of seawater fish, either by increasing the Na^+ efflux for several days, an effect produced typically by cortisol or ACTH, or by decreasing the Na^+ efflux, an effect produced by mammalian prolactin (Maetz *et al.*, 1969b; Mayer, 1970; Potts and Fleming, 1971; Fleming and Ball, 1972; Macfarlane and Maetz, 1974; Dharmamba *et al.*, 1973). Moreover, hypophysectomy is accompanied by a decrease in the Na^+ efflux in all seawater fish so far examined (Maetz *et al.*, 1967a, b; Stanley and Fleming, 1966a; Macfarlane and Maetz, 1974). Interrenal-ectomy, which is possible in the eel, is also accompanied by a decrease in the Na^+ turnover rate. Injection of cortisol restores the Na^+ exchange to the control rate (Mayer *et al.*, 1967) In one species, the flounder, the changes in Na^+ efflux have been found to be paralleled by a depression of the Na^+/K^+ exchange which presumably represents the active component of the Na^+ efflux (Macfarlane and Maetz, 1974). It is probable that in the other species investigated (e.g. *Fundulus heteroclitus*, *Anguilla anguilla*), Na^+ excretion is also affected by hypophysectomy and replacement therapy, since the branchial $Na^+ + K^+$ dependent ATPase activity varied in parallel with the Na^+ efflux (Epstein *et al.*, 1967; Pickford *et al.*, 1970b; Milne *et al.*, 1971). In freshwater yellow eels, long term cortisol treatment increased the $Na^+ + K^+$ dependent ATPase activity in the gills to a level characteristic of seawater-adapted animals. Upon transfer to sea water, the treated animals exhibited an accelerated pattern of Na^+ efflux adjustment, and the plasma electrolyte level did not rise as high as that in control animals abruptly transferred to seawater. Thus cortisol successfully prepares these euryhaline teleosts for combating the osmotic stress of migration into sea water (Epstein *et al.*, 1971; Forrest *et al.*, 1973*a*). Conversely interrenal-ectomy or hypophysectomy results in a slowing down of the Na^+ efflux readjustment accompanying seawater adaptation in the eel. The inter-renalectomized animals did not survive more than a few days in sea water (Mayer *et al.*, 1967; Mayer, 1970), but the hypophysectomized animals were able to adapt (Langford, 1971; Butler and Carmichael, 1973). After 2–5 weeks in seawater the internal Na^+ level was identical to that of control animals, but according to Langford (1971) the total Na^+ efflux and the Na^+/K^+ exchange component were about one-third of the values observed in controls. The $Na^+ + K^+$ dependent ATPase activity in the gill was reduced by nearly 50%. As hypophysectomy is also followed by an impairment of the gut absorption mechanism (see Section III A4), it is probable that the Na^+_{int} remained unchanged because both Na^+ entry in the gut and Na^+ exit in the gill were reduced to the same extent. Similar observations were made by Pickford *et al.* (1970*b*) on *Fundulus heteroclitus*. Injection of cortisol produces a 140% increase in the total Na^+ efflux and the $Na^+ + K^+$ dependent ATPase activity in the gill of the eel, while the Na^+/K^+ exchange component increased more than 4-fold. That this latter effect corresponds to a stimulation of the Na^+ extrusion pump was demonstrated by a simultaneous decrease in the internal Na^+ level. Cortisol presumably acts on both effector organs to increase Na^+ entry via the intestine (see Section III A4) and Na^+ excretion by way of the gill. Obviously, however, it must act more rapidly on the gill, hence the decreased Na^+_{int} level. Table XIV summarizes these data obtained by Langford (1971).

It is interesting to note that the $Na^+ + K^+$ dependent ATPase level attained in the hypophysectomized *Anguilla anguilla* in seawater was almost twice as high as that observed in freshwater animals (Langford, 1971). Butler and Carmichael (1973) made similar observations in *Anguilla rostrata*. They suggest that the increased salt load which accompanies salt water adaptation is responsible for this effect independently of the influence of the hypophysial-interrenal system.

The morphological aspects of endocrine control of gill function have not yet been studied intensively. It would be of interest to examine the fine structure of the epithelium in hypophysectomized or interrenalectomized eels, either adapted to sea water or during freshwater to seawater tran-sition. Very recently Doyle and Epstein (1972) investigated the effects of saline adaptation and cortisol treatment on the ultrastructure of the gill of *Anguilla rostrata*. Cortisol treatment in freshwater eels induces the development of additional chloride cells with a conspicuous branching tubular reticulum. Such cells did not, however, seem to be functional as very few were in contact with the external medium. Thus some additional

TABLE XIV: Effect of hypophysectomy and cortisol treatment on the various components of Na^+ transfer across the gill of the eel and on the $Na^+ + K^+$ dependent ATPase activity.

	n	Na^+_{int} (mMol l^{-1})	J_{out} (in SW)	J_{out} (in FW)	Na^+/K^+ exchange	$Na^+ + K^+$ ATPase
control SW	7	163±1·7	1173±199*	244±64	371±40*	77·5±4·5* (10)
hypophysectomized (2–5 weeks in SW)	9	163±3·6	352±24	136±34	117±43	44·8±3·7 (8)
hypophysectomized + cortisol (SW)	9	148±2·0*	835±102*	254±71	517±70*	107·8±7·6* (8)
control FW	10	142±5·0*	—	—	—	25·5±4·3* (8)

*: P <0·01 Comparison between hypophysectomized and either control (FW or SW) or cortisol treated animals.
J_{out}: efflux measured in sea water and after transfer in fresh water.
Na^+/K^+ exchange and K^+-dependent Na^+ efflux measured after addition of 10 mM K^+ to FW.
J_{out} and Na^+/K^+ exchange in μequiv/100g/h; ATPase activity in μmoles Pi/h.mg protein.
Doses of cortisol as in Table VIII. Taken from Langford, 1971.

factor is required to cause migration of the cell to the surface. Transfer to salt water promptly induces cell migration, allowing the treated animals to adjust to their new surroundings very rapidly. Nevertheless there is a delay before the Na^+ efflux increases, although this time lag is much shorter than in controls. Possibly the decrease in the circulating prolactin level which accompanies salt water adaptation may be the additional hormonal factor involved.

8. COMPARATIVE ASPECTS: OCCURENCE OF EXTRARENAL SALT EXCRETION IN HYPO-OSMOTIC SALTWATER INVERTEBRATES

Among the hypo-osmoregulating crustaceans, the brine shrimp, *Artemia salina*, affords the best evidence for extrarenal-branchial salt excretory mechanisms. Croghan (1958c) was the first to suggest this extraordinary physiological convergence with teleosts living in sea water. The passive uptake of silver ions localized in the first ten pairs of branchiae suggest that the branchial epithelium is highly permeable to Cl^- ions. Animals in which the branchiae had been damaged by a brief exposure to saturated $KMnO_4$ solution lost their ability to osmoregulate and became iso-osmotic to the external medium. Croghan suggested that this isotonicity is not due to increased water permeability, but to a specific destruction of the salt excreting mechanism. This was clearly indicated by alterations in the chemical composition of whole, $KMnO_4$ treated animals kept in 25% seawater and transferred to 50% seawater. Croghan also studied the ontogeny of the regulatory mechanism, showing that in the gill-less nauplii, the dorsal or neck-organ is most probably the effector organ of osmoregulation; when the branchiae develop the dorsal organ degenerates.

Croghan (1958b) also measured the ionic fluxes between hemolymph and external medium using ^{24}Na and ^{82}Br. Extremely high fluxes were found representing about 100% of the salt content per hour. The gut did not participate significantly in this exchange. Animals with ligatures preventing drinking show no significant difference in the ionic turnover rates. The steady state fluxes increased markedly in proportion to the external salinity. In this respect the brine shrimp resembles stenohaline marine teleosts (see Table XII). Rapid transfer to media of lower external salinity was accompanied by a very swift decline of the Na^+ efflux. This rapid readjustment is analogous to that observed in the euryhaline flounder. Croghan suggested that the greatest part of these ionic exchanges was due to exchange-diffusion. Croghan also showed that external K^+ and Na^+ compete for the exchange sites and a one to one relationship was found. Thuet et al. (1968) observed that the variation of the unidirectional Na^+ efflux with external Na^+ concentration could be described by an equation of the Michaelis-Menton type, as would be expected from a carrier-mediated process responsible for a one for one linkage of the Na^+ influx and efflux. They also confirmed the presence of a K^+/Na^+ exchange and studied the specificity of the Na^+ and Cl^- exchange carriers. Smith (1969a) measured the potential difference across the gill epithelium and found it to be 23 mV (inside positive), a value which is nearly identical to that found in the seawater eel. He concluded that, as in teleosts, the chloride exchange and the chloride net extrusion rate across the gill is an active process, and that Na^+ movements are more or less accounted for by the electro-chemical gradient. Measurement of the potential difference in relation to transfer into media with various Na^+, K^+ and Cl^- concentrations, suggests that the gill-potential results from the free-diffusion of the main cations Na^+ and K^+. The permeability ratios with respect to Na^+, are for K^+, 0.6, and for Cl^-, 0.1. When compared with the values found by House and Maetz (1974) for the eel it can be seen that the teleostean gill is far more permeable to K^+. Gill resistance was measured and found to be extremely low. Finally Smith (1969b) studied the effects of transfer into Na^+-free or Cl^--free media on the Na^+ or Cl^- effluxes. Both declined instantaneously and independently on transfer. The decreases in the Na^+ efflux can be explained by changes in the potential difference and diffusional permeability. For Na^+, exchange-diffusion does not occur. For Cl^-, however, 70% of the exchange fluxes are due to exchange-diffusion and most of the remainder is due to active transport. Furthermore, Smith showed that ion efflux curves which can be expressed by Michaelis-Menten kinetics do not constitute evidence for the presence of exchange-diffusion, because curves of a similar shape can be obtained from purely diffusional processes, given a constant electric field across the whole epithelium. As in the seawater-adapted eel, the Na^+/K^+ exchange process is also explained by the potential changes, but it remains to be seen whether the active K^+ transport, which was found to occur across the gill by Smith (1969a), is mediated by a $Na^+ + K^+$ dependent ATPase. The presence of this enzyme in *Artemia* is documented by Augenfeld (1969) who observed that its activity increased with increasing salinity in the adaptation medium. Unfortunately the activity was measured in whole homogenates of the animals. It is probable, however, that the enzyme is located in the "chloride cells", which were described by Copeland (1967) in the branchial metepipodites of the adult. Similar cells are found in the neck-organ of the nauplii (Conte *et al.*, 1972). Moreover Ewing *et al.* (1972) showed that addition of ouabain to the external medium caused a rapid decrease in survival of the nauplii. This effect is dependent on the salinity of the incubation medium and the dosage of ouabain. Furthermore, various inhibitors of protein or nucleic acid syntheses, when added to the incubation media, also caused marked reduction of survival.

All these agents were more effective in higher salinities. Thus, as in the seawater-adapted eel, transcriptional and translation events in relation to salt-dependent macromolecular synthesis and cell differentiation play an important role in the functioning of the gill.

There is very little information about the occurrence and mechanism of extrarenal salt excretion in hypo-osmoregulating crustaceans other than *Artemia*.

The problem of the role of the gut has been discussed previously (see p. 108) with respect to the penaeid shrimp, *Metapenaeus bennettae*, and the grapsid crab, *Metapograpsus gracilipes*. In *Metapenaeus*, Dall (1967) demonstrated that the cephalothoracic region—presumably the gills—is responsible for most of the rapid electrolyte turnover observed in sea water. He argues that it does not necessarily mean that the gills are the site of salt excretion. Indeed, after salt-loading, the Na^+ and Cl^- effluxes from this region decreased, while the abdominal efflux, which presumably includes salt excretion *via* the anus, increased. In *Palaemonetes varians*, Potts and Parry (1964b) also observed very rapid Na^+ and Cl^- turnover fluxes in seawater: 15 000 and 50 000 $\mu Eq/100g/h$ respectively. These steady state fluxes vary in relation to the external salinity. The electric potential was found to be negative inside whereas in marine teleosts, it is positive. The flux-force relationship indicates an active excretion of Na^+ and a passive behaviour of Cl^-. In neither palaemonid nor penaeid shrimps have the morphological or biochemical aspects of gill or gut been studied in relation to their presumed excretory functions.

For the isopod, *Sphaeroma*, more information is available. Harris (1972) observed a negative potential (inside) in seawater-adapted animals, but in contrast to *Palaemonetes*, which exhibits an internal Na^+ and Cl^- concentration about two-thirds of that measured in sea water, *Sphaeroma rugicauda* is nearly iso-osmotic. This species hypo-osmoregulates, however, at higher external salt water concentrations (150% sea water), but the potential across the body surface was unfortunately not measured under such conditions. Both Na^+ and Cl^- move passively across the body surface of animals kept in sea water, but both ions are actively absorbed in lower salinities when the animals are hyper-osmoregulating. The results of Thuet *et al.* (1969) concerning the $Na^+ + K^+$ dependent APTase activity found in the pleopods of *Sphaeroma serratum* have been discussed in the preceding section. It is of interest to note that in 150% sea water and 100% sea water the activity is lower than in 20% sea water. Unfortunately this species does not clearly exhibit hypo-osmoregulating tendencies in higher salinities. Finally copepods, which have recently been recognized as hypo-osmotic regulators (Brand and Bayly, 1971), lack gills. Specialized cells have been observed on the mandibles which strongly resemble "chloride cells". Ong (1969) suggests that these cells may serve osmoregulatory functions.

In insect larvae, which are able to adapt to salt water pools, osmotic regulation is carried out in a variety of ways. In the euryhaline dipterous *Aëdes detritus*, *Ephydra riparia* or *Coelopoda frigida*, for example, preliminary measurements of the rectal fluid osmolality indicate that the excretory malpighian tubules are capable of secreting a highly concentrated fluid (Ramsay, 1950; Sutcliffe, 1960). Osmotic independence of these larvae is achieved by a considerable degree of impermeability to water and salts, and it is significant that they do not possess functional anal papillae characteristic of many freshwater dipterous larvae. Intake of salt water through the gut is likely to occur, and the larvae rely on renal excretory mechanisms to economize water and excrete salt. In *Aëdes campestris* larvae a pattern of osmotic and ionic regulation very similar to that of *A. detritus* is observed, yet this species retains anal papillae with an ultrastructure similar to that described by Copeland in *A. aegypti*. Animals which had their papillae destroyed by 1% $AgNO_3$, when reared in iso-osmotic saline and then transferred to hyper-osmotic lake water, had a significantly greater blood Cl^- level than control larvae. Phillips and Meredith (1969) suggest that the direction of the ionic transport across the anal papillae may be reversed in a manner reminiscent of gill function in euryhaline fish. In the caddisfly larvae of *Philanisus plebeius* which are exceptional in being able to live in seawater rock pools, Leader (1972) found that the rectal fluid was much less concentrated than the external medium and thus the kidney is inadequate to maintain salt balance. It is thus necessary to postulate an extrarenal site of salt excretion. Measurements of the electric potential difference across the body wall indicated that the haemolymph was negative (-5.5 mV) with respect to the outside medium. Active transport of both Na^+ and Cl^- must, therefore, occur across the body wall. It would be of great interest to verify whether, in this trichopteran, chloride cells are present either in anal papillae analogous to those observed in *Philopotamus* or the Glossomatidae, or in specialized abdominal fields analogous to those found in *Anabolia* (see preceding section p. 58).

C. Branchial and body surface water permeability in relation to hypo-osmotic regulation.

In the preceding section the characteristics of the diffusional and osmotic permeabilities of various aquatic animals were considered in relation to external salinity. Data from seawater living animals, whether iso-osmotic regulators (worms, crustaceae) or hypo-osmotic regulators, were included for comparison in Tables IV, V, VI and VII (pp. 66-68, 70-73, 74-75).

1. DIFFUSIONAL WATER PERMEABILITY

Several points concerning the diffusional water permeabilities emerge from a study of Tables IV – VII.

(a) Iso-osmotic regulators exhibit higher HTO turnovers in sea water than hypo-osmotic regulators. Such a difference is well documented in fish by comparing the iso-osmotic elasmobranchs or myxinoids with the hypo-osmotic teleosts, and in crustaceans by comparing *Artemia, Penaeus, Uca* and *Palaemonetes* with the iso-osmotic *Gammarus, Carcinus* and *Macropipus.* The chelicerate, *Limulus,* also comes into this category.

(b) Hypo-osmotic regulators, with the highest degree of osmotic independence from the external medium, exhibit the lowest HTO turnover rate. This is shown by comparing the efficient *Artemia* with *Palaemonetes, Penaeus* and *Uca,* all which are far less efficient osmoregulators. According to Hannan and Evans (1973) the rather low HTO turnovers found in *Uca,* particularly in *Uca rapax,* may be ascribed to the tendency towards terrestrial life and the resistance to desiccation exhibited by this species.

(c) According to Leader (1972), in the hypo-osmoregulating insect larvae there is a considerable difference in the body surface permeability of the diptera larvae, which possess spiracles and are able to gain access to atmospheric air for their respiratory needs, and the trichopteran larvae which depend upon dissolved oxygen and possess trachael gills. Thus, the HTO turnover of the marine mosquito *Opifex fuscus* (Nicholson, un-published data given by Leader, 1972) is very low with a P_d value of 0.15 μsec^{-1} which is 4 to 5 times lower than the values observed in *Artemia salina* or *Opsanus tau,* the latter being one of the most impermeable teleosts (see Table VII). The value for *Opifex* is also 3 times lower than that given by Shaw (1955a) for the cuticle of *Sialis lutraria.* From a study of the rate of penetration of D_2O, Nemenz (1960) concludes that the permeability of another salt water mosquito, *Ephydra cineria,* is similar to that of *Sialis* larvae. However, Shaw and Stobbart (1963) interpret the data presented as in fact showing a much lower D_2O permeability constant, at least 10 times less than that of *Sialis.* Comparative HTO turnover experiments in freshwater and salt water culicines would be of great interest, especially in salt water forms such as *Aëdes detritus* or *A. campestris,* which differ in the development of their anal papillae.

The HTO turnover rate in the hypo-osmoregulating caddisfly larvae of *Philanisus* is indeed relatively high, about 40% of the body weight per hour (Leader, 1972). The permeability coefficient, however, could not be calculated because of the uncertainty in evaluating the body surface and of the branched nature of the tracheal gills.

(d) Euryhaline teleosts show in general a lower diffusional water permeability in sea water than in fresh water. This difference may be considered as an adaptive feature in relation to an osmotic gradient about two to three times higher in seawater than in freshwater. There are, however, many exceptions, including *Anguilla anguilla* (Evans, 1969b; Motais and Isaia, 1972a), *Xiphister atropurpureus* (Evans, 1967b) and *Pholis gunnellus* (Evans, 1969a). When the osmotic permeabilities are considered, these differences in water permeabilities in relation to the external salinities or to the osmotic gradients between external and internal media are confirmed, but complicating factors arise which need to be discussed.

2. OSMOTIC WATER PERMEABILITY OF THE TELEOSTEAN GILL: THE PROBLEM OF SOLUTE-SOLVENT INTERACTION

Various studies have been undertaken on euryhaline and stenohaline teleosts, adapted to fresh water or sea water, to evaluate indirectly the osmotic net flow across the gills (Motais *et al.*, 1969; Motais and Isaia, 1972a; Isaia, 1972). The drinking rate minus the urine flow is taken as a measure of the rate of water loss in the marine environment. The urine flow minus the drinking rate corresponds to the osmotic inflow of water from the freshwater environment. The osmotic net flow is higher in freshwater fish than in seawater ones. This applies to both stenohaline and euryhaline species. Yet the osmotic gradient is about 3 times greater in seawater than in freshwater. Table VI expresses the results in % body weight per osmole gradient per hour. Thus it seems definitely established, at least for Teleosts, that seawater adaptation is characterized by a great reduction in the osmotic permeability of the body surface. There are, however, problems concerning the measurement of the osmotic water permeability coefficient of the body surface.

(1) The osmotic net flow has been determined indirectly from two independent series of measurements (urine flow and drinking rate), while the fish were assumed to be in a steady state with respect to their water balance. Some of these measurements were made on handled fish and handling may result in weight loss in seawater fish and weight gain in freshwater fish (Stevens, 1972). These factors may lead to an underestimation of the net osmotic flow. The amount of water swallowed is also assumed to be completely absorbed. While this is probably true in freshwater fish, it is not the case in seawater fish in which only 60–80% of the water is absorbed (see above and Table IX). The urine flow may also be overestimated as catheterization prevents the bladder from completing its absorptive function.

(2) To calculate P_{os}, it is necessary to take into account not only the difference in osmolalities of the external and internal media, but also σ, the reflexion coefficient which indicates the degree of interaction of solutes

(essentially salt) and solvent. In the calculations given in Tables VI and VII the assumption was made that the membrane separating both media is semi-permeable. This is probably true for the body surfaces of the fresh-water fish, which are rather impermeable to Na^+ and Cl^-. In seawater teleosts, however, the gills are characterized by a high ionic permeability. In the preceeding section, the active and passive nature of the electrolyte exchanges were discussed. The carrier-mediated active transport and exchange-diffusion components may not interact with the passive flow of solvent. Only the passive flux of electrolytes moving by diffusion along the electro-chemical gradient would probably interfere with the water flow, causing the reflexion coefficient to be less than 1. This passive component is obviously important in stenohaline or euryhaline fish, such as *Serranus* and *Fundulus*, which display no "Na^+ or Cl^--free" effect. In the preceding paragraph, we have seen that, even in the eel or the flounder, the mono-valent cation movements may result from diffusional flows, while for Cl^- exchange-diffusion almost certainly take place. Thus even in those fish, cationic movement may interfere with water movement. Furthermore, in both *Serranus* and *Anguilla* after transfer from sea water to fresh water, the addition to the external medium of mannitol, considered to be an im-permeant solute, was shown to interfere with the Na^+ efflux. This finding shows that an experimental change of the osmotic gradient influences electrolyte movement in these fish (Motais *et al.*, 1966; Motais, 1967). It may be assumed that the opposite is also true, i.e. that electrolyte movement influences water movement. Very recently J. Isaia (personal communi-cation) showed that replacement of seawater by an iso-osmotic sucrose solution produces a 5 to 6 fold increase in the water loss in eels which were prevented from drinking by blocking the oesophagus with an inflatable balloon. This confirms that $\sigma \ll 1$ in the seawater eel.

In Table VII, the values calculated for P_{os} in various fish, assuming that $\sigma = 1$, are compared with the P_d value obtained from the HTO turn-over studies. For many seawater teleosts the two permeability coefficients have similar values, which may be taken as an indication that all water movements are diffusive, or else that no movement occurs by way of water-filled channels and bulk flow. Evans (1969b), Potts *et al.* (1970) and Motais *et al.* (1969) have considered this possibility. It must be emphasized that the suggestion only holds if one assumes that there is no solute-solvent interaction.

This problem was recently reconsidered by Motais and Isaia (1972a) and by Isaia (1972). The temperature dependence of diffusional perme-ability to water and Na^+ was compared in freshwater and seawater-adapted eels, in goldfish and in seawater perch. The most interesting observations were obtained on the seawater-adapted eel. The ratio of water to Na^+ permeabilities was approximately 3 at 25°C but fell to 1 at 5°C. This drop was mainly due to a diminution of the water permeability, the temperature coefficients being much higher for water than for Na^+. The relatively independent variations of water and Na^+ permeabilities probably indicate a certain dissociation of water and salt movements. Moreover, the osmotic water permeability declined even faster than the diffusional water permeability so that, if one assumes $\sigma = 1$, the ratio P_{os}/P_d decreased with decreasing temperature. Values as low as 0.28 and 0.17 were given. This considerable discrepancy can only be explained by inferring that the osmotic pressure difference does not represent the true osmotic pressure gradient across the epithelial barrier. The second possibility envisaged by Motais and Isaia (1972a) and Motais and Garcia-Romeu (1972) is that in the chloride cells water absorption occurs against its chemical gradient from the external toward the internal medium, and is probably linked with solute movement. The net solute movement, however, is in the opposite (outside) direction. The authors, therefore, make the further assumption that a recycling of the ions takes place in a manner analogous to that proposed by Phillips (1970) for the rectal pad of insects. If indeed drinking rates in the eel in unstressed conditions are very low, or even if the gut degenerates during seaward migration in such animals, some extra-intestinal means of up-hill water intake must then develop to compensate for osmotic and renal water losses. The problem is whether such mechanisms are indeed necessary in the majority of seawater living teleosts, for which the gut represents the obvious site of "uphill" water intake and the gill the obvious site for extra-renal compensatory salt excretion.

3. EVOLUTION OF THE WATER PERMEABILITY OF THE GILL AFTER TRANSFER FROM
   FRESH WATER TO SEA WATER

Comparison of the reversed osmotic flows of gills of euryhaline fishes in fresh water and in sea water suggests that the mucosal to serosal permeability is higher than that in the opposite direction. Such a difference in permeability with reversed flows was first described by Bentley (1964) in the toad bladder and called rectification of flow. It must be emphasized, however, that the osmotic flow across the gill is measured in long term adapted fish and it is possible that the observed differences of the permeabilities are the result of slow adaptive mechanisms modifying the epithelial permeability.

A kinetic study of the changes in drinking rate, intestinal absorption and renal excretion of water, together with changes of body weight and water content in *Anguilla japonica* in relation to transfer from fresh water to sea water, has shown that during the first day of transfer the rate of water loss across the gill is three times higher than after 3 days adaptation (Oide and Utida, 1968). This strongly suggests changes in osmotic water permeability. Kamiya (1967) arrived at the same conclusion from her investigations on incubated gill preparations. Excised freshwater eel gills lost about twice as much water as seawater eel gills when incubated in seawater. Gills excised from eels after 6–12 h adaptation to sea water already lost significantly less water than freshwater gills. After 24 h adaptation the gill behaved as a seawater gill.

Another striking adaptive feature of the gill during fresh water to sea water-adaptation is its progressive augmentation of permeability to Na$^+$ and Cl$^-$ ions, which has been discussed in the preceding section. It is quite possible that the resulting increase in "leakiness" of the gill is responsible for the reduction in the effective osmotic gradient or reflexion coefficient. This explanation of the osmotic behaviour of the gill is more likely than the one which postulates a disappearance of the water-filled channels which are then replaced by molecular pores. Differences in the Ca^{2+} content of fresh water and sea water have recently been suggested to be of importance in the increase of diffusional permeability which accompanies fresh water-adaptation. Potts and Fleming (1970) observed that removal of the Ca^{2+} from fresh water or sea water is followed by a 100% increase in the HTO turnover rate in *Fundulus kansae*. The Mg^{2+} which is found in high concentrations in sea water may also play a role. Addition of Ca^{2+} and Mg^{2+} salts to fresh water significantly reduces the HTO turnover rate. Calcium ions tend to reduce the hydration of polar organic molecules by reducing the repulsive forces between fixed anions of the cell membrane, thus allowing for closer packing of the molecules. It is likely that the effect of Ca^{2+} removal does not interfere with the stability of the inter-cellular cements, as there is probably enough Ca^{2+} in the internal body fluids to maintain these structures.

Whether these changes in diffusional water permeability are paralleled by changes in osmotic water permeability remains to be investigated. However, according to Potts and Fleming (1970) it is probable that Ca^{2+}, by reducing the hydration of the polar organic groups of the cell membrane also decreases membrane porosity.

The time course of the Ca^{+2} effect also requires further investigation. Preliminary experiments by Potts and Fleming (1970) suggest that permeabilities continue to increase during the second day after transfer from high to low Ca^{2+} concentrations. Adaptation to high Ca^{2+} is probably more rapid. Cuthbert and Maetz (1972) and Bornancin *et al.* (1972) observed that the reduction of branchial permeability to Na$^+$ after addition of Ca^{2+} to the external medium (deionized water or Ca^{2+}-free sea water) occurs within an hour.

Adaptation from fresh water to sea water is accompanied by relatively slow changes in water permeability: 3 days in the *in vivo* studies by Oide and Utida (1968) and 1 day in the *in vitro* study by Kamiya (1967). More recently Macfarlane and Maetz (1974) have observed that the increase in HTO turnover rate which accompanies freshwater adaptation in the flounder occurs only during the second week of transfer. It thus seems that permeability changes are controlled in part by exogenous factors (Ca^{2+}, Mg^{2+}) and in part by endogenous factors (hormones).

The role of endocrines, especially prolactin, has been discussed in the preceding section. Prolactin is characterized by its relatively slow mechanism of action and its long term effect. Another hormone which is characterized by a very rapid action is adrenalin. Pic *et al.* (1973) have recently shown that high doses of adrenalin injected intraperitoneally into the seawater-adapted mullet produced a loss of weight caused by an increased osmotic flow across the gill. Within minutes of injection a 100% increase of the diffusional water efflux was observed. This effect was mediated by β receptors since propranolol alleviates the effects of adrenalin. Adrenalin was previously shown by Rankin and Maetz (1971) to produce haemodynamic changes in the gill blood flow (also mediated by β receptors). The relative importance of such alterations in the branchial surface available for water diffusion, as well as of true changes of the epithelial water permeability, remains to be studied. In any case alterations in the circulating adrenalin level are probably responsible for the effects of handling on the water balance of the fish.

4. OSMOTIC WATER PERMEABILITY: COMPARATIVE ASPECTS

Investigations have been made on the osmotic permeability of the body surface in hypo-osmotic regulating crustaceans.

The most complete information concerns the brine shrimp. Smith (1969*b*) compared the diffusional water permeability as deduced from HTO turnover experiments (see above) with the osmotic water permeability calculated from the drinking rate, neglecting urine loss from the maxillary gland. The value obtained, *i.e.* 3.3% body weight/h/osmole gradient, is much larger than those recorded from seawater teleosts (see Table VI). The ratio P_{os}/P_d is accordingly very high (13.2). Thus *Artemia* differs from the hypo-osmoregulating teleosts. It may be noted that solute-solvent flow interaction may be of importance in *Artemia*, because of the high rate of ionic exchange and the high permeability of the gill to electrolytes. Smith (1969*b*) suggests that, at least for the cations, the movements are diffusive in nature. Thuet *et al.* (1968) compared the Na^+ and Cl^- efflux of *Artemia* after transfer to fresh water or to fresh water made iso-osmotic by the addition of mannitol. No significant difference was observed in the presence or absence of an osmotic flow. They concluded that solute-solvent interactions are probably of little importance.

Osmotic permeabilities of the body surfaces of various other hypo-osmotic crustaceans, such as the penaeid shrimp, *Metapenaeus bennettae*, or the grapsid crab, *Metapograpsus gracilipes*, may be calculated from the data given by Dall (1967), taking into account the drinking rate and ignoring the urinary output. These animals are more permeable than *Artemia* (see Table VI), but the osmotic gradient is much less. The HTO permeabilities are also higher than in *Artemia* (see Table V).

In the hypo-osmoregulating Tricoptera larvae, which depend on dissolved oxygen and possess tracheal gills, the osmotic permeability is also very high. In the most recently investigated euryhaline caddisfly larvae, *Philanisus*, Leader (1972) observed a high drinking rate of about 1% of the bodyweight per hour, which is indicative of a high permeability. High values are also suggested by the rate of shrinking in sea water, and the rate of swelling in fresh water of larvae with the mouth and anus ligatured to prevent drinking and urine loss. The values, 8–10%/day given by Leader (1972), are almost identical with those obtained by Sutcliffe (1961*b*) for the freshwater caddisfly larvae, *Limnophilus stigma* and *Anabolia nervosa*, and for the more euryhaline *L. affinis*. Thus, in caddisfly larvae the capacity to adapt to saline media is not accompanied by a reduction of the osmotic permeability coefficient of the body wall.

Other freshwater insect larvae are characterized by either a lower or a higher osmotic permeability than that found in the caddisfly larvae. Values 3–4 times higher are found in *Aëdes aegypti* or *Chaoborus plumicornis* (see review by Shaw and Stobbart, 1953) which are characterized by the development of anal papillae. A value 50% lower is found in the alderfly larvae, *Sialis lutaria* (Shaw, 1955*a*).

IV. Concluding remarks: Evolutionary aspects of hyper and hypo-osmotic regulation

Conservation of water is an important problem of life because water provides the physical framework in which energy transformations characteristic of life take place. The presence of salt in the cellular fluids suggests that life originated in the sea, and in all organisms water metabolism is closely connected with that of salts. It is probable that the cells of primitive organisms, like those of most marine invertebrates today, were iso-osmotic to the external medium, although the ionic composition of the cellular fluids may have been different from that of sea water.

A critical moment of evolution must have been the invasion of brackish waters, followed later by the invasion of fresh water and dry land. In brackish water the body fluids of some organisms remain iso-osmotic with the environment. The adjustment of the cellular osmotic pressure is obtained by non-electrolytes (generally amino acids) as osmotic "fillers". Well established invaders of low salinity media, however, whether invertebrates or vertebrates, used extracellular fluids as a buffer compartment allowing for the maintenance of a constant cellular environment. The regulation of hetero-osmotic body fluids was achieved by the development of specialized membranes separating the external and internal media and controlling salt and water input and output. The present review demonstrates that similar mechanisms of active salt uptake have evolved independently in the various invertebrate and vertebrate phyla which invaded low salinity media, whether they were primary aquatic animals, such as teleosts and crustaceans, or secondary aquatic animals reinvading freshwater from dry land, such as the pulmonate snails and the various insect larvae.

Reinvasion of brackish waters and the seas by invertebrates and vertebrates has occured at various geological times. Most organisms readapted to salt water by developing ingenious devices to acquire water and to eliminate excess salt, while remaining hypo-osmotic to the environment. The ability of the intestinal tract to absorb water by absolute-linked water flow mechanism was used by most species, allowing for "up-hill" water movement in the most successful ones. Except in mammals and some insect larvae, the kidney remained unsuitable for the purpose of salt excretion. Extra-renal salt excretion evolved independently in primary and secondary aquatic animals. In the primary aquatic animals, which are the main object of this review, and in some insect larvae, the same effector organ which specialized in salt absorption in low salinity media became capable of salt excretion in high salinity media. Whether this "switch-over" corresponds to a true reversal of the functional polarity of the same epithelial cells has been discussed. In most hypo-osmotic crustaceans, however, the site of extra-renal salt excretion remains in doubt.

From an evolutionary point of view, the case of the "ureo-osmotic" regulators is of special interest. In these animals, as in hypo-osmotic regulators, the extracellular fluids have a lower concentration of NaCl than does sea water. Iso-osmolarity between external and internal media is achieved by urea retention, accompanied by urea accumulation in the tissues. The use of urea as an "osmotic filler" has evolved independently in three groups of vertebrates: elasmobranchs, coelacanths and amphibians.

In the amphibians reinvasion of brackish waters was undoubtedly made relatively recently by ancestral freshwater species (Gordon *et al.* 1961). In the two other groups, however, freshwater ancestry is still being debated. The coelacanths are a side-line of the crossopterygians, appearing in the Upper-Devonian period in both fresh water and salt water deposits. An early freshwater stage may, therefore, be responsible for the low salt content characterising the body fluids of the recently discovered *Latimeria* (Pickford and Grant, 1967). The elasmobranchs, in contrast, seem to have been ocean-dwellers since their appearance in the geological record, although side-lines of the main evolutionary branch in the Upper-Devonian were freshwater species (Romer, 1946). The observation that a blood urea concentration approximately one-third of that of marine species is found in estuarine freshwater elasmobranchs from localities not very far from the sea, is suggestive of a marine or brackish-water origin for this group rather than a freshwater origin (Thorson, 1967). The recent discovery of *Potamotrygon*, a sting-ray occuring in places as far as 5000 km from the sea, with an urea content lower than that of many higher vertebrates, poses many questions especially as these animals, when placed in salt water, were found to be unable to add urea to their body fluids to increase their internal osmolarity (Thorson *et al.*, 1967; Thorson, 1970). Thorson concludes that "It is . . . impossible to state that the rays under study were derived from marine ancestors that possessed mechanisms for retaining urea. The possibility exists, although remotely, that they descended from ancestors that did not leave fresh water . . . and thus never developed urea retention". Payan and Maetz (1973) have recently summarized the physiological evidence showing that marine elasmobranchs display branchial electrolyte exchange mechanisms which are rather similar to those characterizing freshwater teleosts. The statement of Homer Smith (1953) that the class Chondrichthyes derived from freshwater ancestors may have to be reconsidered.

The question of the habitat of primitive vertebrates has recently been surveyed by Robertson (1957), and a freshwater origin was rejected on paleontological and physiological grounds. Paleontologically, the remains of the earliest vertebrates are found in Ordovician and Silurian sediments associated with typical marine invertebrates. Only in the Devonian period are ostracoderms and bony fishes represented in both marine and fresh-water deposits. Physiologically, the development and widespread distribution of the filtration-type kidney does not necessarily provide evidence that vertebrates evolved in fresh water. Glomerular kidneys "are neither uniquely vertebrate no exclusively fresh water" according to Kirschner (1967). A glomerular kidney is found in the primitive myxinoids, which must have inherited their internal medium of high salt concentration from marine chordate ancestors, and in marine crustaceans and molluscs.

V. References

Aceves, J. and Erlij, D. (1971). *J. Physiol., Lond.* **212**, 195–210.
Adams, B. L., Zaugg, W. S. and McLain, L. R. (1973). *Comp. Biochem. Physiol.* **44A**, 1333–1339.
Alvarado, R. H. and Dietz, T. H. (1970). *Comp. Biochem. Physiol.* **33**, 93–110.
Alvarado, R. H. and Johnson, S. R. (1965). *Comp. Biochem. Physiol.* **16**, 531–546.
Alvarado, R. H. and Moody, A. (1970). *Am. J. Physiol.* **218**, 1510–1516.
Alvarado, R. H. and Stiffler, D. F. (1970). *Comp. Biochem. Physiol.* **33**, 209–212.
Andersen, B. and Ussing, H. H. (1957). *Acta physiol. scand.* **39**, 228–239.
Augenfeld, J. M. (1969). *Life Sci.* **8**, 973–978.
Balinsky, J. B. and Baldwin, E. (1961). *J. exp. Biol.* **38**, 695–705.
Ball, J. N., Chester Jones, I., Forster, M. E., Hargreaves, G., Hawkins, E. F. and Milne, K. P. (1971). *J. Endocr.* **50**, 75–96.
Bank, N. and Schwartz, W. B. (1960). *J. appl. Physiol.* **15**, 125–127.
Bates, R. B. and Pinching, G. D. (1949). *J. Res. natn. Bur. Stan.* **42**, 419–430.
Beadle, L. C. (1939). *J. exp. Biol.* **16**, 346–362.
Bellamy, D. (1961). *Comp. Biochem. Physiol.* **3**, 125–135.
Bentley, P. J. (1962). *Comp. Biochem. Physiol.* **6**, 95–98.
Bentley, P. J. (1964). *Life Sci.* **4**, 133–140.
Bentley, P. J. (1971a). "Endocrines and Osmoregulation. A Comparative Account of the Regulation of Water and Salt in Vertebrates". Springer Verlag, Berlin.
Bentley, P. J. (1971b). *J. Endocr.* **49**, 183–184.
Bentley, P. J. and Heller, H. (1964). *J. Physiol. Lond.* **711**, 434–453.
Bentley, P. J. and Main, A. R. (1972). *Am. J. Physiol.* **223**, 361–363.
Bergmiller, E. and Bielawski, J. (1970). *Comp. Biochem. Physiol.* **37**, 85–91.
Bethe, A. (1934). *Pflügers Arch. ges. Physiol.* **234**, 629–644.
Biber, T. U. L. (1971). *J. gen. Physiol.* **58**, 131–144.
Biber, T. U. L. and Curran, P. F. (1970). *J. gen. Physiol.* **56**, 83–99.
Biber, T. U. L., Aceves, J. and Mandel, L. J. (1972). *Am. J. Physiol.* **222**, 1366–1373.
Bielawski, J. (1964). *Comp. Biochem. Physiol.* **13**, 423–432.
Bielawski, J. (1971). *Protoplasma* **73**, 177–190.
Bierther, M. (1970). *Z. Zellforsch. mikrosk. Anat.* **107**, 421–446.
Bliss, D. E. (1968). *Am. Zool.* **8**, 355–392.
Bliss, D. E., Wang, S. M. E. and Martinez, E. A. (1966). *Am. Zool.* **6**, 197–212.
Bonting, S. L. (1971). *In* "Membranes and ion Transport" (E. E. Bittar, Ed.), Vol. 1, pp. 257–363, Wiley-Interscience, London and New York.
Boonkoom, V. and Alvarado, R. H. (1971). *Am. J. Physiol.* **220**, 1820–1824.
Bornancin, M. and de Renzis, G. (1972). *Comp. Biochem. Physiol.* **43A**, 577–591.
Bornancin, M., Cuthbert, A. W. and Maetz, J. (1972). *J. Physiol. Lond.* **222**, 487–496.
Brand, G. W. and Bayly, I. A. (1971). *Comp. Biochem. Physiol.* **38B**, 361–371.
Britton, H. G. (1965). *J. theor. Biol.* **10**, 28–52.
Britton, H. G. (1970). *Nature, Lond.* **225**, 746–747.

Bromberg, P. A., Robin, E. D. and Forkner, C. E. (1960). *J. clin. Invest.* **39,** 382–341.

Brown, A. C. (1962). *J. cell. comp. Physiol.* **60,** 263–270.

Bryan, G. W. (1960*a*). *J. exp. Biol.* **37,** 83–99.

Bryan, G. W. (1960*b*). *J. exp. Biol.* **37,** 100–112.

Bryan, G. W. (1960*c*). *J. exp. Biol.* **37,** 113–128.

Bursey, C. R. and Lane, C. E. (1971). *Comp. Biochem. Physiol.* **39A,** 483–493.

Burton, R. F. (1973). *Biol. Rev.* **48,** 195–231.

Butler, D. G. and Carmichael, F. J. (1972). *Gen. comp. Endocr.* **19,** 421–427.

Cameron, J. N. and Randall, D. J. (1972). *J. exp. Biol.* **57,** 673–680.

Capen, R. L. (1972). *J. exp. Zool.* **182,** 307–320.

Carasso, N., Favard, P., Jard, S. and Rajerison, R. M. (1971). *J. Microscopie.* **10,** 315–330.

Carlisle, D. B. (1956). *Pubbl. Staz. zool. Napoli* **27,** 227–231.

Carrier, J. C. and Evans, D. H. (1972). *Comp. Biochem. Physiol.* **41A,** 761–764.

Carrier, J. C. and Evans, D. H. (1973). *Comp. Biochem. Physiol.* **45A,** 667–670.

Chaisemartin, C. (1967). "Contribution à l'étude de l'économie calcique chez les Astacidae. Influence du milieu de vie". Thèse de Doctorat d'Etat, Univ. Poitiers.

Chaisemartin, C. (1968*a*). *C.r. Séanc. Soc. Biol.* **162,** 1193–1195.

Chaisemartin, C. (1968*b*). *C.r. Séanc. Soc. Biol.* **162,** 1194–1198.

Chaisemartin, C. (1969). *C.r. Séanc. Soc. Biol.* **163,** 2422–2425.

Chaisemartin, C., Martin, P. N. and Bernard, M. (1968). *C.r. Séanc. Soc. Biol.* **162,** 523–526.

Chaisemartin, C., Martin, P. N. and Bernard, M. (1970). *C.r. Séanc. Soc. Biol.* **164,** 877–880.

Charmentier, G. (1971). *C.r. hebd. Séanc. Acad. Sci., Paris* **272,** 444–446.

Chester Jones, I. and Bellamy, D. (1964). *Symp. Soc. exp. Biol.* **18,** 195–236.

Chester Jones, I., Chan, D. K. O., Henderson, I. W. and Ball, J. N. (1969). *In* "Fish Physiology" (Hoar, W. S. and Randall, D. J., Eds.) Vol. 2., pp. 321–376. Academic. Press. New York.

Conner, R. L. (1967). *In* "Chemical Zoology" (Florkin, M. and Scheer, B. T., Eds.) Vol. 1., "Protozoa" (Kidder, G. W., Ed.), pp. 309–350. Academic Press, New York and London.

Conte, F. P. (1965). *Comp. Biochem. Physiol.* **15,** 293–302.

Conte, F. P. (1969). *In* "Fish Physiology" (Hoar, W. S. and Randall, D. J., Eds.), Vol. 1. pp. 241–292. Academic Press, New York.

Conte, F. P. and Lin, D. H. Y. (1967). *Comp. Biochem. Physiol.* **23,** 945–957.

Conte, F. P., Hootman, S. R. and Harris, P. J. (1972). *J. Comp. Physiol.* **80,** 239–246.

Copeland, D. E. (1948). *J. Morph.* **82,** 201–227.

Copeland, D. E. (1950). *J. Morph.* **87,** 369–379.

Copeland, D. E. (1964). *J. cell. Biol.* **23,** 253–263.

Copeland, D. E. (1967). *Protoplasma* **63,** 363–384.

Copeland, D. E. (1968). *Am. Zool.* **8,** 417–433.

Copeland, D. E. and Fitzjarrell, A. T. (1968). *Z. Zellforsch. mikrosk. Anat.* **92,** 1–23.

Crabbé, J. (1964). *Endocrinology* **75,** 809–811.

Crabbé, J. and de Weer, P. (1964). *Nature, Lond.* **202,** 298–299.

Crabbé, J. and de Weer, P. (1969). *Pflügers Arch. ges. Physiol.* **313,** 197–221.

Croghan, P. C. (1958a). *J. exp. Biol.* **35,** 219–233.

Croghan, P. C. (1958b). *J. exp. Biol.* **35,** 234–242.

Croghan, P. C. (1958c). *J. exp. Biol.* **35,** 243–249.

Croghan, P. C. (1958d). *J. exp. Biol.* **35,** 425–436.

Croghan, P. C. and Lockwood, A. P. M. (1968). *J. exp. Biol.* **48,** 141–158.

Curran, P. F. (1960). *J. gen. Physiol.* **43,** 1137–1148.

Curran, P. F. and Macintosh, J. R. (1962). *Nature, Lond.* **193,** 347–348.

Cuthbert, A. W. and Maetz, J. (1972). *J. Physiol. Lond.* **221,** 633–643.

Cuthbert, A. W. and Pic, P. (1973). *Br. J. Pharmac.* **49,** 134–137.

Dainty, J. (1965). *Symp. Soc. exp. Biol.* **19,** 75–85.

Dainty, J. and House, C. R. (1966a). *J. Physiol. Lond.* **182,** 66–78.

Dainty, J. and House, C. R. (1966b). *J. Physiol. Lond.* **185,** 172–184.

Dall, W. (1965). *Aust. J. mar. Freshwat. Res.* **16,** 181–203.

Dall, W. (1967). *Comp. Biochem. Physiol.* **21,** 653–679.

Dall, W. and Milward, N. E. (1969). *Comp. Biochem. Physiol.* **30,** 247–260.

Dejours, P. (1969). *J. Physiol. Lond.* **202,** 113P–114P.

Dejours, P., Armand, J. and Verriest, G. (1968). *Respiration Physiol.* **5,** 23–33.

Denne, L. B. (1968). *Comp. Biochem. Physiol.* **26,** 17–31.

Deyrup, I. J. (1964). *In* "Physiology of the Amphibia" (Moore, J. A., Ed.), pp. 251–328. Academic Press, New York and London.

Dharmamba, M. and Maetz, J. (1972). *Gen. Comp. Endocrin.* **19,** 175–183.

Dharmamba, M., Mayer-Gostan, N., Maetz, J. and Bern, H. A. (1973) **21,** 179–187.

Diamond, J. M. (1964a). *J. gen. Physiol.* **48,** 2–14.

Diamond, J. M. (1964b). *J. gen. Physiol.* **48,** 15–42.

Diamond, J. M. (1965). *Symp. Soc. exp. Biol.* **19,** 329–347.

Diamond, J. M. (1971). *Fedn. Proc. Fedn. Am. Socs exp. Biol.* **30,** 6–13.

Dick, D. A. T. (1966). "Cell Water" *in* Molecular biology and medicine series (Bittar, E. E., Ed.) Butterworths, London.

Dicker, S. E. and Elliott, A. B. (1970). *J. Physiol. Lond.* **207,** 119–132.

Dietz, T. H. and Alvarado, R. H. (1970). *Biol. Bull.* **138,** 247–261.

Dietz, T. H., Kirschner, L. B. and Porter, D. (1967). *J. exp. Biol.* **46,** 85–96.

Downing, K. M. and Merkens, J. C. (1955). *Ann. appl. Biol.* **43,** 243–246.

Doyle, W. L. and Epstein, F. H. (1972). *Cytobiology* **6,** 58–73.

Doyle, W. L. and Gorecki, D. (1961). *Physiol. Zool.* **34,** 81–85.

Dyck, M. J., van (1966). *Verh. Kongr. Vlaamse Acad. Kl. Wet.* **28,** nr 91, 99 pp.

Edelman, I. S. (1968). *In* "Functions of the Adrenal Cortex" (McKerns, K.W., Ed.), Vol. 1, pp. 79–133. North Holland, Amsterdam.

Edelman, I. S. and Fanestil, D. D. (1970). *In* "Biochemical Actions of Hormones" (Litwak, G., Ed.), Vol. 1, 324–331. Academic Press, New York and London.

Ehrenfeld, J. (1972). "Les mécanismes de transport ionique à travers la peau de *Rana esculenta in vivo*". Thèse de doctorat de Specialité, Univ. Nice.

Ellory, J. C., Lahlou, B. and Smith, M. W. (1972). *J. Physiol. Lond.* **222**, 487-509.

Ensor, D. M. and Ball, J. N. (1972). *Fedn. Am. Socs exp. Biol.* **31**, 1615-1623.

Epstein, F. H., Katz, A. I. and Pickford, G. E. (1967). *Science* **516**, 1245-1247.

Epstein, F. H., Cynamon, M. and McKay, W. (1971). *Gen. Comp. Endocrin.* **16**, 323-328.

Epstein, F. H., Maetz, J. and de Renzis, G. (1973). *Am. J. Physiol.* **224**, 1295-1299.

Erlij, D. (1971). *Phil. Trans. R. Soc. Ser. B.* **262**, 153-161.

Erlij, D. and Smith, M. W. (1973). *J. Physiol. Lond.* **228**, 221-239.

Ernst, S. A. (1972a). *J. Histochem. Cytochem.* **20**, 13-22.

Ernst, S. A. (1972b). *J. Histochem. Cytochem.* **20**, 23-38.

Ernst, S. A. and Philpott, C. W. (1970). *J. Histochem. Cytochem.* **18**, 251-263.

Evans, D. H. (1967a). *J. exp. Biol.* **47**, 519-524.

Evans, D. H. (1967b). *J. exp. Biol.* **47**, 525-534.

Evans, D. H. (1968). *Comp. Biochem. Physiol.* **25**, 751-753.

Evans, D. H. (1969a). *J. exp. Biol.* **50**, 179-190.

Evans, D. H. (1969b). *J. exp. Biol.* **50**, 689-705.

Evans, D. H. (1973). *Comp. Biochem. Physiol.* **45A**, 843-850.

Ewer, D. W. and Hattingh, I. (1952). *Nature Lond.* **169**, 460.

Ewing, R. D., Peterson, G. L. and Conte, F. P. (1972). *J. comp. Physiol.* **80**, 247-254.

Fanelli, G. M. and Goldstein, L. (1964). *Comp. Biochem. Physiol.* **13**, 193-204.

Farquhar, M. G. and Palade, G. E. (1964). *Proc. natn. Acad. Sci. U.S.A.* **51**, 569-577.

Farquhar, M. G. and Palade, H. E. (1966). *J. cell. Biol.* **30**, 359-379.

Ferreira, H. G. and Jesus, C. H. (1973). *J. Physiol. Lond.* **228**, 583-600.

Fleming, W. R. and Ball, J. N. (1972). *Z. vergl. Physiol.* **76**, 125-134.

Forrest, J. N., Jr., Cohen, A. D. and Epstein, F. H. (1971). *Bull. Mt. Desert Isl. Biol. Lab.* **11**, 18-21.

Forrest, J. N., Jr., Cohen, A. D., Schon, D. A. and Epstein, F. H. (1973a). *Am. J. Physiol.* **224**, 709-713.

Forrest, J. N., Jr., MacKay, W., Gallagher, B. and Epstein, F. H. (1973b). *Am. J. Physiol.* **224**, 714-717.

Foster, M. A. (1969). *Comp. Biochem. Physiol.* **30**, 751-759.

Fromm, P. O. (1968). *Comp. Biochem. Physiol.* **27**, 865-869.

Fromm, P. O. and Gillette, J. R. (1968). *Comp. Biochem. Physiol.* **26**, 887-896.

Gaitskell, R. E. and Chester Jones, I. (1970). *Gen. comp. Endocrin.* **15**, 491-493.

Gaitskell, R. E. and Chester Jones, I. (1971). *Gen. comp. Endocrin.* **16**, 478-483.

Garcia-Romeu, F. (1971). *Phil. Trans. R. Soc. Ser. B.* **262**, 163-174.

Garcia-Romeu, F. and Ehrenfeld, J. (1972). *In* "Role of Membranes in Secretory processes" (Bolis, L., Keynes, R. D. and Wilbrandt, W., Eds.), pp. 264-278. North Holland, Amsterdam.

Garcia-Romeu, F. and Maetz, J. (1964). *J. gen. Physiol.* **47**, 1195-1207.

Garcia-Romeu, F. and Motais, R. (1966). *Comp. Biochem. Physiol.* **17,** 1201–1204.
Garcia-Romeu, F. and Salibian, A. (1968). *Life Sci.* **7,** 465–471.
Garcia-Romeu, F., Salibian, A. and Pezzani-Hernandez, S. (1969). *J. gen. Physiol.* **53,** 816–835.
Getman, H. C. (1960). *Biol. Bull.* **99,** 439–445.
Gicklhorn, J. and Keller, R. (1925). *Biol. Zbl.* **45,** 154–169.
Gordon, M. S., Schmidt-Nielsen, K. and Kelly, H. M. (1961). *J. exp. Biol.* **38,** 659–678.
Gray, I. E. (1954). *Biol. Bull.* **107,** 219–225.
Gray, I. E. (1957). *Biol. Bull.* **112,** 34–42.
Greenaway, P. (1970). *J. exp. Biol.* **53,** 147–163.
Greenaway, P. (1971). *J. exp. Biol.* **54,** 199–214.
Greenaway, P. (1972). *J. exp. Biol.* **57,** 417–487.
Greenwald, L. (1971). *Physiol. Zool.* **44,** 149–161.
Greenwald, L. (1972). *Physiol. Zool.* **45,** 229–237.
Greenwald, L. and Kirschner, L. B. (1971). *Am. Zool.* **11,** 664 (abstract).
Grigera, J. R. and Cereijido, M. (1971). *J. Membrane Biol.* **4,** 148–155.
Handler, J. S., Preston, A. S. and Orloff, J. (1972). *Am. J. Physiol.* **222,** 1071–1074.
Hannan, J. V. and Evans, D. H. (1973). *Comp. Biochem. Physiol.* **44A,** 1199–1213.
Harris, R. R. (1970). *Comp. Biochem. Physiol.* **32,** 763–773.
Harris, R. R. (1972). *Mar. Biol.* **12,** 18–27.
Hays, R. M. (1968). *J. gen. Physiol.* **51,** 385–398.
Hays, R. M. and Franki, N. (1970). *J. Membrane Biol.* **2,** 263–276.
Hays, R. M., Franki, N. and Soberman, R. (1971). *J. Clin. Invest.* **50,** 1016–1018.
Hazelwood, D. H., Potts, W. T. W. and Fleming, W. R. (1970). *Z. vergl. Physiol.* **67,** 186–191.
Henderson, I. W. and Chester Jones, I. (1967). *J. Endocrin.* **37,** 319–327.
Henderson, I. W. and Chester Jones, I. (1972). *Annales Inst. Michel Pacha* **5,** 69–235.
Hickman, C. P. Jr. (1968). *Can. J. Zool.* **46,** 457–467.
Hickman, C. P. Jr. and Trump, B. F. (1969). *In* "Fish Physiology" (Hoar, W. S. and Randall, D. J., Eds.), Vol. 1, pp. 91–239. Academic Press, New York and London.
Hill, J. H., Cortas, N. and Walser, M. (1973). *J. Clin. Invest.* **52,** 185–189.
Hirano, T. (1967). *Proc. Japan Acad.* **43,** 793–797.
Hirano, T. (1974) (in preparation).
Hirano, T. and Utida, S. (1968). *Gen. Comp. Endocrin.* **11,** 373–381.
Hirano, T. and Utida, S. (1971). *Endocr. jap.* **18,** 47–52.
Hirano, T., Kamiya, M., Saishu, S. and Utida, S. (1967). *Endocr. jap.* **14,** 182–186.
Hirano, T., Johnson, D. W. and Bern, H. A. (1971). *Nature, Lond.* **230,** 469–471.
Hoar, W. S. and Randall, D. J. (1969). "Fish Physiology" Vol. 2. "The Endocrine System". Academic Press, New York and London.
Hodgkin, A. L. and Katz, B. (1949). *J. Physiol. Lond.* **108,** 37–77.
Holeton, G. F. and Randall, D. J. (1967). *J. exp. Biol.* **46,** 317–327.

Holmes, W. N. and Stainer, I. M. (1966). *J. exp. Biol.* **44**, 33–46.

Hornby, R. and Thomas, S. (1969). *J. Physiol. Lond.* **200**, 321–345.

Horne, F. R. (1967). *Comp. Biochem. Physiol.* **21**, 525–533.

Horne, F. R. (1968). *Crustaceana* **14**, 271–274.

Horne, R. A. (1972). "Water and Aqueous Solutions. Structure, Thermodynamics and Transport processes." Wiley-Interscience, New York and London.

House, C. R. (1963). *J. exp. Biol.* **40**, 87–104.

House, C. R. and Green, K. (1965). *J. exp. Biol.* **42**, 177–189.

House, C. R. and Maetz, J. (1974). *Comp. Biochem. Physiol.* **47A**, 917–924.

Huang, K. C. and Chen, T. S. T. (1971). *Am. J. Physiol.* **220**, 1734–1738.

Huf, E. G. (1972). *Acta physiol. scand.* **84**, 366–381.

Hughes, G. M. (1963). "Comparative Physiology of Vertebrate Respiration". Heinemann, London.

Hughes, G. M. (1966). *J. exp. Biol.* **45**, 177–195.

Hviid Larsen, E. (1971). *Gen. Comp. Endocrin.* **17**, 543–553.

Isaia, J. (1972). *J. exp. Biol.* **57**, 359–366.

Istin, M. and Kirschner, L. B. (1968). *J. gen. Physiol.* **51**, 478–496.

Istin, M. and Masoni, A. (1973). *Calc. Tiss. Res.* **11**, 151–162.

Jampol, L. M. and Epstein, F. H. (1970). *Am. J. Physiol.* **218**, 607–611.

Jard, S. (1958). "Recherches sur l'adaptation de grenouilles vertes à des milieux de salinité diverse et sur son déterminisme endocrinien". Diplôme d'Etudes Supérieures, Université de Paris.

Johnson, D. W., Hirano, T., Bern, H. A. and Conte, F. P. (1972). *Gen. Comp. Endocrin.* **19**, 115–128.

Jørgensen, Barker, C. (1949). *Acta physiol. scand.* **18**, 171–180.

Jørgensen, Barker, C. (1954). *Acta physiol. scand.* **30**, 171–177.

Jørgensen, Barker, C. and Dales, R. P. (1957). *Physiologia. comp. Oecol.* **4**, 357–374.

Jørgensen, Barker, C. and Larsen, J. D. (1964). *Gen. Comp. Endocrin.* **4**, 389–400.

Jørgensen, Barker, C. and Rosenkilde, P. (1956). *Biol. Bull.* 300–305.

Jørgensen, Barker, C., Levi, H. and Zerahn, K. (1954). *Acta physiol. scand.* **30**, 178–190.

Kado, Y. and Momo, Y. (1971). *J. Sci. Hiroshima Univ. B Div.* 1 **23**, 215–228.

Kamemoto, F. I. (1964). *Gen. Comp. Endocrin.* **4**, 420–426.

Kamemoto, F. I. and Ono, J. K. (1969). *Comp. Biochem. Physiol.* **29**, 393–401.

Kamemoto, F. I. and Tullis, R. E. (1972). *Gen. Comp. Endocrin.* Suppl. 3, 299–307.

Kamemoto, F. I., Kato, K. N. and Tucker, L. E. (1966). *Am. Zool.* **6**, 213–219.

Kamiya, M. (1967). *Ann. Zool. Japon.* **40**, 123–129.

Kamiya, M. (1972). *Comp. Biochem. Physiol.* **43B**, 611–617.

Kamiya, M. and Utida, S. (1968). *Comp. Biochem. Physiol.* **26**, 675–685.

Kamiya, M. and Utida, S. (1969). *Comp. Biochem. Physiol.* **31**, 671–674.

Katchalsky, A. and Curran, P. F. (1965). "Nonequilibrium thermodynamics in Biophysics". Harvard University Press, Cambridge, Mass.

Kato, K. N. and Kamemoto, F. I. (1969). *Comp. Biochem. Physiol.* **28**, 665–674.

Katz, U. and Lindemann, B. (1972). *Israel J. med. Sci.* **8**, 10.

Kawada, J., Taylor, R. E. and Barker, S. B. (1969). *Comp. Biochem. Physiol.* **30**, 965–975.

Kerstetter, T. H. and Kirschner, L. B. (1972). *J. exp. Biol.* **56**, 263–272.

Kerstetter, T. H. Kirschner, L. B. and Rafuse, D. D. (1970). *J. gen. Physiol.* **56**, 342–359.

Kessel, R. G. and Beams, H. W. (1962). *J. Ultrastruct. Res.* **6**, 77–87.

Keys, A. (1931*a*). *Z. vergl. Physiol.* **15**, 352–363.

Keys, A. (1931*b*). *Z. vergl. Physiol.* **15**, 364–388.

Keys, A. (1933). *Proc. Roy. Soc. Ser. B* **112**, 184–199.

Keys, A. and Willmer, E. N. (1932). *J. Physiol. Lond.* **76**, 368–378.

King, E. N. and Schoffeniels, E. (1969). *Archs. int. Physiol. Biochim.* **77**, 105–112.

Kirsch, R. (1971). "Echanges d'eau, de chlorures et de sodium au niveau des différents effecteurs de l'osmorégulation chez l'anguille (*Anguilla anguilla* L.) en eau douce et au cours de l'adaptation d'eau douce à l'eau de mer." Thèse de Doctorat d'Etat. Université de Strasbourg.

Kirsch, R. (1972*a*). *J. exp. Biol.* **57**, 489–512.

Kirsch, R. (1972*b*). *J. Physiol. Paris* **65**, 428A (abstract).

Kirsch, R. and Mayer-Gostan, N. (1973). *J. exp. Biol.* **58**, 105–121.

Kirschner, L. B. (1955). *J. cell. comp. Physiol.* **45**, 61–88.

Kirschner, L. B. (1967). *Ann. Rev. Physiol.* **29**, 169–196.

Kirschner, L. B. (1969). *Am. J. Physiol.* **217**, 596–604.

Kirschner, L. B. (1970). *Am. Zool.* **10**, 365–376.

Kirschner, L. B., Kerstetter, T. H., Porter, D. and Alvarado, R. H. (1971). *Am. J. Physiol.* **220**, 1814–1819.

Kirschner, L. B., Greenwald, L. and Kerstetter, T. H. (1973). *Am. J. Physiol.* **224**, 832–837.

Kirsten, E., Kirsten, R., Leaf, A. and Sharp, G. W. G. (1968). *Pflügers Arch ges. Physiol.* **300**, 213–225.

Kitching, J. A. (1967). *In* "Research in Protozoology" (Tze-Tuan Chen, Ed.) Vol. 1, pp. 307–336. Pergamon Press, Oxford.

Koch, A. R. (1970). *Am. Zool.* **10**, 331–346.

Koch, H. (1934). *Annls Soc. r. Sci. méd. nat. Brux. Ser. B* **54**, 346–361.

Koch, H. (1938). *J. exp. Biol.* **15**, 152–160.

Koch, H., Evans, J. and Schicks, E. (1954). *Meded. vlaam. Acad. Kl. Wet.* **16**, 3–16.

Koefoed-Johnsen, V. and Ussing, H. H. (1953). *Acta Physiol. Scand.* **28**, 60–76.

Koefoed-Johnsen, V. and Ussing, H. H. (1958). *Acta Physiol. Scand*, **42**, 298–308.

Koefoed-Johnsen, V. and Ussing, H. H. (1960). *In* "Mineral Metabolism" (Comar, D. F. and Bronner, F., Eds.), Vol. I, part A, pp. 169–203. Academic Press, New York and London.

Koefoed-Johnsen, V., Levi, H. and Ussing, H. H. (1952). *Acta Physiol. Scand.* **25**, 150–163.

Komnick, H., Rhees, R. W. and Abel, J. H. Jr. (1972). *Cytobiology* **5**, 65–82.

Kristensen, P. (1972). *Acta Physiol. Scand.* **84**, 338–346.

Krogh, A. (1937). *Z. vergl. Physiol.* **24**, 656–666.

Krogh, A. (1938). *Z. vergl. Physiol.* **25**, 335–350.

Krogh, A. (1939). "Osmotic Regulation in Aquatic Animals". Cambridge University Press.

Lacanilao, F. (1972a). *Gen. Comp. Endocrin.* **19**, 405–412.

Lacanilao, F. (1972b). *Gen. Comp. Endocrin.* **19**, 413–420.

Lahlou, B. (1970). *Bull. Inf. Scient. Tech. Commt. Energ. Atom.* **144**, 17–52.

Lahlou, B. and Giordan, A. (1970). *Gen. Comp. Endocrin.* **14**, 491–509.

Lahlou, B. and Sawyer, W. H. (1969a). *Gen. Comp. Endocrin.* **12**, 370–377.

Lahlou, B. and Sawyer, W. H. (1969b). *C.r. hebd. Séanc. Acad. Sci., Paris* **268**, 725–728.

Lahlou, B. and Sawyer, W. H. (1969c). *J. Physiol. Paris*, **61**, 143 (abstract).

Lahlou, B. and Sawyer, W. H. (1969d). *Am. J. Physiol.* **216**, 1273–1278.

Lahlou, B., Henderson, I. W. and Sawyer, W. H. (1969). *Comp. Biochem. Physiol.* **28**, 1427–1433.

Lam, T. J. (1969). *Comp. Biochem. Physiol.* **31**, 909–913.

Lam, T. J. (1972). *Gen. Comp. Endocrin. Supp.* **3**, 328–338.

Lange, R. (1972). *Oceanogr. Mar. Biol. Ann. Rev.* **10**, 97–137.

Langford, R. W. (1971). "Relationship between Sodium-potassium Adenosine triphosphatase and Sodium Fluxes in the European Eel *Anguilla anguilla.*" Ph.D. University of Oregon.

Lasker, R. and Threadgold, L. T. (1968). *Expl. Cell Res.* **52**, 582–590.

Lasserre, P. (1971). *Life Sci.* 10 part II, 113–119.

Lauer, G. J. (1969). *Physiol. Zoöl.* **42**, 381–387.

Leader, J. P. (1972). *J. exp. Biol.* **57**, 821–838.

Leaf, A. and Macknight, A. D. C. (1972). *J. Ster. Biochem.* **3**, 237–245.

Leblanc, G. (1972). *Pflügers Arch. ges. Physiol.* **337**, 1–18.

Leersnyder, M., de (1967). *Cah. Biol. mar.* **8**, 295–321.

Leiner, M. (1938). *Z. vergl. Physiol.* **26**, 416–466.

Lennep, E. W., van (1968). *J. Ultrastr. Res.* **25**, 94–108.

Lennep, E. W., van and Komnick, H. (1971). *Cytobiology* **3**, 137–151.

Lennep, E. W., van and Lanzing, W. J. R. (1967). *J. Ultrastr. Res.* **18**, 333–344.

Lever, J., Jansen, J. and Vlieger, T. A., de (1961). *Proc. K. ned. Akad. Wet. C* **64**, 532–542.

Lilly, S. J. (1955). *J. exp. Biol.* **32**, 423–439.

Little, C. (1965). *J. exp. Biol.* **43**, 39–54.

Lloyd, R. and Herbert, D. W. M. (1960). *Ann. appl. Biol.* **48**, 399–404.

Lloyd, R. and White, W. R. (1967). *Nature, Lond.* **216**, 1341–1342.

Lockwood, A. P. M. (1959). *J. exp. Biol.* **37**, 614–630.

Lockwood, A. P. M. (1961). *J. exp. Biol.* **38**, 647–658.

Lockwood, A. P. M. (1962). *Biol. Rev.* **37**, 257–305.

Lockwood, A. P. M. (1965). *J. exp. Biol.* **42**, 59–69.

Lockwood, A. P. M. (1968). "Aspects of the Physiology of Crustacea". Oliver and Boyd, Edinburgh and London.

Lockwood, A. P. M. (1970). *J. exp. Biol.* **53**, 737–751.

Lockwood, A. P. M. and Andrews, W. R. H. (1969). *J. exp. Biol.* **51**, 591–605.

Lockwood, A. P. M. and Inman, C. B. E. (1973). *J. exp. Biol.* **58**, 149–163.
Lockwood, A. P. M., Inman, C. B. E. and Courtenay, T. H. (1973). *J. exp. Biol.* **58**, 137–148.
Lotan, R. (1969). *Z. vergl. Physiol.* **65**, 455–462.
Lotan, R. (1971). *Z. vergl. Physiol.* **75**, 383–387.
Lotan, R. and Skadhauge, E. (1972). *Comp. Biochem. Physiol.* **42A**, 303–310.
Lubet, P. and Pujol, J. P. (1963). *C.r. hebd. Séanc. Acad. Sci. Paris* **257**, 4032–4034.
Lubet, P. and Pujol, J. P. (1965). *Rapp. Comm. Int. Explor. Sci. Mediterr.* **18**, 149–154.
Macfarlane, N. A. A. and Maetz, J. (1974). *Gen. Comp. Endocrin.* **22**, 77–89.
Macklin, M. and Josephson, R. K. (1971). *Biol. Bull.* **141**, 299–318.
Macklin, M., Roma, T. and Drake, K. (1973). *Science* **179**, 194–195.
Maetz, J. (1946). *Bull. Inst. Oceanog. Monaco* **43n**, 899, 1–20.
Maetz, J. (1956*a*). *Bull. Biol. Fr. Belg. Suppl.* **40**, 1–129.
Maetz, J. (1956*b*). *J. Physiol. Paris* **48**, 1085–1099.
Maetz, J. (1959). *In* "La méthode des indicateurs nucléaires dans l'étude des transports actifs d'ions." (Coursaget, J., Ed.), pp. 185–196. Pergamon Press, Oxford.
Maetz, J. (1964). *Bull. Inf. scient. tech. Commt. Energ. atom.* **86**, 11–70.
Maetz, J. (1968). *In* "Perspectives in Endocrinology. Hormones in the Lives of Lower Vertebrates" (Barrington, E. J. W. and Jørgensen, Barker C., Eds.), pp. 47–162, Academic Press, London and New York.
Maetz, J. (1969). *Science N.Y.* **166**, 613–615.
Maetz, J. (1970). *Bull. Infs. Scient. tech. Commt Energ. atom.* **146**, 21–43.
Maetz, J. (1971). *Phil. Trans. R. Soc. Ser. B* **262**, 209–251.
Maetz, J. (1972*a*). *In* "Nitrogen Metabolism and the Environment" (Campbell, J. W. and Goldstein, L., Eds.), pp. 105–154. Academic Press, London and New York.
Maetz, J. (1972*b*). *J. exp. Biol.* **56**, 601–620.
Maetz, J. (1973*a*). *J. exp. Biol.* **58**, 255–275.
Maetz, J. (1973*b*). Alfred Benzon Symp. no. 5. "Transport Mechanisms in Epithelia" (Ussing H. H. and Thorn N. A., Eds), pp. 427–441. Munskaard, Copenhagen.
Maetz, J. and Campanini, G. (1966). *J. Physiol. Paris* **58**, 248 (abstract).
Maetz, J. and Evans, D. H. (1972). *J. exp. Biol.* **56**, 565–585.
Maetz, J. and Garcia-Romeu, F. (1964). *J. gen. Physiol.* **47**, 1209–1227.
Maetz, J. and Lahlou, B. (1974). *In* "Handbook of Physiology, Section on Endorinology, Vol. on Hypothalamo-hypophysical System" (Knobil, E. and Sawyer, W. H., Eds.), American Physiological Society, Washington, D.C.
Maetz, J. and Skadhauge, E. (1968). *Nature, Lond.* **217**, 371–373.
Maetz, J., Jard, S. and Morel, F. (1958). *C.r. hebd. Séanc. Acad. Sci. Paris* **247**, 516–518.
Maetz, J., Sawyer, W. H., Pickford, G. E. and Mayer, N. (1967*a*). *Gen. Comp. Endocrin.* **8**, 163–176.

Maetz, J., Mayer, N. and Chartier-Baraduc, M. M. (1967b). *Gen. Comp. Endocrin.* **8**, 177–188.

Maetz, J., Nibelle, J., Bornancin, M. and Motais, R. (1969a). *Comp. Biochem. Physiol.* **30**, 1125–1151.

Maetz, J., Motais, R. and Mayer, N. (1969b). *Excerpta med. Int. Congr. Series* **184**, 225–232.

Mantel, L. H. (1967). *Comp. Biochem. Physiol.* **20**, 743–753.

Mantel, L. H. (1968). *Am. Zool.* **8**, 433–442.

Maren, T. H. (1967a). *Physiol. Rev.* **47**, 595–781.

Maren, T. H. (1967b). *Fedn. Proc. Fedn. Am. Socs exp. Biol.* **26**, 1097–1104.

Martin, D. W. and Curran, P. F. (1966). *J. cell. Physiol.* **67**, 367–374.

Martinez-Palomo, A., Erlij, D. and Bracho, H. (1971). *J. cell. Biol.* **50**, 277–287.

Mashiko, K. and Josuka, K. (1964). *Annotnes. zool. jap.* **37**, 41–50.

Mattheij, J. A. M. and Stroband, H. W. (1971). *Z. Zellforsch. mikrosk. Anat.* **121**, 93–101.

Mayer, N. (1969). *Comp. Biochem. Physiol.* **29**, 27–50.

Mayer, N. (1970). *Bull. Inf. Scient. tech. Commt Energ. atom.* **146**, 45–75.

Mayer, N. and Maetz, J. (1967). *C.r. hebd. Séanc. Acad. Sci. Paris* **264**, 1632–1635.

Mayer, N. and Nibelle, J. (1970). *Comp. Biochem. Physiol.* **35**, 553–566.

Mayer, N., Maetz, J., Chan, D. K. O., Foster, M. and Chester Jones, I. (1967). *Nature, Lond.* **214**, 1118–1120.

Milne, K. P., Ball, J. N. and Chester Jones, I. (1971). *J. Endocrin.* **49**, 177–178.

Morris, R. (1956). *J. exp. Biol.* **33**, 235–248.

Morris, R. (1957). *Q. Jl. microsc. Sci.* **98**, 473–485.

Morris, R. (1958). *J. exp. Biol.* **35**, 649–665.

Morris, R. (1965). *J. exp. Biol.* **42**, 359–371.

Morris, R. and Bull, J. M. (1970). *J. exp. Biol.* **52**, 275–290.

Mossberg, S. M. (1967). *Am. J. Physiol.* **213**, 1327–1330.

Motais, R. (1961a). *C.r. hebd. Séanc. Acad. Sci. Paris* **253**, 724–726.

Motais, R. (1961b). *C.r. hebd. Séanc. Acad. Sci. Paris* **253**, 2609–2611.

Motais, R. (1967). *Annals. Inst. Oceanog. Monaco* **45**, 1–84.

Motais, R. (1970a). *Bull. Inf. Scient. techn. Commt. Energ. atom.* **146**, 3–19.

Motais, R. (1970b). *Comp. Biochem. Physiol.* **34**, 497–501.

Motais, R. and Garcia-Romeu, F. (1972). *Ann. Rev. Physiol.* **34**, 141–176.

Motais, R. and Isaia, J. (1972a). *J. exp. Biol.* **56**, 587–600.

Motais, R. and Isaia, J. (1972b). *J. exp. Biol.* **57**, 367–373.

Motais, R. and Maetz, J. (1964). *Gen. Comp. Endocrin.* **4**, 210–224.

Motais, R. and Maetz, J. (1967). *J. Physiol. Paris* **59**, 271 (abstract).

Motais, R., Garcia-Romeu, F. and Maetz, J. (1966). *J. gen. Physiol.* **50**, 391–422.

Motais, R., Isaia, J., Rankin, J. C. and Maetz, J. (1969). *J. exp. Biol.* **51**, 529–546.

Myers, R. M., Bishop, W. R. and Scheer, B. T. (1961). *Am. J. Physiol.* **200**, 444–450.

Nagel, H. (1934). *Z. vergl. Physiol.* **21**, 468–491.

Nemenz, H. (1960). *J. Insect Physiol.* **4**, 38–44.

Newstead, J. D. (1971). *Z. Zellforsch. mikrosk. Anat.* **116**, 1–6.

Nielsen, R. (1969). *Acta Physiol. Scand.* **77**, 85–94.
Nielsen, R. and Tomilson, R. W. S. (1970). *Acta Physiol. Scand,* **79**, 238–243.
Nüske, H. and Wichard, W. (1971). *Cytobiology* **4**, 480–486.
Nüske, H. and Wichard, W. (1972). *Cytobiology* **6**, 243–249.
Ogawa, M., Yagasaki, M. and Yamazaki, F. (1973). *Comp. Biochem. Physiol.* **44A**, 1177–1183.
Oglesby, L. C. (1969). *In* "Chemical Zoology" (Florkin, M. and Scheer, B. T., Eds.), Vol. 4, pp. 211–311. Academic Press, New York and London.
Oglesby, L. C. (1972). *Comp. Biochem. Physiol.* **41A**, 756–790.
Oide, M. (1967). *Annotnes zool. jap.* **40**, 130–135.
Oide, M. (1970). *Comp. Biochem. Physiol.* **36**, 241–252.
Oide, M. and Utida, S. (1967). *Mar. Biol.* **1**, 102–106.
Oide, H. and Utida, S. (1968). *Mar. Biol.* **1**, 172–177.
Olivereau, M. (1970). *C.r. Séanc. Soc. Biol.* **164**, 1951–1955.
Olivereau, M. (1971). *C.r. Séanc. Soc. Biol.* **165**, 1009–1013.
Ong, J. E. (1969). *Z. Zellforsch. mikrosk. Anat.* **97**, 178–195.
Packer, R. K. and Dunson, W. A. (1970). *J. exp. Zool.* **174**, 65–72.
Parry, G. (1955). *J. exp. Biol.* **32**, 408–422.
Parry, G. (1957). *J. exp. Biol.* **34**, 417–423.
Parry, G. (1966). *Biol. Rev.* **41**, 392–444.
Parry, G. and Potts, W. T. W. (1965). *J. exp. Biol.* **42**, 415–422.
Patlak, C. S., Goldstein, D. A. and Hoffman, J. F. (1963). *J. theor. Biol.* **5**, 426–442.
Payan, P. and Maetz, J. (1970). *Bull. Infs. Scient. tech. Commt Energ. atom.* **146**, 77–96.
Payan, P. and Maetz, J. (1971). *Gen. Comp. Endocrin.* **16**, 535–554.
Payan, P. and Maetz, J. (1973). *J. exp. Biol.* **58**, 487–502.
Payan, P., Goldstein, L. and Forster, R. P. (1973). *Am. J. Physiol.* **224**, 367–372.
Pequin, L. (1967). *Archs. Sci. physiol.* **21**, 193–203.
Petrik, P. (1968). *Z. Zellforsch. mikrosk. Anat.* **92**, 422–427.
Petrik, P. and Bucher, O. (1969). *Z. Zellforsch. mikrosk. Anat.* **96**, 66–75.
Pettengill, O. and Copeland, D. E. (1948). *J. exp. Zool.* **108**, 235–242.
Pfeiler, E. and Kirschner, L. B. (1972). *Biochem. Biophys. Acta* **282**, 301–310.
Philippot, J., Thuet, M. and Thuet, P. (1972). *Comp. Biochem. Physiol.* **41B**, 231–243.
Phillips, J. E. (1970). *Am. Zool.* **10**, 413–436.
Phillips, J. E. and Meredith, J. (1969). *Nature, Lond.* **222**, 168–169.
Philpott, C. W. (1965). *Protoplasma* **60**, 7–23.
Philpott, C. W. (1967). *J. cell. Biol.* **35**, 104A (abstract).
Philpott, C. W. and Copeland, D. E. (1963). *J. cell. Biol.* **18**, 389–404.
Pic, P. (1972). *C.r. Séanc. Soc. Biol.* **166**, 131–136.
Pic, P., Mayer-Gostan, N. and Maetz, J. (1973). *In* "Comparative Physiology: Locomotion, Respiration, Transport and Blood" (Bolis, L., Schmidt-Nielsen, K. and Maddrell, S. H. P., Eds.), pp. 293–322. North-Holland, Amsterdam.
Pickering, A. D. and Dockray, G. J. (1972). *Comp. Biochem. Physiol.* **41A**, 139–147.

Pickering, A. D. and Morris, R. (1970). *J. exp. Biol.* **53**, 231–243.
Pickering, A. D. and Morris, R. (1973). *J. exp. Biol.* **58**, 165–176.
Pickford, G. E. and Grant, F. B. (1967). *Science N.Y.* **155**, 568–570.
Pickford, G. E., Griffith, R. W., Torretti, J., Hendler, E. and Epstein, F. H. (1970a). *Nature, Lond.* **228**, 378–379.
Pickford, G. E., Pang, P. K. T., Weinstein, E., Torretti, E., Hendler, E. and Epstein, F. H. (1970b). *Gen. Comp. Endocrin.* **14**, 524–534.
Pickford, G. E., Srivastava, A. K., Slicher, A. M. and Pang, P. K. T. (1971a). *J. exp. Zool.* **177**, 89–96.
Pickford, G. E., Srivastava, A. K., Slicher, A. M. and Pang, P. K. T. (1971b). *J. exp. Zool.* **177**, 97–108.
Pickford, G. E., Srivastava, A. K., Slicher, A. M. and Pang, P. K. T. (1971c). *J. exp. Zool.* **177**, 109–117.
Potts, W. T. W. (1967). *Biol. Rev.* **42**, 1–41.
Potts, W. T. W. (1968). *Ann. Rev. Physiol.* **30**, 73–104.
Potts, W. T. W. and Evans, D. H. (1966). *Biol. Bull.* **131**, 362–368.
Potts, W. T. W. and Evans, D. H. (1967). *Biol. Bull.* **133**, 411–425.
Potts, W. T. W. and Fleming, W. R. (1970). *J. exp. Biol.* **53**, 317–327.
Potts, W. T. W. and Fleming, W. R. (1971). *J. exp. Biol.* **54**, 63–75.
Potts, W. T. W. and Parry, G. (1964a). "Osmotic and Ionic Regulation in Animals". Pergamon Press, Oxford.
Potts, W. T. W. and Parry, G. (1964b). *J. exp. Biol.* **41**, 591–602.
Potts, W. T. W., Foster, M. A., Rudy, P. P. and Parry-Howells, G. (1967). *J. exp. Biol.* **47**, 461–470.
Potts, W. T. W., Foster, M. A. and Stather, J. W. (1970). *J. exp. Biol.* **52**, 553–564.
Price, J. W. (1931). *Stud. Ohio State Univ.* **4**, 1–46.
Quinn, D. J. and Lane, C. E. (1966). *Comp. Biochem. Physiol.* **19**, 533–543.
Ramamurthi, R. and Scheer, B. T. (1967). *Life Sci.* **6**, 2171–2175.
Ramsay, J. A. (1949). *J. exp. Biol.* **26**, 46–56.
Ramsay, J. A. (1950). *J. exp. Biol.* **27**, 145–157.
Ramsay, J. A. (1953). *J. exp. Biol.* **30**, 79–89.
Ramsay, J. A. (1971). *Phil. Trans. R. Soc. Ser. B* **262**, 251–260.
Rankin, J. C. and Maetz, J. (1971). *J. Endocrin.* **51**, 621–635.
Rawlins, F., Mateu, L., Fragachan, F. and Whittembury, G. (1970). *Pflügers Arch. ges. Physiol.* **316**, 64–80.
Renzis, G., de and Maetz, J. (1973). *J. exp. Biol.* **59**, 339–358.
Richards, B. D. and Fromm, P. O. (1970). *Comp. Biochem. Physiol.* **33**, 303–310.
Ritch, R. and Philpott, C. W. (1969). *Expl. Cell. Res.* **55**, 17–24.
Robertson, J. D. (1957). *Biol. Rev* **32**, 156–187.
Romer, A. S. (1946). *Q. Rev. Biol.* **21**, 33–69.
Rudy, P. P. Jr. (1967). *Comp. Biochem. Physiol.* **22**, 581–589.
Rudy, P. P. Jr. and Wagner, R. C. (1970). *Comp. Biochem. Physiol.* **34**, 399–403.
Salibian, A., Pezzani-Hernandez, S. and Garcia-Romeu, F. (1968). *Comp. Biochem. Physiol.* **25**, 311–317.
Satir, P. and Gilula, N. B. (1970). *J. cell. Biol.* **47**, 468–487.

164   J. MAETZ

Schafer, J. A. and Andreoli, R. E. (1972). *Archs intern. Med.* **129**, 279–291.
Schlieper, C. (1933a). *Z. vergl. Physiol.* **18**, 682–695.
Schlieper, C. (1933b). *Z. vergl. Physiol.* **19**, 68–83.
Schmidt-Nielsen, B. (1971). *Fedn. Proc. Fedn. Am. Socs exp. Biol.* **30**, 3–5.
Schoffenliels, E. (1973). *In* "Comparative Physiology: Locomotion, Respiration, Transport and Blood". (Bolis, L., Schmidt-Nielsen, K., Maddrell, S. H. P., Eds), in pp. 353–385. North-Holland, Amsterdam.
Schoffeniels, E. and Gilles, R. (1972). *In* "Chemical Zoology" (Florkin, M. and Scheer, B. T., Eds), Vol. 7, pp. 393–420. Academic Press, New York and London.
Scholles, W. (1933). *Z. vergl. Physiol.* **19**, 522–554.
Schultz, S. G. and Zalusky, R. (1964a). *J. gen. Physiol.* **47**, 567–584.
Schultz, S. G. and Zalusky, R. (1964b). *J. gen. Physiol.* **47**, 1043–1058.
Schwabe, E. (1933). *Z. vergl. Physiol.* **19**, 183–236.
Scudamore, H. H. (1947). *Physiol. Zool.* **20**, 187–208.
Sharp, G. W. G. and Leaf, A. (1966). *Recent Prog. Horm. Res.* **22**, 431–471.
Sharratt, B. M., Bellamy, D. and Chester Jones, I. (1964). *Comp. Biochem. Physiol.* **11**, 19–30.
Shaw, J. (1955a). *J. exp. Biol.* **32**, 330–352.
Shaw, J. (1955b). *J. exp. Biol.* **32**, 353–382.
Shaw, J. (1959a). *J. exp. Biol.* **36**, 126–144.
Shaw, J. (1959b). *J. exp. Biol.* **36**, 157–716.
Shaw, J. (1960a). *J. exp. Biol.* **37**, 534–547.
Shaw, J. (1960b). *J. exp. Biol.* **37**, 548–556.
Shaw, J. (1960c). *J. exp. Biol.* **37**, 557–572.
Shaw, J. (1961a). *J. exp. Biol.* **38**, 135–152.
Shaw, J. (1961b). *J. exp. Biol.* **38**, 153–162.
Shaw, J. (1963). *In* "Viewpoints in Biology" (Carthy, J. D. and Duddington, C. L., Eds.), Vol. 2, pp. 163–201. Butterworths, London.
Shaw, J. (1964). *Symp. Soc. exp. Biol.* **18**, 237–254.
Shaw, J. and Stobbart, R. H. (1963). *Adv. Insect Physiol.* **1**, 315–399.
Shaw, J. and Sutcliffe, D. W. (1961). *J. exp. Biol.* **38**, 1–15.
Shehadeh, Z. H. and Gordon, M. S. (1969). *Comp. Biochem. Physiol.* **30**, 397–419.
Shirai, N. (1972). *J. Fac. Sci. Univ.* (*Tokyo Section*). **6, 12**, 385–403.
Shirai, N. and Utida, S. (1970). *Z. Zellforsch. mikrosk. Anat.* **103**, 247–264.
Shuttelworth, T. J. (1972). *Comp. Biochem. Physiol.* **43A**, 59–64.
Skadhauge, E. (1969). *J. Physiol. Lond.* **204**, 135–158.
Slayman, C. L. (1970). *Am. Zool.* **10**, 377–392.
Smith, D. S. and Linton, J. R. (1971). *Comp. Biochem. Physiol.* **39A**, 367–378.
Smith, H. W. (1929). *J. Biol. Chem.* **81**, 727–742.
Smith, H. W. (1930). *Am. J. Physiol.* **93**, 480–505.
Smith, H. W. (1953). "From Fish to Philosopher" Little Brown, Boston.
Smith, M. W. (1964). *J. Physiol. Lond.* **175**, 38–49.
Smith, M. W. (1966). *J. Physiol. Lond.* **182**, 559–573.
Smith, P. G. (1969a). *J. exp. Biol.* **51**, 727–738.

Smith, P. G. (1969*b*). *J. exp. Biol.* **51**, 739–757.
Smith, R. I. (1964). *Biol. Bull.* **126**, 142–149.
Smith, R. I. (1967). *Biol. Bull.* **133**, 643–659.
Smith, R. I. (1970*a*). *J. exp. Biol.* **53**, 75–92.
Smith, R. I. (1970*b*). *J. exp. Biol.* **53**, 93–100.
Smith, R. I. (1970*c*). *J. exp. Biol.* **53**, 101–108.
Smith, R. I. (1970*d*). *Biol. Bull.* **139**, 351–362.
Smith, R. I. and Rudy, P. P. (1972). *Biol. Bull.* **143**, 234–246.
Sohal, R. S. and Copeland, D. E. (1966). *J. Insect Physiol.* **12**, 429–439.
Staddon, B. W. (1955). *J. exp. Biol.* **32**, 84–94.
Staddon, B. W. (1959). *J. exp. Biol.* **36**, 566–574.
Staverman, A. J. (1951). *Recl. Trav. chim. Pays-Bas* **70**, 344–352.
Stanley, J. G. and Fleming, W. R. (1966*a*). *Biol. Bull.* **131**, 155–165.
Stanley, J. G. and Fleming, W. R. (1966*b*). *Biol. Bull.* **130**, 430–441.
Stanley, J. G. and Fleming, W. R. (1967). *Comp. Biochem. Physiol.* **20**, 489–497.
Steen, J. B. and Kruysse, A. (1964). *Comp. Biochem. Physiol.* **12**, 127–142.
Stein, W. D. (1967). "The Movement of Molecules across Cell Membranes" Academic Press, New York and London.
Stevens, E., Don (1972). *J. Fish. Res. Bd Can.* **29**, 202–203.
Stobbart, R. H. (1959). *J. exp. Biol.* **36**, 641–653.
Stobbart, R. H. (1960). *J. exp. Biol.* **37**, 594–608.
Stobbart, R. H. (1965). *J. exp. Biol.* **42**, 29–43.
Stobbart, R. H. (1967). *J. exp. Biol.* **47**, 35–57.
Stobbart, R. H. (1971*a*). *J. exp. Biol.* **54**, 19–27.
Stobbart, R. H. (1971*b*). *J. exp. Biol.* **54**, 29–66.
Stobbart, R. H. (1971*c*). *J. exp. Biol.* **54**, 67–82.
Stobbart, R. H. and Shaw, J. (1964). *In* "The Physiology of Insecta" (Rockstein, M., Ed.) Vol. 3, pp. 189–258. Academic Press, New York and London.
Sutcliffe, D. W. (1960). *Nature, Lond.* **187**, 331–332.
Sutcliffe, D. W. (1961*a*). *J. exp. Biol.* **38**, 501–519.
Sutcliffe, D. W. (1961*b*). *J. exp. Biol.* **38**, 521–530.
Sutcliffe, D. W. (1962). *J. exp. Biol.* **39**, 141–160.
Sutcliffe, D. W. (1967*a*). *J. exp. Biol.* **46**, 499–518.
Sutcliffe, D. W. (1967*b*). *J. exp. Biol.* **46**, 529–550.
Sutcliffe, D. W. (1968). *J. exp. Biol.* **48**, 359–380.
Sutcliffe, D. W. (1970). *Nature, Lond.* **228**, 875–876.
Sutcliffe, D. W. (1971). *J. exp. Biol.* **54**, 255–268.
Sutcliffe, D. W. and Shaw, J. (1967). *J. exp. Biol.* **46**, 519–529.
Sutcliffe, D. W. and Shaw, J. (1968). *J. exp. Biol.* **48**, 339–359.
Taylor, R. E. and Barker, S. B. (1965). *Science*, 1612–1613.
Thompson, L. C. (1967). *Am. Zool.* **7**, 736 (abstract).
Thorson, T. B. (1967). *In* "Sharks, Skates and Rays" (Gilbert, P. W., Matheson, R. F. and Rall, R. P., Eds.), pp. 265–270. John Hopkins, Baltimore.
Thorson, T. B. (1970). *Life Sci.* **9**, 893–900.
Thorson, T. B., Cowan, C. M. and Watson, D. E. (1967). *Science* 375–377.

Threadgold, L. T. and Houston, A. H. (1964). *Expl. Cell Res.* **34**, 1–23.
Thuet, P., Motais, R. and Maetz, J. (1968). *Comp. Biochem. Physiol.* **26**, 793–818.
Thuet, M., Thuet, P. and Philipott, J. (1969). *C.r. hebd. Séanc. Acad. Sci.*, *Paris* **269**, 233–236.
Tosteson, D. C. (1962). *Bull. Mt. Desert Isl. Biol. Lab.* *4* **82** (abstract).
Treherne, J. E. (1954). *J. exp. Biol.* **31**, 386–404.
Turnberg, L. A., Bieberdorf, F. A., Morawski, S. G. and Fordtran, J. S. (1970). *J. clin. Invest.* **49**, 557–567.
Ussing, H. H. (1960). *In* "The Alkali Metal Ions in Isolated Systems and Tissues", Handbuch Exp. Pharmakologie 13, 1–195. Springer Verlag, Berlin.
Ussing, H. H. (1971). *Phil. Trans. R. Soc. Ser. B* **262**, 85–90.
Utida, S., Isono, N. and Hirano, T. (1967a). *Zool. Mag.* **76**, 203–204.
Utida, S., Oide, M., Saishu, S. and Kamiya, M. (1967b). *C.r. Séanc. Soc. Biol.* 1201–1204.
Utida, S., Oide, M. and Oide, H. (1968). *Comp. Biochem. Physiol.* **27**, 239–249.
Utida, S., Kamiya, M. and Shirai, N. (1971). *Comp. Biochem. Physiol.* **38A**, 443–447.
Utida, S., Hirano, T., Oide, H., Ando, M., Johnson, D. W. and Bern, H. A. (1972). *Gen. Comp. Endocrin. Suppl.* **3**, 317–327.
Vancura, P., Sharp, G. W. G. and Malt, R. A. (1971). *J. clin. Invest.* **50**, 543–551.
Vickers, J. A. (1961). *Q. Jl. microsc. Sci.* **102**, 507–518.
Vicente, N. (1969). *Recl. Trav. St. Mar. Endoume* **46**, 13–121.
Vooys, G. G. N., de (1968). *Archs int. Physiol. Biochim.* **76**, 268–273.
Voûte, C. L. and Ussing, H. H. (1968). *J. cell. Biol.* **36**, 625–638.
Voûte, C. L., Hanni, S. and Ammann, E. (1972). *J. steroid Chem.* **3**, 161–165.
Wang, J. W., Robinson, C. V. and Edelman, I. S. (1953). *J. Am. chem. Soc.* **75**, 466–470.
Webb, D. A. (1940). *Proc. R. Soc. Ser. B* **129**, 107–136.
Weel, P. B., van (1957). *Z. vergl. Physiol.* **39**, 492–506.
Weendelaar Bonga, S. E. (1972). *Gen. Comp. Endocrin.* suppl. 3, 308–316.
Whitlock, R. T. and Wheeler, H. D. (1964). *J. clin. Invest.* **43**, 2249–2265.
Wichard, W. and Komnick, H. (1971). *Cytobiology* 3, 215–228.
Wichard, W. and Komnick, H. (1973). *Z. Zellforsch. mikrosk. Anat.* **136**, 579–590.
Wiebelhaus, V. D., Sung, C. P., Helander, H. F., Shah, G., Blum, A. L. and Sachs, G. (1971). *Biochim. Biophys. Acta* **241**, 49–56.
Wigglesworth, V. B. (1933a). *J. exp. Biol.* **10**, 1–15.
Wigglesworth, V. B. (1933b). *J. exp. Biol.* **10**, 16–26.
Wigglesworth, V. B. (1933c). *J. exp. Biol.* **10**, 27–37.
Wikgren, B. (1953). *Acta zool. fenn.* **71**, 1–102.
Windhager, E. E. (1968). "Micropuncture Techniques and Nephron Function." Butterworths, London.
Wolbach, R. A., Heinemann, H. O. and Fishman, A. P. (1959). *Bull. Mt. Desert Isl. Biol. Lab.* **4**, 56–57 (abstract).

Wood, C. M. and Randall, D. J. (1973). *J. comp. Physiol.* **82**, 235–256.
Whurmann, K. von and Woker, H. (1948). *Schweiz. Z. Hydrol.* **11**, 210-244.
Whurmann, K. von and Woker, H. (1953). *Schweiz. Z. Hydrol.* **15**, 235–259.
Zadunaisky, J. A. and Fisch, F. W., de. (1964). *Am. J. Physiol.* **207**, 1010–1014.
Zadunaisky, J. A., Candia, O. A. and Chiarandini, D. J. (1963). *J. gen. Physiol.* **47**, 393–402.
Zaugg, W. S. and McLain, L. R. (1970). *Comp. Biochem. Physiol.* **35**, 587–596.
Zaugg, W. S. and McLain, L. R. (1972). *J. Fish. Res. Bd Can.* **29**, 167–171.
Zaugg, W. S., Adams, B. L. and McLain, L. R. (1972). *Science N.Y.* **176**, 415–416.

Biochromes. Occurrence, Distribution and Comparative Biochemistry of Prominent Natural Pigments in the Marine World

DENIS L. FOX

Division of Marine Biology, Scripps Institution of Oceanography,
University of California, San Diego, La Jolla, California

I. Introduction

". . . mention shall be made of corals, . . . of pearls; for
the price of wisdom is above rubies". Job—xxviii, 18.

Panoramic vistas, even through the naked (or aided) eyes of generalists or
amateurs, spanning the two million or more known species of living
organisms, will bring sharply into notice not only the conspicuous differences in size, form and movement, but great contrasts in color or color-
combinations and patterns. Yet a few moments of reflection will
remind the inquiring observer that, despite the incalculable number of
different proteins involved, constituting the basis of species differentiation,
the various entities conferring upon living organisms the property of
coloration reside in but a few molecular types or polymolecular structural
configurations. There are but those few basic kinds to go round.

Countless organisms share the ability to synthesize their own supply of
some colored molecules *de novo*, e.g. chlorophylls and carotenoids are
manufactured by all green plants, carotenoids indeed also by many fungal
and bacterial species; dark melanins and bright tetrapyrroles (including
hemoporphyrins and bilichromes) by many animals; and the list could be
extended.

In this chapter, following some discussion with illustrative examples of
schemochromic (physically evoked) manifestations of conspicuous colors,
we shall consider four or five of the more common, widely distributed
classes of true biochromes, or colored molecules manifested by organisms
of the hydrosphere. These examples will include notably the yellow, orange
or red polyenes (carotenoids) accompanying chlorophylls in all algae and
exhibited also by many animal species; the polyhydroxy naphthoquinones
and anthraquinones, of limited distribution; indolic pigments, including
colorful indigoids that occur in certain marine snails, as well as dark,
endogenous melanins of very wide distribution; and some of the cyclic and
open-chained members of the tetrapyrrole class, i.e. porphyrins and bilichromes. These and certain less common or less noticeable molecular
types including the ill-defined chromolipids or lipofuscines, flavones,
flavins, pterins, and some miscellaneous kinds, have been considered in
greater detail elsewhere (e.g. Fox, 1953, revised, 1974). Save for an occasional exception, the less common types will not be dealt with here.

In many instances, references may be limited to books, chapters, and
other reviews which already will have covered extensively some of the
original articles and reports scattered through the literature.

Colors are evoked in nature through the reflection or transmission of
incident light, from or through surfaces or systems characterized by:
(1) the presence of entities involving multiple ultra-fine physical dimensions

(giving optical colors or schemochromes); (2) certain molecules (bio-chromes) endowed with chemical resonance of frequencies residing within the wavelength span of the visible spectrum (400 to 750 nm); (3) resultants of certain combinations of both biochromes and schemochromes. In any treatment of natural chromogenesis, the schemochromic entities deserve passing consideration in order to set them recognizably apart from the biochromes, or chemical pigments.

II. Schemochromes

A. Blues, from light scattering

Blue colors of the cloudless, smokeless sky arise from Rayleigh-scattering of incident sunlight by atmospheric gaseous molecules in thick layers. Thus the color of pure, mote-free air is a soft blue, as are likewise many other gases, as well as many clear liquids, including freshly distilled, mote-free water. This Rayleigh-scattering can be seen and demonstrated in two ways. Firstly, by viewing a light-source through pure gaseous or pure liquid systems in long columns or thick layers; and secondly by focusing white light laterally through a lens at the center of a column of the fluid within a blackened, non-reflecting tube, and by then viewing from a 90° angle the blue, luminous cone through a window in the end of the tube (Wood, 1934). Due largely to Rayleigh-scattering, wherein the longer light-rays fail to penetrate so far, the ocean, notably at its deepest, least populated areas, is blue in its own right, besides constituting a surface-reflector of the blue sky above. Marine divers, descending to substantial depths in clear water areas, observe blueness in the environment generally, and on looking upward, e.g., at white or light-colored objects above them, even on the surface, see these as pale blue in color. Indeed, bodies of clear oceanic water, or of clear mountain lakes, e.g., Crater Lake in California, reflect the blue colors of Rayleigh-scattering even under overcast skies. Moreover, if one remains afloat in a prone position, with one's face barely submerged while viewing downward through the clear glass of a face-mask, the color thus observed will be of the same deep blue. Rayleigh-scattering thus exerts its effect independently of the fact that the blue fraction of white sunlight is the portion that penetrates farthest into water bodies.

Most blue colors of solid objects or of disperse systems, however, are generated through Tyndall-scattering, i.e., by finely divided particles of submicroscopic, colloidal dimensions, dispersed in translucent media. Common examples include freshly generated wood-smoke in air, very fine white suspensoids in aqueous media, and the microvesicles of air (ca 0.6μ or less in diameter) within the solid keratin-proteins overlying melanin in feather-barbs, rendering them blue. Another example lies in the solid-in-liquid or liquid-in-liquid colloidal systems characteristic of blue eye-iris and of blue skin areas of numerous fishes, reptiles, birds, and some primates, such as in the muzzle and buttocks of male mandrill baboons.

White scattering results from reflection of all light-waves equally and completely by dispersed systems whose constituent microparticles exceed in size the small dimensions capable of selectively reflecting the shortest rays (blue and violet) of incident whole light (Fox, 1953).

B. Changeable, polychromic colors, from interference

Another outstanding example of schemochromic manifestation resides in the iridescent colors evoked by *interference* between incident (entering) and reflected (returning) light rays passing through systems involving multiple layers of extremely fine dimensions (again, about 0.6 to 0.7μ or less in thickness). On land, wing-scales of butterfly and hard wing-covers of beetle members of the great insect class, as well as feathers of numerous birds such as peafowl, hummingbirds, mallard ducks, pigeons and many others, advertise strikingly their possession of ultra-thinly laminated materials of alternating composition.

In the marine world as well, interference coloration has many representatives, e.g. pearly or nacreous deposits of alternating calcium carbonate and water laminations in (or from) shells of gastropod, lamellibranch or nautiloid molluscs; proteinaceous layers in bristles or setae of some marine worms (e.g. *Aphrodite*); cuticular body-coverings of some worms; the thin carapaces of a few pelagic or burrowing crustaceans; and scales of some fishes (Fox, 1953).

III. Carotenoids

" . . . thus . . . will the multitudinous seas incarnadine,
making the green one red". (MacBeth, Act II Scene II)

A. General

The colors of the hydrosphere, such as, in areas of the oceans and in many lakes, are changeable depending on relative seasonal populations of pigment-bearing, microscopic phyto- and zooplanktonic organisms. Mid-oceanic waters, characterized by the greatest depths and by the lowest concentrations of nutrient chemicals, hence of minimal numbers of phytoplankton and other organisms, exhibit as we have seen the deepest schemochromic blue colors. In contrast extended areas of the sea bordering or close to land masses assume seasonal greenish, yellow, or even reddish or deeper colors due to high populations of diatoms, dinoflagellates, or both, which bear chlorophylls and often rich stores of yellow or orange carotenoids.

Some coastal waters of North American Pacific, Atlantic, or Gulf regions harbor resurgent populations of dinoflagellates such as *Prorocentrum, Gonyaulax, Gymnodinium, Peridinium* or other species. These plankters may at certain seasons achieve such enormous populations as to confer yellow, orange or red colors upon extensive masses of off-shore and near-shore waters. Such intense populations actually can take on a kind of deep brownish or chocolate aspect at some sites, and may exhibit lines, streaks or masses of small surface-bubbles from their photosynthetic activity. Some planktonic species are also luminescent, exhibiting brilliant nocturnal flashes of bluish light when stimulated by objects moving rapidly through the water, or noticeably at the impact of long, incoming waves breaking on or near a beach.

Adverse effects may arise as a consequence of these resurgent, so-called "red tides", or more generally "red-water" incidences, due partly to the excessive consumption of dissolved oxygen during night hours, and even in daylight by cell populations located beneath the compensation depth imposed through light-screening by the portion of the colony present in the waters above. More serious lethal effects arise, however, as the microscopic plant cells pass their maximal reproductive period, then die and decompose by countless millions. As a result not only is still more oxygen consumed, but toxic products of bacterially implemented putrefaction are released into the water, killing many resident invertebrates and fishes, while the pigmented plastids expelled from the ruptured cells continue to color the water for relatively short periods. The processes involved in the rotting of organic matter give rise to sundry odors, wherefore "red water" has at times been accorded the alternative term "stinking water".

Pigmentary yellow, orange or red colors may characterize natural waters for varying periods. Still other colors may arise from the combined effects of Tyndall or Rayleigh blue with biochromic yellow resulting in green; or if with biochromic red a purplish coloration. These are all based upon the presence of fat-soluble polyenes or carotenoids synthesized by all green plants and by many bacteria and fungi, and present also through nutrition in the tissues of nearly all animal organisms, including many of the colorful zooplankters.

The phytoplankton of the oceans, representing the most productive biochemical factory in the world, synthesize some 4×10^{10} metric tons of (dryweight) organic matter annually. Of this, about 0.10% or several hundred million tons constitute newly manufactured carotenoids, chiefly

fucoxanthin and other typical algal xanthophylls, with carotene hydrocarbon types themselves in minor proportion amounting, on the average,
to perhaps 10% of total carotenoids.

The pigmentary coloration of natural waters is contributed not solely by
carotenoids. Photosynthetically active chlorophylls exceed by some 50-fold
the average concentration of intracellular carotenoids in algal organic
matter. Thus greenish colors may arise from chlorophylls as such (as well
as from yellow plus structural blue, cited above). Complementary brownish
colors are generated through the concerted screening of reflected or
transmitted light by both yellow or orange carotenoids and green chlorophylls, or their darker colored degradation products such as phaeophytins
(lacking the magnesium component) or phaeophorbides (without either the
magnesium or phytol moities).

The so-called polyenes, lipochromes, or more properly carotenoids are
the most widely and conspicuously displayed biochromes throughout the
animal kingdom, especially in the marine world. They are conspicuous
also in many species of bacteria, fungi and even algae, although in the latter
the green chlorophyll colors usually predominate, masking the carotenoids.
Carotenoids manifest their yellow, orange or red colors whether dissolved
in various common, so-called fat-solvents such as ethanol, acetone, hexane,
chloroform, etc., or when present in tissues, usually in lipid deposits.
They are of high molecular weight, e.g., $536\cdot85$ for carotene hydrocarbons,
and greater for the oxygen-bearing derivatives of these ^{40}C compounds.
They manifest high melting points and display characteristic absorption
maxima in the visible spectrum, i.e., one, more often two or three, and
sometimes more peaks in violet, blue, blue-green or green regions.

It should be pointed out here, since reference to this will follow, that
certain polar carotenoid compounds may conjugate chemically with
proteins, giving blue, green, purple, red or other colors to the outer surfaces or eggs of various crustaceans, echinoderms, nudibranch molluscs
and other invertebrate animals; or with calcareous skeletal material,
contributing purple, pink, orange or other colors to some coelenterates of
the coral class; or actually with chitin of certain crustacean carapaces
(Fox and Wilkie, 1970; Fox, 1972b, 1973; McBeth, 1972a, b).

Colors of compounds are evoked by chemical resonance of the molecules
arising from vibrational frequencies as modified by the presence of certain
double-bond sequences. In the instance of carotenoids, these unsaturated
sequences take the form of alternating, or so-called conjugated, carbon-to-
carbon double bonds within the long chromophoric chain (Fig. 1).Caroten-
oids notably in solution *in vitro* are sensitive to the presence of oxygen,
acidic substances, light, or combinations of these, and this sensitivity is
enhanced by elevations in temperature.

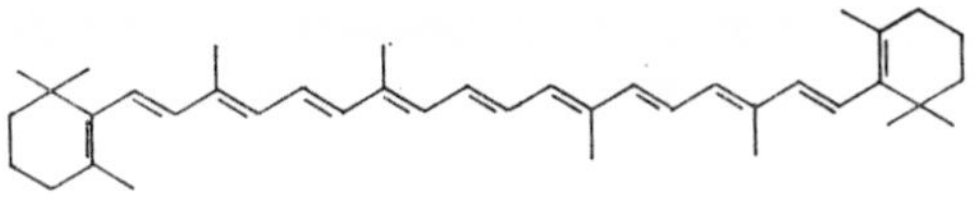

β-carotene $C_{40}H_{56}$

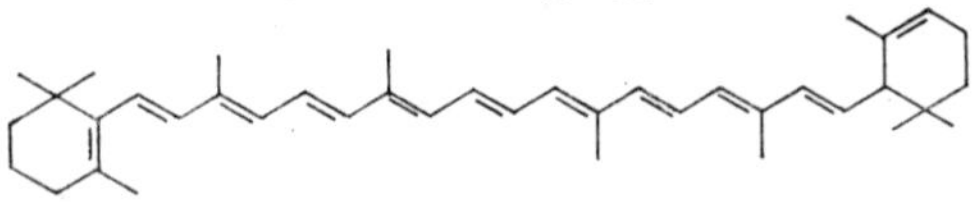

α-carotene $C_{40}H_{56}$

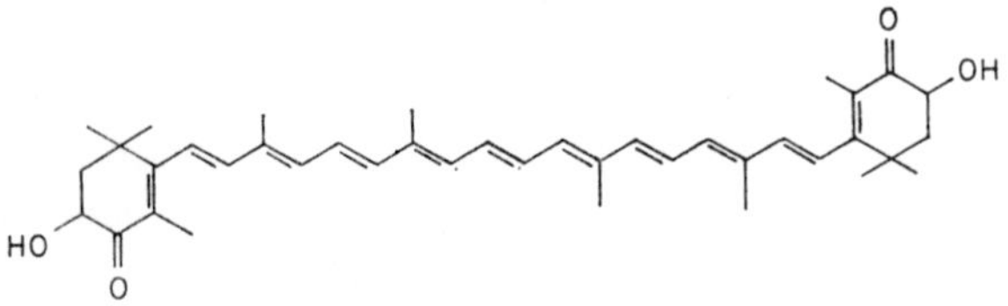

astaxanthin $(= 3,3'$ dihydroxy-$4,4'$dioxo-β-carotene$)$

$C_{40}H_{52}O_4$

FIG. 1. Some common carotenoid molecules.

In certain deep marine sediments, however, carotenoids may remain buried for many thousands of years with other detritus from plant and animal organisms (see further). Hence they are extractable by washing the damp (or vacuum-dried), dark, muddy, silty or other sediments with water-miscible fat-solvents, such as alcohol or acetone. Many bottom-dwelling, detritus-feeding animals doubtless sustain their supplies of body-carotenoids along with their food, from such sources.

In metabolism typical of marine animals, with few exceptions, carotenoids consumed with the food are selectively mobilized through the intestine (or coelenteron), via the circulatory fluid, not only to glandular sites such as liver (or hepatopancreas) and ripening gonads, but often in very rich supply to the integument, thus rendering many species brightly colored and conspicuous, at least in lighted waters and to human viewers.

B. Fishes and Birds

Many species of teleost fishes, notably inhabitants of tropical tide-pools and other near-shore waters, exhibit the bright yellow, orange and red colors of integumentary carotenoids over the body surface, whether as solid hues or in bars, spots, patches or other interrupted patterns.

The Garibaldi or marine goldfish, *Hypsypops rubicunda,* of Southern Californian rocky shore waters exhibits brilliant, specular blue schemochromic skin-patches when young and small, but in adulthood displays bright orange-yellow integumentary xanthophylls over the general body surface. The juvenile blue spots are generated by interference-reflection of light by thin epidermal multilayers of flat, flaky, crystalline guanine, a colorless purine, overlying deposits of dark melanin.

Some fishes inhabiting unlighted depths possess not only black or other dark, melanistic skin pigmentation, but also (sometimes instead) the bright red colors of keto-carotenoids. Red coloration, at considerable depth where the color of the light is blue, confers upon the animal a dark appearance; hence the animal is visually inconspicuous to either predators or prey.

Anadromous salmon, migrating upstream into rivers toward their hatching sites to spawn, following their years of maturation in the ocean, exhibit the rich red integumentary pigmentation of red carotenoid, chiefly astaxanthin, derived from their extensive consumption of crustaceans. This red coloration is more conspicuous in the males than in females, which deposit much of the astaxanthin in ovaries and developing eggs.

Even more striking examples of this kind of sexual dichromatism appear in numerous other fishes, e.g., the sheep-head *Pimelometopon pulcher*. The male sheephead exhibits a broad red band across the mid-portion of the body, while head and caudal area have black, melanistic pigment covering the red. The female exhibits a fairly uniform pink color over her whole body skin.

Certain fishes are capable of alternatively diffusing or densely concentrating the pigment particles within the chromatophore cells of their scales, thus rendering their bodies respectively brightly attractive or pale and inconspicuous. The physiological control of visible coloration involves the tawny, brown, sometimes ruddy, or more commonly black melanins within the so-called melanophores (see further), more than it does the carotenoids contained within the smaller, less numerous xanthophores or erythrophores.

Some white or dark-feathered birds of the open marine waters or of shore areas exhibit carotenoid pigmentation in parts of bare skin. Certain gulls, for example, display yellow pigmentation over the exposed tarsal and toe-skin, while others exhibit red colors instead, both in those sites and about the face and bill. Flamingos, roseate spoonbills and scarlet ibis, inhabiting swampy, brackish, or shore water places, and feeding there, display rich red carotenoid pigmentation extravagantly in their feathers and in naked tarsal and digital skin, and in some patches about the face.

Flamingos, equipped with specialized filter-feeding equipment within the mandible and over the lateral margins of the tongue, are able to ingest great quantities of planktonic algae, pelagic crustaceans, larval fishes and other small animals, as well as much organically rich detrital mud from the floors of lagoons, shore areas and river mouths. Undoubtedly, it is the algal material which contributes by far the major quantities of carotenoid pigment wherefrom the birds acquire the raw material for adornment of their naked skin parts and bright plumage. Indeed it has been shown that *Phoenicopterus ruber*, the so-called American, West Indian, or Caribbean flamingo, most brightly pigmented of today's six living species, oxidizes

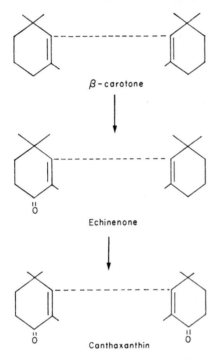

Fig. 2 Possible metabolic oxidation of β-carotene in flamingo (cf. Davies *et al.*, 1970).

β-carotene in its diet to ketonic derivatives: orange echinenone (4-mono-keto-β-carotene), and hence the red derivative canthaxanthin (4, 4'-diketo-β-carotene) Fox *et al.*, 1969*b*) (Fig. 2). Phoenicoxanthin (3-hydroxy-canthaxanthin) and astaxanthin (3, 3'-dihydroxy-canthaxanthin), the familar crustacean carotenoid, also arise as ensuing red derivatives deposited in skin and feathers (Fox and Hopkins, 1966*a*).

Parenthetically, it may be remarked that zoo-keepers or bird fanciers who fail to provide rich sources of carotenoids in the diets of yellow, pink or red-feathered birds observe a gradual fading of skin-pigmentation and a loss of plumage color with each seasonal molt. They have long recognized

178 DENIS L. FOX

the advantage of including in the diet generous supplies of yellow plant or red animal carotenoids for realizing maximal coloration of certain birds. Similarly, it has been demonstrated by Schmidt and Baker (1969), that the pink to red pigmentation of salmon and trout flesh may be materially augmented by feeding such fishes on a canthaxanthin-rich diet.

There are exceptional numbers of brightly colored carotenoid-displaying species among the invertebrates of the sea whether they be pelagic or benthic, beneath waters of greatly varying depth, or living between high and low tide levels. A few typical examples from each of several more representative animals should serve for present purpose of general comparison.

C. Tunicates (Ascidians or Sea-squirts)

These, strictly speaking, are classed as primitive chordates, however, they lose the notochord as they develop toward adulthood, when they become sac-like and usually sedentary, but some species are free-swimming animals, bearing one inhalent and one excurrent siphon and invested within a covering cloak or tunic. Bright colors are evident in a few free-swimming or planktonic members, but are far more common in the attached species.

Some special biochemical claims that this interesting, rather aberrant group exercise upon our attention include their ability, unique in animals, to lay down an animal-cellulose, tunacin, as the material composing the tunic or over-cloak. Also a number of the species carry small, greenish, rounded cells (vanadocytes) in the body fluid, which contain relatively high concentrations of ionic vanadium linked to a tetrapyrrolic biochrome, and which bear also striking concentrations of sulfuric acid amounting to ca 1.8 N or some 9%.

Conspicuous among the brightly pigmented tunicates are the carmine-red and orange-pigmented simple ascidian *Styela grossularia*, and such colonial forms as *Amaroucium stellatum*, often called sea pork due to the pink or pale bluish colors of the common gelatinous tunic enclosing the massive colonies. *Botryllus schlosseri*, *Maxilla mammilaris* and others are present in this group. The dark red tunic of *Halocynthia papillosa*, an ascidian related to the sea peach *H. pyriformis*, was found to yield rich amounts of astaxanthin with a second xanthophyllic carotenoid and minor quantities of α- and β-carotenes. The violet-red or roseate social tunicate *Dendrodoa grossularia* likewise yields astaxanthin in considerable quantity (Lederer, 1938).

A number of tunicate species store assortments of both the hydrocarbon (carotene) and alcoholic or the ketonic (xanthophyllic) type of carotenoid, accompanied by some chlorophylloid compounds, derived from a diet rich in phytoplankton and finely particulate algal detritus all readily screened from the water by these ciliary-mucous-feeding animals.

D. Crustaceans

Under this class of arthropods there are five sub-classes, of which the first three (branchiopods, ostracods and copepods) are small animals, even microscopic in many orders; a fourth, the cirripeds or barnacles are mostly sessile and less conspicuous than the fifth subclass, Malacostraca, which includes the larger, more conspicuous and colorful members, e.g., lobsters, crayfishes, crabs, shrimps, krill, etc.

Many copepod species, predominant organisms of the zooplankton, exhibit red or orange pigmentation notably in the viscera, fat-body, or ripe ova. And while reproductive eruptions of some copepod species may redden extended patches of ocean waters in certain seasons (a sign regarded favorably by commercial fishermen), the incarnadine effects thus generated are of less frequency, covering less extensive areas, of shorter duration, and usually not so deeply colored in comparison with the respective effects of phytoplankton blooms. The small, pelagic or planktonic crustaceans' supplies of carotenoids are derived, in any event, from their consumption of microscopic algae and finely particulate organic detritus or leptopel. This applies likewise to the euphausiids, including red krill, consumed in vast numbers by the toothless, filter-feeding blue (baleen) whales. Indeed there are reports of occasional specimens of the blue whale exhibiting red colors in its skin and tissues, from some aberrant metabolic processing of the astaxanthin consumed with the vast number of krill (Fox, 1953).

The bright blue, green, purple, red or other colors of common crustaceans are familiar to most of us, as is the spectacular change to the bright red color of the carotenoid, usually astaxanthin, when the protein moiety of the colored complex has been denatured, whether suddenly as by exposure to boiling water, or more gradually as when immersed in alcohol. Similarly, numerous such species carry eggs of colors like those of the epicuticle or other tissues (e.g., the pelagic, blue goose-barnacle *Lepas fascicularis* (Fox *et al.*, 1967*b*). There also are occasional instances of mutational crustacean specimens, e.g., of the American lobster *Homarus americanus*, which instead of displaying the common overall dark green with interspersed red colors, may be characterized by blue, red, creamy or calico patterns. Specimens have been reported and photographed showing discontinuous splotches of green and red, and even occasionally a striking harlequin, or pied lobster showing a dark-green color on one longitudinal half of its carapace, including the chela, walking legs and telson, and on the other half, bright orange over the body shell and appendages (Atz, 1939). The striped shore-crab *Pachygrapsus crassipes* exhibits broad purple stripes on claws and legs, and wide green bands across the dorsal surface of the carapace. Heat-denaturation of the protein moiety unmasks the uniform red color of the heretofore conjugated carotenoid.

The conjugated carotenoids of certain crustaceans may be naturally red, as in some freshwater and marine crayfish, e.g. *Astacus* and *Panulirus* respectively, and in some marine shrimps and prawns.

In at least one instance, i.e., the southern red kelp-crab *Taliepus nuttallii*, common in and near kelp beds along the coast of Southern California, four keto-carotenoids, viz: echinenone, canthaxanthin, phoenicoxanthin, and astaxanthin, are firmly anchored chemically to the animal's chitinous armor. It is logical to suppose that the linkage is between the terminal amino groups of chitin, an acetylaminopolysaccaride, and the ketonic radicals of the carotenoid molecules. This is the kind of conjugating linkage believed to hold likewise for the binding of keto-carotenoids to terminal amino groups of proteins (Cheeseman, 1967), whether thus affording colloidally soluble chromoproteins or hard, insoluble, stable complexes, typified by the keratin-proteins of bright red feathers, as in flamingos, scarlet ibis, or roseate spoonbill. The keto-cum-amino linkages are believed to be effected through the formation of carbonylamino or perhaps protonated Schiff's base configurations, as represented by the following equilibrial equations:

$$
\begin{array}{ccccc}
\diagdown\!\diagup & & \diagdown\!\diagup & & \diagdown\!\diagup \\
\text{C} & & \text{C} & & \text{C} \\
\| & & \diagup\ \diagdown & & \| \\
\text{O} & & \text{HN} \quad \text{OH} & & \text{N} \\
+ & \rightleftharpoons & | & \rightleftharpoons & | \qquad\quad + \text{H}_2\text{O}\\
\text{NH}_2 & & \text{—C—} & & \text{—C—} \\
| & & | & & | \\
\text{—C—} & & \text{R} & & \text{R} \\
| \\
\text{R} & & \text{Carbonylamino} & & \text{Schiff's} \\
& & \text{linkage} & & \text{base} \\
& & & & \text{bonding}
\end{array}
$$

The reverse of each (rightward indicated) reaction, e.g., by brief exposure to elevated temperatures or to acidic, saline media, would hardly be expected to occur when insoluble substrates, such as keratin-protein or chitin, were involved, or if carotenoid linkages were being forged. In any event, carotenoids thus conjugated to hard, insoluble, relatively inert materials exhibit exceptional chemical stability against relative excesses of environmental light, temperature, atmospheric oxygen, and certain other chemical factors (Fox, 1973).

The brine-inhabiting branchiopod or fairy-shrimp *Artemia salina*, is capable of oxidizing dietary β-carotene to produce echinenone and canthaxanthin (Davies *et al.*, 1965; 1970) as has since been shown for flamingos (Fox *et al.*, 1969*b*). In all likelihood, other crustacean species likewise are capable of this and similar metabolic oxidations, whence their rich stores of red astaxanthin must derive. This is very likely, for example, in the above-cited kelp-eating crab *Taliepus*, which obtains from its prominent algal food adequate supplies of carotenoids, including β-carotene, whence it consequently must generate the four keto-carotenes mentioned, conjugating them with the chitin of its exoskeleton.

The time is now at hand for beginnings to be made on a search for the site, nature and reaction-rate of oxidative enzyme systems responsible for introducing O-atoms, each one in place of two H-atoms, onto the 4-C positions of β-cyclohexenyl rings of carotenoid molecules. Materials for a promising beginning in this kind of research should be available in certain invertebrate animals exhibiting comparatively rich stores of keto-carotenoids. Thus, such a survey might include certain coelenterates (e.g. *Metridium*, the plumose anemone is currently under such study in this laboratory), asteroid echinoderms (*q.v. infra*), and particularly, one might believe, among the Crustaceae (see Goodwin's extensive survey, 1971). Special attention ought to be given to any differences in the animals' metabolism of hydrocarbon *versus* alcoholic or other xanthophyllic types of carotenoid consumed, as has been done with *Artemia*, flamingos, and earlier with certain species of fish (Davies *et al.*, 1970; Fox *et al.*, 1969*b*; Sumner and Fox, 1935).

E. Molluscs

In this diversified phylum, carotenoid pigmentation is most richly advertized by members of the gastropod class (snails, slugs, etc.), while some of the pelecypods (Lamellibranchia or Bivalvia, including clams, mussels, and oysters) may exhibit similar bright pigmentation. In most such instances, the conspicuous coloration is chiefly in the integument, brightly colored molluscan shells owing their adornment usually to the presence of tetrapyrrolic pigments (see below). The overall body surfaces of many brilliant sea-slugs of the nudibranch order, and notably also exposed areas of mantle in some marine snails and bivalves, such as the common mussel *Mytilus*, afford conspicuous examples of carotenoid storage. These and many other allied forms carry rich stores of such pigments also in the "liver" organ or hepatopancreas, and deposit them thence into the yolky part of their ripening eggs.

The great majority of marine animals including molluscs, crustaceans, fishes and members of many other phyla are, like the domestic fowl, chiefly selectors and assimilators of the xanthophyll (or hydroxy-carotene) type, unlike the horse, which selectively assimilates and stores carotenes, discarding the xanthophylls. There are a few marine exceptions to this wide generalization, such as the intertidal, sand-dwelling polychaete worm *Thoracophelia* (*Euzonus*) *mucronata*, which, consuming marine organic detrital materials in and adsorbed to the sand, eliminate all xanthophylls, despite the 10-fold preponderance thereof over the carotenes, which they assimilate. Moreover this assimilation is unusually selective, since these worms seem to store only the β-isomer (Fox *et al.*, 1948).

Molluscs exhibiting the most brilliant and beautiful integumentary colors and patterns are the nudibranch slugs of the chromodorid type, which display light or dark blue, purple, violet and delicate pink pigments, extractable in aqueous solvents, and magenta-red or other colors, in some instances with orange or lemon-yellow stripes, spots or other patches not leached out by watery media.

The water-extractable blue, purple, pink, or violet pigments are carotenoids chemically conjugated with proteins, and exhibit red or orange colors once the protein constituent has been denatured with alcohol. The red, orange and yellow skin-pigments represent carotenoids associated with lipids, or perhaps merely dissolved in the lipid portions of lipoproteins, and hence remain untouched by water systems.

Curiously enough, a number of brilliantly pigmented nudibranch species studied in this laboratory seem to display a conspicuous lack in the metabolic originality that characterizes many other animals, such as coelenterates, crustaceans, echinoderms, and even some bird species. That is, rather than chemically modifying relatively conventional, yellow dietary carotenoids to yield colorful ketone derivatives, these slugs merely assimilate specific carotenoids already present in their prey, including sponges, bryozoans or hydroids.

A couple of examples among the nudibranchs are *Hopkinsia rosacea*, the delicate pink-bodied, pink-gilled slug, and the purple-skinned, orange-gilled *Flabellinopsis iodinea*, both encountered in California coastal waters. The carotenoid-rich *Hopkinsia* yields, among its several pigment fractions, 70% of their total as hopkinsiaxanthin, an orange apo-carotenoid bearing oxygen groups (one alcoholic, one ketonic) on its single cyclohexenyl terminal ring, an acetylenic (triple-bonded) link between C-atoms 7 and 8 of the chromophore chain, and a ketonic oxygen atom on the terminal carbon (Fig. 3). But this same compound also is the prominent carotenoid

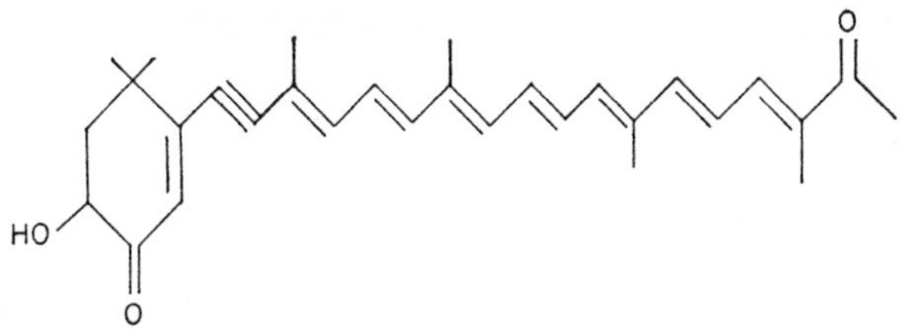

Fig. 3. Hopkinsiaxanthin.

in the slug's principal prey, i.e., the pink, encrusting bryozoan *Eurystomella bilabiata* (McBeth, 1972a, b).

Flabellinopsis iodinea was indeed singular, in that it yielded but one carotenoid, astaxanthin, from its body-parts, i.e., one free and two esterified fractions from the reddish rhinophores (a pair of fleshy, tentacle-like extensions from the head. Likewise from the orange-colored cerata (external, finger-like gills over the dorsal body surface) there were recovered free astaxanthin (80% of the total there present) and two esters thereof. The ripe, pink egg-masses, visible through the blue-violet skin, yielded only unesterified astaxanthin, as its usual status in eggs, and finally the blue-violet color of the integument was due to the presence of a conjugated astaxanthin-protein. This common marine slug feeds upon a hydroid, *Eudendrium ramosum*, which stores in its bright orange-colored gastrozooids free astaxanthin and five esters thereof. It has been assumed, accordingly, that this nudibranch predator, rather than chemically modifying a carotenoid precursor, derives its astaxanthin directly from its victim's tissues (McBeth, 1972a, b).

Other molluscan classes, including the cephalopods (octopus, squid) and many common bivalves, such as oysters and clams, store no observable carotenoids in skin, mantle, or other external surfaces, and only moderate amounts derived directly from the diet in the "liver", or digestive diverticulum. (re: octopus, see Fox and Crane, 1942).

Ovorubin, a glycoprotein of red color due to its chemical conjugation with the tautomeric carotenoid astaxanthin, occurs in the egg-jelly of an herbivorous, freshwater snail, *Pomacea canaliculata australis*. This complex showed a molecular weight of 330 000 and absorption maxima at 510 and 545 nm in aqueous systems. It was relatively highly stable against denaturation by heat (below 100°C), by cold alkali, or by adsorption to solid surfaces. Moreover, this chromoprotein was not readily attacked by trypsin, which it indeed inhibited, nor by heavy populations of bacteria and molds during the course of a year at pH 6 and 5°C. The carbohydrate moiety, constituting about 20% of the ovorubin molecule, involved some

5% of amino sugars, determined in terms of glucosamine, and the other 15% as non-amino sugars. Glucosamine, mannose and galactose were among the carbohydrate fractions recognized.

Cheesman (1958), who carried out this research, pointed out certain manifest ecological advantages of ovorubin's general chemical stability in relation to the natural habits and ecology of the snail, which deposits its eggs in exposed sites. The chromoprotein's resistance against thermal, chemical and adsorptive denaturation, and against attack by bacterial, fungal or other enzymes, as well as its apparent role in the conservation of imbibed water during the eggs' exposure to air at elevated temperatures, all suggest a considerable degree of survival value for the species.

F. Echinoderms

Within this completely marine animal phylum, conspicuous carotenoid pigmentation is exhibited by three classes, namely the asteroids (sea-stars), ophiuroids (brittlestars or serpent stars) and a few holothurians (sea-cucumbers). The two other classes, i.e., the echinoids (sea-urchins and sand-dollars) and crinoids (sea-lilies) may display instead purple, red, or other pigmentation due to their storage of certain quinones (see below), but not commonly carotenoids, which may, however, be carried in ripe or maturing eggs, and occluded in soft somatic tissues.

The asteroid echinoderms recall the nudibranch molluscs, in that both of these invertebrate types are richly colored with carotenoids, often carotenoid-protein complexes. Since both are largely if not exclusively carnivorous, they may derive their various pigment fractions directly from their prey.

As one example, the large, so-called ochre star, *Pisister ochraceus*, most abundant of the Pacific shore asteroids, displays skin coloration varying from yellow, through orange and brown even to purple, and may store, *inter alia*, considerable proportions of mytiloxanthin, a xanthophyll elaborated by *Mytilus californianus*, the sea mussel, upon which the sea-star preys (Fox and Scheer, 1941).

One of the so-called webbed sea-stars *Patiria miniata*, a low-tide species encountered along the Pacific coast from Alaska to Lower California, occurs as yellow, orange, red, brown and purple variants. Extracts of the integument yield correspondingly increasing proportions of astaxanthin, compared with another xanthophyll present but not yet identified. Thus the ratios of astaxanthin to the congener xanthophyll have been measured as 51/40, 66/25, 72/21, and 80/19 for the respective orange, red, brown and purple genotypes (Wells, cited by Fox and Hopkins, 1966*b*).

The majority of sea-stars investigated store varying amounts and proportions of carotenes (always in lowest ratio), hydroxy-carotenes and keto-carotenes (Fox and Hopkins, 1966b).

G. Coelenterates

It is in this phylum, notably among the sea anemones, certain hydrocorals, some gorgonians and other alcyonarians, and to some extent in the Scyphozoa (jellyfishes) that we encounter perhaps the greatest variety and brilliance of animal biochromy. Indeed, rich pigmentation characterizes not only the thin, fleshy parts of many species, but the solid, calcareous, skeletal parts of numerous hydrocorals, gorgonians (sea-fans) and other alcyonarians as well (Fox and Pantin, 1944; Fox et al., 1969a; Fox and Wilkie, 1970; Fox, 1972b). There are numerous species of anemone and coral which harbor symbiotic unicellular algae (zooxanthellae or zoochlorellae) within their tissues thus exhibiting greenish, yellowish or brown colors in such sites.

Among the non-commensal-bearing anemones and corals, the bright pigments of the coenenchyme or fleshy parts belong, for the most part, to the carotenoid class. The older work relating to such instances was reviewed by Fox and Pantin (1944) and Fox (1953), while later researches have afforded additional examples, of which only a few will be cited here.

Among countless sea anemone species which exhibit brilliant carotenoid hues, whether over the whole body or in various patterns, is the plumose form, *Metridium senile*, common along the shores of Britain and of Europe in general. Its sub-specific relative *M. s. fimbriatum* occurs along the Pacific Coast. The species is polychromic, occurring in a number of genotypic colors or color combinations, including white, red, brown, grey, red-with-brown, red-with-grey, brown-with-white or with grey, red-with-brown-with-grey and other combinations (Fox and Pantin, 1941; Fox et al., 1967a).

Moreover, superimposed upon these contrasting colors and color patterns, there occurs a limited, incidental and seasonal instance of sexual dichromatism in that the white female genotype, yielding negligible carotenoids when unripe, stores rich quantities of red carotenoids in her eggs as these ripen within the coelenteron. When in the expanded condition, such a specimen presents a beautiful cloudy-pink aspect through the mid-scapus, in contrast with the snow-white male. And when dissected, the gravid white female genotype presents an appearance reminiscent of raspberries served over ice-cream. The carotenoid of these ripe eggs is preponderantly astaxanthin accompanied by lesser fractions of zeaxanthin-like esters and some unfamiliar ketones (Fox et al., 1967a).

The red variant yields epiphasic and secondary fractions of ambiphasic astaxanthin esters from somatic tissues, and from ripe ovaries nearly all epiphasic astaxanthin esters, with traces of esterified zeaxanthin-like material and some unfamiliar ketones. As in ovarian tissues from the simple white and in somatic or ovarian tissues of simple red variants, the tan, brown, or red-brown specimens also yielded principally astaxanthin esters from both somatic and gonadal tissues, but usually in lower concentration (Fox *et al.*, 1967a).

As we concluded in our joint publication more than 30 years ago, "the vivid colours of the anemones may signify freedom from certain environmental restraints, rather than any functional adaptation" (Fox and Pantin, 1941). There remains the fact, however, of widely distributed carotenoid pigmentation of ovarian tissues and ripe eggs, and the decrease of pigment with embryonic development, suggesting the value of continued inquiry concerning the possible roles of carotenoids, not necessarily limited to the A-vitamins, in development and growth among widely differing phyla.

There are many other anemone species that exhibit equally or perhaps more vivid carotenoid pigmentation than do the colored variants of *Metridium*. Some consideration was given to comparative coelenterate pigments by Fox and Pantin in 1944 (*op cit*).

Numerous free-floating or swimming coelenterates, i.e., siphonophores and jellyfishes, display arresting colors and/or color-patterns. By no means does all of such pigmentation involve carotenoids. The venomous, so-called purple jellyfish, *Pelagia noctiluca panopyra* (= *colorata*), is a conspicuous example along the California coast and far seaward. Its blue, brown, purple and magenta integumentary spots and stripes involve no carotenoids, but suspected precursors of the endogenous, dark pigment melanin, of which more will be related below (Fox and Millott, 1954b). There are other scyphozoans like this one, exhibiting in some instances homogeneous blue coloration over the whole exterior, which may or may not involve carotenoid-protein complexes, according to species.

Among the siphonophores, however, there are several species which manifest blue coloration indeed involving carotenoid chromoprotein conjugants (Fox and Pantin, 1944). A conspicuous example in Pacific waters is the so-called by-the-wind-sailor, *Velella lata*, which occurs occasionally in extravagant population-explosions in the oceanic areas off western North America. And when an ill wind blows shoreward for several successive days, these sailed colonies are propelled in and deposited by countless millions upon our beaches, lying in vast, blue windrows along the sand or in piles in more sequestered sites. The blue pigment of the mantle exhibits in clear aqueous media a single, rounded maximum at 585 to 588 nm., whereas a similar preparation of tentacular blue pigment absorbs maximally at *ca* 610 nm. These findings suggest differences in the protein moieties involved, or perhaps in the manner of chemical conjugation between the protein and astaxanthin, the red carotenoid involved in both instances, which is promptly unmasked when the protein portion has been denatured by heat or with alcohol.

This colonial animal form carries vast numbers of zooxanthellae—small, yellow, ovate algal cells within the endodermal tissues of its digestive equipment, as well as in the medusa buds of the gonozoids. The commensal algal cells are readily capable of carrying on photosynthesis in daylight, the blue mantle integument serving as a sufficiently translucent window while efficiently shading the plant cells against excessive sunlight incident upon the ocean surface. In return for safe harborage and transport, and doubtless continuous supplies of metabolic waste chemicals, useful as nutrient materials, the commensal plant organisms produce oxygen in excess of the respiratory needs of the partnership, such that the gas finds its way into the air-chamber canals lying above the plant colonies and directly beneath the horny cover supporting the blue mantle. The animal association thus is enabled to better assure its sustained support at the water's surface, where microscopic organisms, planktonic eggs, and finely particulate organic matter are more concentrated, and thus available as food (Fox and Haxo, 1958).

Leaving for now these few examples of coelenterate tissue-carotenoids, we should next consider briefly the recent discovery of carotenoids conjugated in a heretofore unrecognized manner, i.e., to the calcareous (aragonitic) skeletal material of at least one gorgonian and of several hydrocoral species. The sea-fan *Eugorgia ampla* bears, in the calcareous microspicules within its coenenchyme or fleshy parts, a pale yellow carotenoid "eugorgiaenoic acid", firmly bonded to the inorganic carbonaceous material, whence it is detachable only with acidic reagents such as acetic acid, oxalic acid, or di-sodium EDTA (ethylene-diamino-tetra-acetate) which, at pH *ca* 5, detach the calcium, releasing CO_2 and the carotenoid, in the latter solvent as a sodium soap (Fox *et al.*, 1969a). Similarly several species of hydrocoral store astaxanthin, and in some instances even neutral xanthophylls, as calcareous complex esters of carbonate and the carotenoid, bound through terminal enolic or alcoholic -OH radicals (Fig. 4).

Herein, as in the gorgonian example, none of the pigment is extractable with cold or warm, neutral, organic fat-solvents, such as alcohol, acetone or chloroform, even from the finely ground aragonitic material. Treatment with acetic acid or with Na_2EDTA, always preferably with mild warming and continual stirring, is most effective and safest, since even dilute mineral acids such as HCl can be too drastic for some carotenoids. The Na_2EDTA,

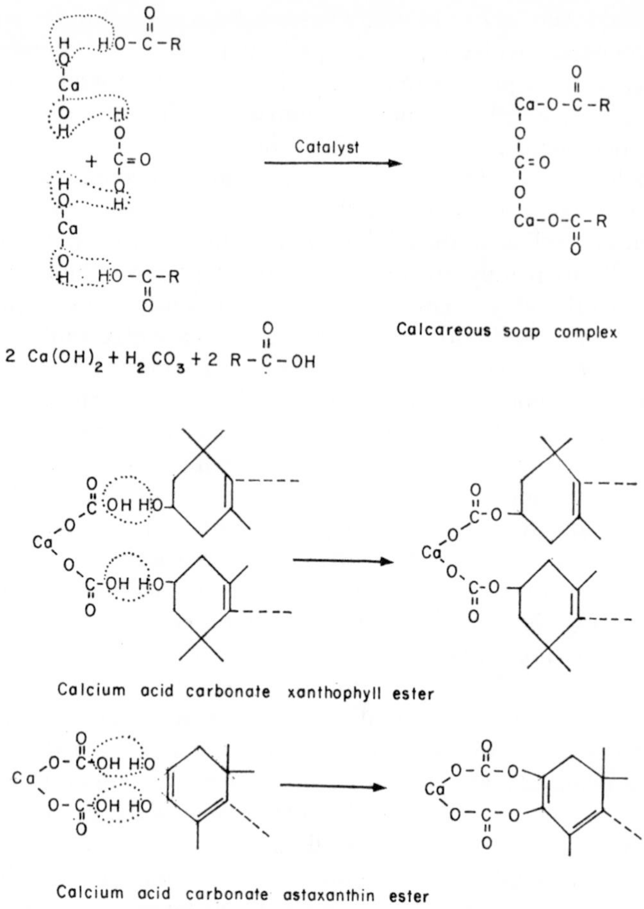

Calcium acid carbonate xanthophyll ester

Calcium acid carbonate astaxanthin ester

FIG. 4. Proposed linkages: between calcareous solids and carboxylic acids forming complex soaps; and with hydroxyl and enolic groups, yielding complex esters.

affording a mildly acidic solution (*ca* pH 5.0) dissolves the calcium to form a soluble complex, and releases the carotenoid as its soap or perhaps as a carbonyl ester, while neutral xanthophylls are released also in a colloidal state.

The purple hydrocoral *Allopora californica* and the vermilion-red species, *Distichopora violacea*, are two examples which yielded astaxanthin only, on such treatment (Fox and Wilkie, 1970). Two additional *Distichopora* species, i.e., *D. coccinea*, also with a red skeleton, and *D. nitida*, whose skeleton is pale orange, likewise yielded astaxanthin as the only recognized carotenoid. Of three *Stylaster* species, *S. roseus* (with a purplish skeleton) also yielded astaxanthin as the only detected skeletal carotenoid, while skeletons of *S. elegans* (pink and orange) and *S. sanguineus* (pale pink) yielded the same enolically acidogenic carotenoid, but accompanied by small amounts of neutral xanthophyllic material as well (Fox, 1972*b*).

By no means are all colored coral skeletons examples of anchored carotenoids. The deep red skeleton of the Organ-Pipe coral *Tubipora musica*, and several bright gorgonaceans, including *Corallium rubrum*, the deep red Precious Coral, yielded no detectable traces of carotenoid on treatment with the usual methods applied to other specimens (Fox, 1972*b*). These and many other examples, notably among the sea-fans, thus pose enigmatic challenges to the naturalist and biochemist. In the words of the young instructor, observing a beautiful pair of eyes in his new class: they invite looking into.

H. Porifera (Sponges)

No account of vivid biochromy should omit some reference to at least a few of the sponges, the most primitive animal phylum above the Protozoa. Like numerous other invertebrate species many sponges are black, grey or other drab, inconspicuous colors; but a considerable number exhibit brilliant yellow, orange, red or purple pigmentation. Like the coelenterates, the bright species of the sponge phylum have long arrested the attention and curiosity of naturalists and others.

Astaxanthin and kindred red carotenoids were recognized in the red sponge *Axinella crista-galli*, by Karrer and Solmssen (1935), while Lederer (1938) detected red carotenoids similar to lycopene, torulene or gamma-carotene in other red sponges, e.g., *Suberites domuncula* and *Ficulina ficus*. Drumm and O'Connor (1940) and Drumm *et al.* (1945) reported in the red sponge *Hymeniacidon sanguineum*, no astaxanthin, but both echinenone and gamma-carotene.

In more recent years Japanese workers have reported their findings of aromatic carotenoid hydrocarbons, renieratene, isorenieratene, and renierapurpurin in a marine sponge *Reniera japonica* (Yamaguchi, 1957, 1958, 1959, 1960).

V. E. Smith, who cited the Japanese work and completed his doctoral research in this laboratory of the Scripps Institution of Oceanography, dealt with the "Comparative Cytology and Biochemistry of Two Marine Sponges" (1968). Working with the two red, silicious sponges *Cyamon neon* and *Trikentrion helium*, both common on the sea floor off La Jolla, and San Diego, California, Smith described each as having a Scarlet to Brazil Red color (Ridgeway's classification), and reported that both yielded essentially the same mixture of carotenoids, as identified by thin-layer chromatography and spectral absorption profiles of all fractions.

No fewer than twelve fractions were recoverable by thin-layer chromatographic resolution, but some of these doubtless represented contiguous isomers. Smith encountered unusual proportions of hydrocarbons in both species, as compared with proportions in the great majority of marine animals. These included β-carotene, 3, 4-dehydro-β-carotene, and red fractions resembling the aromatic compound β-isorenieratene. One substantial fraction was recognized as being possibly identical with a mono-hydroxy-β-isorenieratene, while about 40% of the total carotenoid yield was a previously unfamiliar dihydroxy-carotene, which appeared to be partly in esterified condition and which exhibited some very unusual properties. Smith believed it to be a dihydroxy-bis-dehydro-β-carotene. There was no evidence that the sponges gained their carotenoids unchanged from their finely particulate diet. The whole colorful sponge phylum wants further biochromic research.

I. Sediments

In having accorded the carotenoids a major share of relative importance and corresponding space in this treatment, we hardly shall have gone astray. For not only are these physiologically significant biochromes among the most conspicuous, and certainly the most widely distributed through living organisms, but they are further distinguished by their presence in non-living matter of biological origin as well. Examples are to be found not only in solid skeletal calcareous materials, in keratinaceous proteins of feathers, insect elytra, etc., and in hard chitinous armor, but also in colloidal or other finely divided organic matter both in suspension in ocean waters as so-called leptopel, and in bottom sediments, partly in the thin carpets of sapropel, and in part buried in mud or silt.

Accordingly, filter-feeders obtain their carotenoid supplies largely by screening the suspended leptopel from the water, and benthic forms by sweeping, scraping or shovelling material from the sea floor into their digestive tracts. Thus the carotenoids are ubiquitous in all biologically tolerable environments, not omitting even some deep freshwater caves where bacterially and fungally produced organic matter is essential for the support of the unusual fauna living there, and into which carotenoid-containing detritus is introduced by underground streams, thus supplying cave fauna such as the salamander *Proteus anguineus* with carotenes and xanthophylls (Beatty, 1941, 1949).

In a more vital connection, it should be remembered that carotenoids have long been identified in deeply buried marine muds, at depths more than 2000 m, from the water surface. Some of these exhibited typical conventional properties like those of algal carotenoids (chlorophyll-degradation products were also present), while others were more reminiscent of less common carotenoids recoverable from bacteria and fungi (Fox, 1937, 1944; Fox and Anderson, 1941; Fox et al., 1944).

Whereas Gulf of California sediment cores studied by Fox et al. (1944) dated back as far as 7000 years (estimated), older sediments (e.g., ca 11 000 years) recovered from deep bottoms of Canadian lakes and pools also yielded identifiable carotenoids (Vallentyne, 1956).

Such findings offer assurance of a continuing and ample supply of carotenoids for deep-dwelling, benthic, sediment-feeding marine and other aquatic fauna.

Of other natural biochromes meriting attention in this selected treatment, we shall recall the rather limited quinones, whose fundamental aromatic ring skeletons are most likely synthesized de novo by plants, as are the carotenoids. In contrast, the variously colored cyclic and acyclic tetrapyrroles are elaborated by both plants and animals, and the ruddy, brown or dark melanins are more characteristic of animal metabolism. These will now be given some consideration in sequence.

IV. Quinones

Within this class are the ubiquinones, a group of pale yellow p-benzoquinones of wide occurrence, as their name suggests, but whose very low concentrations leave them invisible in situ. Earlier called coenzyme Q_{6-10}, referring to various lengths of the side chains, these benzoquinones are recognized as biocatalysts of metabolic importance in regulating respiratory cellular oxidations. This group will not be treated in the present discussion, since its members play no direct role in coloration proper and since their biochemical significance has been discussed elsewhere (Thomson, 1957).

The other two principal kinds of natural quinones (apart from those constituting biochemical steps in melanogenesis) are the naphthoquinones, wherein one benzene ring is condensed with a p-benzoquinone ring, and the anthraquinones, characterized by a benzene ring condensed to each side of the p-benzoquinone structure.

A. Naphthoquinones

Outstanding representatives of the naphthoquinone compounds displayed conspicuously by marine animals are the polyhydroxy derivatives exemplified by echinochrome A and by several recognized spinochromes (Fig. 5) which occur in tissues, and notably also in the calcareous spines and tests of many echinoid species (sand-dollars and sea-urchins). This

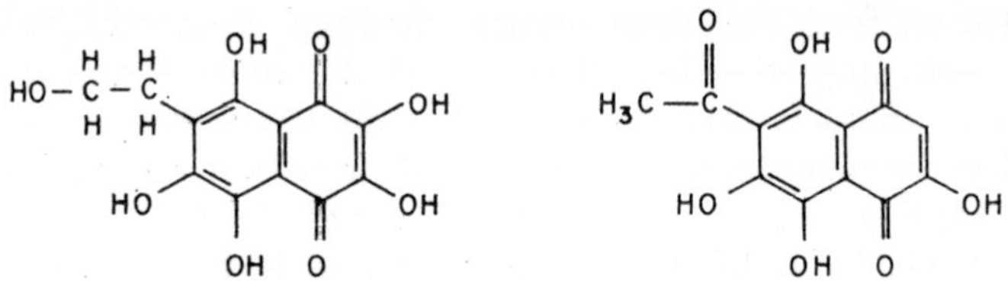

Echinochrome A
$C_{12}H_{10}O_8$

Spinochrome A
$C_{12}H_8O_7$

FIG. 5. Two naturally occurring naphthoquinones from echinoids.

pigment occurs also in the body-wall of a small, deep purple holothurian or sea-cucumber *Polycheira rufescens*, wherein it constitutes the prosthetic group of a chromoprotein (Mukai, 1958, 1960) and also in protein aggregates of connective tissue and ovaries of a crinoid or sea-lily *Antedon bifida* (Dimelow, 1958).

In echinoids these biochromes appear in the gastrular embryonic stage (Chaffee and Mazia, 1963) and later are found in ectodermal and endodermal tissues, perivisceral fluid and contained amoebocytic corpuscles (elaeocytes), wherein they commonly exhibit red to purple colors according to lower or more elevated pH values respectively. The commonest color of skeletally deposited polyhydroxynaphthoquinones is purple, although red, greenish, green-brown, brown-violet and violet colors are manifested by some echinoid species.

Beyond an incidental resemblance in molecular structure to the anti-hemorrhagic or coagulating factors (K-vitamins) biochromes of the echinochrome class have not been accorded a seriously suspected physiological role, e.g., in respiratory augmentation in eggs, sperm or other cells (Fox, 1953). Nor, to the best of my knowledge, have any of these polyhydroxynaphthoquinones been proven to exercise any other biocatalytic function.

The close chemical alliance of these biochromes with calcareous skeletal materials may likely reflect their status as calcareous esters, i.e. very like the bonding believed to prevail in the combining of enolic and even neutral hydroxy-carotenoids with aragonite in the skeletons of some hydrocorals referred to above.

In that regard, it is to be recalled that naphthoquinone biochromes turn up in skeletal structures of at least two marine vertebrate species which feed extensively on sea-urchins. These are the sea otter *Enhydra lutris*, which owes the pink and purple colors in skull, teeth and bones to the chemical binding of one or more polyhydroxynaphthoquinones, presumably complexed with the calcium phosphate (Fox, 1953). This applies also to the teeth, entire calcareous oral equipment and dorsal spine of the horned shark, *Heterodontus francisci*, which frequently manifests bright pink colors from eating the same diet of urchins. Moreover, red sea-urchins (*Strongylocentrotus franciscanus*) finely comminuted, mixed with peanut butter, and fed for several months to young, growing mice, effected a purple coloration to the consumers' skeletons (Yadon, 1968).

B. Anthraquinones

Like the naphthoquinones, the anthraquinones show a frequency of occurrence in (land) plants, but in very few animal species. These conspicuously brilliant compounds have been applied extensively as dyes and as chemical indicators of acid or alkaline conditions. Like the naphthoquinones they are readily reducible chemically.

In terrestrial environments the red dyes cochineal (a potassium salt of carminic acid from fat-body cells of female cochineal insects), kermesic acid from the female kermes insect, and laccaic acid recovered from the solid, resinous exudate covering the female bodies of certain lac insects, have long been known commercially. These are instances of anthraquinones derived from plant-hosts, e.g., certain cacti, oaks and others, by their scale-insect parasites (cited by Fox, 1953).

There are far fewer known examples of anthraquinones occurring in the marine environment. The instances currently known all refer to crinoid members of the echinoderm plylum. A prominent example

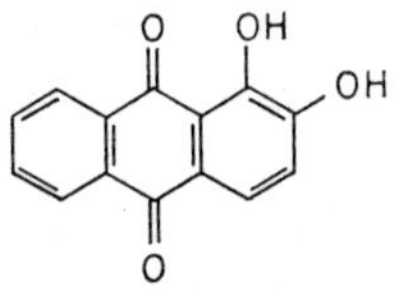

Alizarin (1, 2-dihydroxy-9,10-anthraquinone)

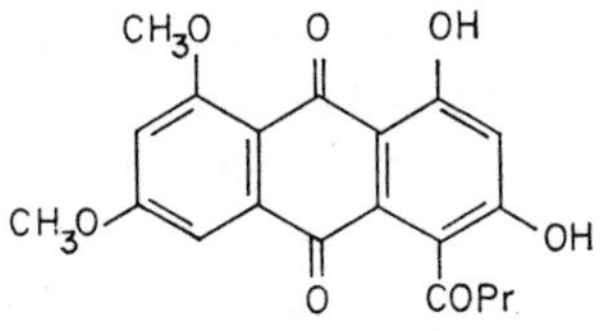

Rhodocomatulin dimethyl ether.

FIG. 6. Two anthraquinones encountered in nature: alizarin from madder root and rhodocomatalin dimethyl ether from crinoids.

involves rhodocomatulin (Fig. 6), a 4-butyryl-1,3,6,8-tetrahydroxy-anthra-quinone, present actually as its 6-methyl and 6, 8-dimethyl ethers, and accompanied by a monomethyl ether of rubrocomatulin (= 8-butyryl-1, 4, 5, 7-tetrahydroxy-2-methoxyanthraquinone). The bright or dark red or purple, 10-armed, free-swimming crinoids *Comatula pectinata* and *C. cratura* owe their pigmentation to the presence of such compounds which, like the carotenoids, are yielded readily to organic solvents such as acetone (Sutherland and Wells, 1959, 1967; Powell *et al.* 1967).

Other free-swimming crinoids from Australian coastal waters, investigated by Powell and Sutherland (1967), include the purple "passion flower", *Ptilometra australis* and another ten-armed comatulid *Tropiometra afra*, which in some instances is yellow, yellowish-brown or brown, or more frequently appears to be black *in situ*, but on close examination is found to be very deep purple. *Ptilometra* yielded three anthraquinone derivatives exhibiting red colors in alkalized aqueous solution. These have been called rhodoptilometrin, isorhodoptilometrin, and ptilometric acid, and were identified respectively with three 1, 6, 8-trihydroxy-3-propyl derivatives as follows: the 1-hydroxypropyl, 2-hydroxypropyl and 3-hydroxypropyl anthraquinones, the latter compound bearing also a 2-carboxylic acid group. The same latter anthraquinone-carboxylic acid compound was recovered also from both yellow and "black" specimens of *Tropiometra afra*.

There seems to be no evidence to date bearing on the possible original source, whether external or within the animal, of the anthraquinone biochromes. Nor have any specific functions of the compounds been recognized for the animal's metabolic economy.

While no examples of anthraquinones in the stalked class of crinoids come to mind, these do bear colored compounds which call for study.

The Australian workers seemed to be inclined toward a theory that the crinoids are able to synthesize the anthraquinones that they store, as do certain fungi, lichens and higher plants of terrestrial habitats. Animal forms ingesting, in their plant diet or detrital food, adventitious components possessing the fundamental skeleton of anthraquinones or other polycyclic character, might well be conceived to effect chemical innovations thereto, e.g., as echinoids may do when forming colored naphthoquinones from colorless precursors during embryonic development, or as when ketonic or other oxygen-functions are introduced at sites on the terminal hexenyl rings of carotenes by some animals. But the whole question concerning the ultimate origins of naturally occurring animal anthraquinones calls for careful, objective study before one might assign such a role, or a part thereof, to animal organisms.

We have thus far been considering the distribution of a few classes of conspicuous animal pigments composed only of C, H, and O atoms (save that the carotenes themselves lack the O). Of these classes, the carotenoids are known to be synthesized *de novo* only in green and in many achlorophyllous plant organisms. The conspicuous quinone pigments, known to occur in both plants and animals may well be supposed, in view of their condensed phenyl-ring structures, to owe the primary synthesis of the basic skeleton at any rate, to elaboration by plants. Certain restricted groups of echinoderms may be able to elaborate echinochromic pigments from colorless precursors, and can assimilate, store and display the colored molecules in tissues and in skeletal parts.

We must now turn to some different molecular species of conspicuous biochromes, in that they are synthesized in both animals and plants; furthermore, they involve heterocyclic N as well as the other named atoms. These include the indolic class, notably the dark melanins, and the variously colored tetrapyrroles, to be considered in that order.

V. Indolic Pigments

A. Indigoids

These compounds of limited marine distribution arise in general from the metabolic degradation of the common amino acid tryptophan, and are represented by colorless heterocyclic end-products such as indole, skatole (= 3-methylindole), indoxyl (= 3-hydroxy-indole), indoxylsulfuric acid, and indican, all characteristic of excreted matter.

Uncombined indole and skatole are encountered in feces, while indoxyl, skatoxyl and their sulfate salts represent detoxicated excretory components of urine from numerous animals, notably herbivores. These compounds are present in more pronounced concentrations in cases of hepatic cancer and other pathological conditions.

Indican, the potassium salt of indoxylsulfuric acid, is converted to indigo-blue (indogotin) through oxidation, e.g., with chlorate or ferric salts added to an acidulated specimen of pathologic urine. Shaking with a little chloroform then extracts the blue, oxidized end-product, indigo (Fig. 7).

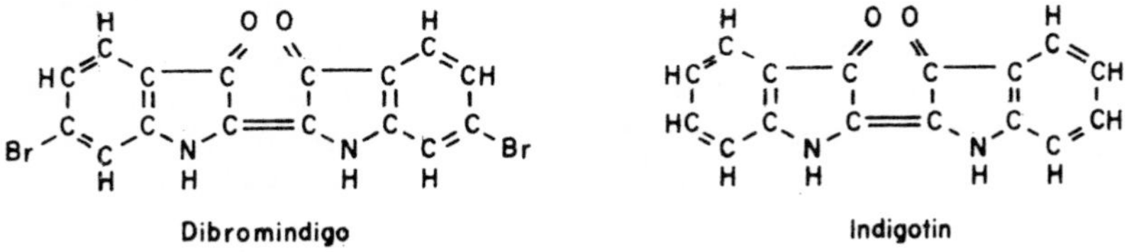

Dibromindigo

Indigotin

FIG. 7. Indigo pigment types.

Indigo pigments may appear as such in certain pathological urines, notably after standing, when decomposition may occur through the agency of bacterial contaminants (Fox, 1953).

The occurrence of indigoid biochromes in the natural marine environment is relatively limited, quite in contrast to the wide incidence of dark melanins discussed below. The indigoids, inconspicuous in the intact organisms, characterize the secretions of certain gastropod molluscs e.g., in the *Mitra*, *Murex*, and *Purpura* genera. Indeed these pigments enjoyed wide popularity in ancient times, were known as Tyrian Purple, Purple of the Ancients, purpurin or punacin, and included as a prominent example the purple dye, 6,6'-dibromindigo (Fig. 7), recovered from several species. This presumably was the product recovered and marketed by one of Paul's early converts, "a certain woman named Lydia, a seller of purple, of the city of Thyatira . . . " (Acts 16, 14).

The prochromogens, present in the pale yellowish hypobranchial or so-called adrectal gland of *Murex* and other related marine snails, are released when the whole tissues are mashed. On exposure to air and sunlight the chromogen is oxidized through the agency of a liberated enzyme, yielding the final colored product dibromindigo, used prior to the invention of modern dyes for applying fast colors to woolen fabrics. "And the purple tint of the shellfish is united . . . with the body of the wool, yet it cannot be separated . . . not . . . if the whole sea should strive to wash it out with all its waves" (Lucretius VI, lines 1074–1078).

Chemical and spectral properties of some indigoid biochromes have been discussed earlier (Fox, 1953, 1966). The colored compounds themselves have not been shown to fulfill any physiological role but would appear to represent merely excretory products of tryptophan breakdown.

B. Melanins

These entities are not stoichiometric compounds, but colored polymers arising from the progressive oxidative degeneration of the amino acid tyrosine, with formation of increasingly complex decomposition products, to yield variously colored dirty yellow, murky orange, reddish, tan, ruddy, dark-brown or black material encountered in widely different biological sites. Omitting more than passing reference to the occurrence of melanins even among higher plants, such as at cut surfaces of apples, potato-tubers, and certain stem-saps of *Rhus* and other genera, as well as in some dark fungal and bacterial species, we shall pass over also the many land animals that conspicuously exemplify melanogenesis, including myriads of insects which exhibit blackness or other dark colors of body-covering, whether as homogeneous surfaces or occurring in stripes, round or other shaped patches on body or wings (elytra). All are familiar, similarly, with dark colors in the scaly covering of reptiles, the feathers and some exposed skin-areas of countless birds, the hair, fur or wool of numerous mammals, and the various dark shades of skin exhibited by many animals including the human race.

Melanins occur not as crystalloid solutes but as minute, often colloidal particles, whether ellipsoids (0.25–0.45 by 0.25–0.40μ), rodlets (0.1 by 0.4 to 0.18 by 0.60μ), spherules (0.20 to 0.30μ diameter) or other rounded shapes within animals' tissues (Mason *et al.*, 1947).

Melanization, from the oxidation of tyrosine and products thereof, is

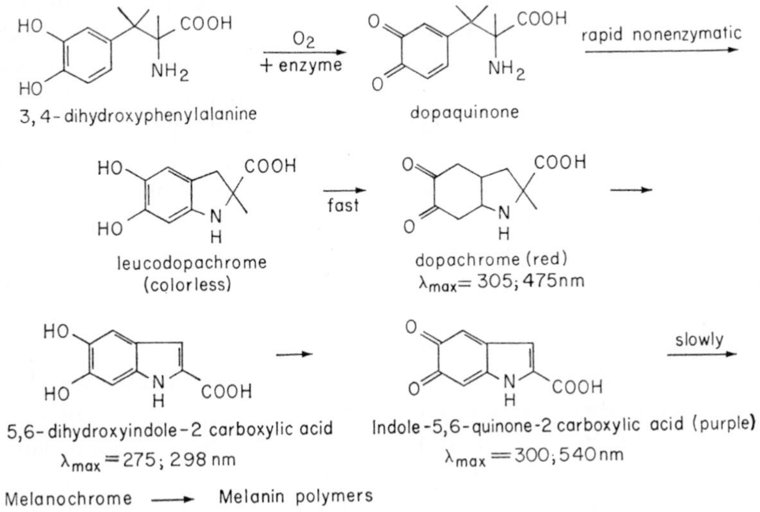

FIG. 8. Abbreviated path of melanin biosynthesis.

catalyzed by the presence of a copper-containing enzyme tyrosinase, which manifestly occurs very widely in the animal kingdom (Fig. 8). Dry, ash-free melanin reportedly has the average chemical composition (by percentage): C, 52.6; H, 7.28; N, 13.42; S, 1.33 and O (by difference), 25.37, these values agreeing closely with the ratios given earlier by Gortner as an empirical formula: $C_{105}H_{173}N_{23}SO_{38}$. Melanins are reversibly dissolved (colloidally) in aqueous alkali, precipitated by dilute acids and bleached by peroxides and other oxidizing agents. Their properties and analytical assay have been outlined elsewhere (Fox, 1953).

A great many fish species exhibit melanism in various degrees and patterns. Some, inhabiting great ocean depths, e.g., the black sea-devil, a deep-sea angler, *Melanocetus johnsoni*, are completely black, while others may have external stripes or patches of black or other dark colors across or along the length of the body. Such patterning may in many instances afford the animals an appreciable degree of camouflage or disruptive protective coloration, thus rendering them less perceptible to predators or prey. Certain species of deep-sea fish may have their skin red with carotenoids, thus rendering them as invisible as melanistic species in their world of soft blue light.

There are indeed certain fish species capable of generating and storing pigment, notably melanin, in the changeable, so-called melanophores of their skin. Within these special cells the microgranules, which Sumner called melanosomes, are capable of being dispersed through multibranched, dendritic channels, giving the whole animal a dark appearance, or alternatively being withdrawn and aggregated into a small, inconspicuous spot in the cell's center, thus evoking a generally pale aspect, i.e., changeable in response to the respectively shaded or illuminated aspect of the animals' background. These adaptive, concealing responses, and even modifications in numbers of melanophores and thus in total integumentary melanin, can be evoked and quantitated by the experimenter (Sumner, 1940).

Black or brown coloration is encountered in some feathers of numerous marine birds (relatively large numbers of which bear also white plumes), including penguins, cormorants, sooty albatross, brown pelicans, juvenile members of some gulls and others.

Dark pigmentation is of common occurrence also in many invertebrates, e.g., even in certain sponges, coelenterates such as dark variants of the anemone *Metridium* (Fox and Pantin. *op. cit.*), some holothurian and ophiuroid echinoderms, many species of worms, a considerable number of gastropod and certain bivalve molluscs, the large, dark, epidermal melanophores of the spined urchin *Diadema antillarum* (Millott and Jacobson, 1951), and the ink of cephalopods. Melanoid substances were also found to be responsible for the reddish, purple and brown stripes of the large, cosmopolitan, venomous jellyfish *Pelagia colorata* (= *noctiluca*) (Fox and Millott, 1954b). *Octopus bimaculatus*, the two-spotted octopus of the North American Pacific coast, stores dark brown melanin or melanoprotein in its glandular ink-sac, attached laterally to the hepatopancreas. From this sac the ink is forcibly ejected via the special duct opening next to the anus. There is ground for believing that this secretion of cephalopods may serve more as an olfactory distraction than as a visually occluding cloud. Thus it may attract the attention of a predator such as the moray eel *Gymnothorax mordax* to search vigorously the area of ejection while the octopus flees the scene, rather than serving merely to screen the animal from sight. The inky colour could be of no usefulness at night, nor at dark marine sites;

indeed some deep-living cephalopods secrete a pale, luminescent fluid instead of a true, darkly colored ink.

Moreover, the mere presence of the ink-secretion, diluted with sea water and introduced into an aquarium harboring the octopus-predator *Gymnothorax*, serves to excite the eel into a febrile search for its prey; and if a previously declined morsel, such as a piece of mussel-flesh, or even a small sea-star's arm, be re-offered, after dipping into octopus ink, it will then be seized avidly. *Sepia* or *Octopus* ink has a seminal or musky odor, which may be the exciting chemical factor, while the included melanin may be merely one of the by-products in a general turn-over of protein, of which tyrosine is a critical component.

There are scattered instances of melanistic or other secretory discharges even among fish species. An outstanding example is a crestfish species of *Lophotus*, which stores copious melanin in a special glandular sac situated contiguously between the two branches of the air-bladder. Viscid, sticky, musty-odored ink, very like that of octopus, is forcibly expelled through a post-anal aperture by muscular contractions against the two sides of the air-bladder (cited by Fox, 1957).

Instances are recognized wherein integumentary melanization occurs concurrently with phases of growth, with sexual maturation, or following spawning. Reference has been made to the dark plume-colors of some juvenile gulls, a phase which is later replaced by generalized whiteness of plumage on maturation. A much more conspicuous instance applies to young flamingo chicks, e.g., the Caribbean or American species *Phoenicopterus ruber*, which has raised young in the San Diego Zoo. Freshly hatched chicks, bearing white or grey, fluffy down over the body, exhibit bright red colors of hemoglobin through the thin skin of the pudgy tarsals and toes. But by about the ninth day, the chick no longer is so continuously shaded from sunlight, whether by sitting on its legs or being sheltered by a hovering parent, and is preparing to make excursions away from the nest. At this juncture the tarsals have become longer, thinner, and increasingly deeply pigmented through dark red, brown-red, to genuine black, with increasing melanogenesis. No carotenoid is yet stored in the tarsal skin, and none occurs there until after a considerable time, when there ensues a passing, half-grown phase of black-and-red, marbled coloration over the tarsal skin. The black iris of the chick likewise does not turn to the pale yellow-orange color until closer to maturity.

Among fishes, pronounced melanogenesis has long been recognized as associated with maturation, or with certain diseased conditions, or with moribundity. The goldfish *Carassius auratus*, however, is dark when juvenile, acquiring its red, golden or yellow coloration with maturation. There occur also jet-black, black-and-yellow or black-and-gold genotypes in this species. There have been reviews of genetically transferred melanotic tumors, arising from clusters of aberrant macro-melanophores in some species of fishes (Gordon, 1948).

Blackburn (1950), studying the Tasmanian whitebait *Lovettia seali*, an anadromous fish species which ascends rivers for spring-spawning, discussed the rapid, nearly explosive integumentary blackening, more extensive in males, in the post-spawning, moribund state. There are conspicuous increases in numbers of melanophores as sexual maturation progresses, and post-spawned individuals, males especially, exhibit large, dark areas, or, while dying, overall blackening of the skin.

Such observations emphasize the likelihood of direct biochemical association between melanogenetic degradation of tyrosine and accelerated protein overturn, whether leading to rapid proliferation of gonadal or of neoplastic cells or toward general cellular breakdown (Fox, 1957).

Integumentary melanism thus may be of genetic origin, or associated with advanced reproductive phases, with pathological or moribund conditions, or may be induced, e.g., by exposure to bright sunlight (tanning in humans) or to dark backgrounds (dark adaptation in fishes). There also remain the widely occurring incidences of Tyndall-scattering of blue light, e.g., by colloidal materials overlying basal layers of melanin. This is commonly seen over the dorsal surface or other blue areas of oceanic fishes, while the ventral skin often is matte white, or silvery in some parts, due to leucocytes heavily packed with microcrystals of the purine compound guanine. If guanine platelets overlie melanized strata, there result some beautiful blue colors of interference, as seen conspicuously in the skin of young Garibaldi, *Hypsypops rubicunda*.

There remains yet another instance of reversible melanistic disposition, quite distinct from the previous examples wherein fishes aggregate or disperse their intra-chromatophoral melanosomes in response to pale or dark backgrounds respectively. This example is the unique, reversibly occlusible, choroidal, guanine-laden tapetum lucidum in the eyes of elasmobranch fishes, i.e., many sharks and some rays. Here, instead of serving a concealing response the melanin fulfills a kind of dynamic function of ocular adjustment, presumably of direct aid to the animal's vision. The tapetum, underlying the retina, remains unoccluded in dim light and so reflects incoming light traversing the retina, thus supposedly increasing visual actuity and enhancing sensitivity while evoking eye-shine as any incident light is reflected. However, in bright light, individual strands or extended processes advance from a basal, choroid layer of dark melanin, moving forward to form a dark, compound curtain covering the tapetal plates and thus shielding the retina against dazzle (Nicol, 1961, 1964; Denton and Nicol, 1964; Gilbert, 1963; Fox and Kuchnow, 1965).

VI. Tetrapyrroles

A. Porphyrins

This nitrogenous class of heterocylcic compounds is exemplified most extravagantly by the green, photosynthetic chlorophylls of algae, and indeed of all green plants, and by red hemoglobins and myoglobins of many animal species.

Extensive greenish areas of surface or near-shore ocean waters afford striking evidence of locally current, maximal photosynthetic activity. This, as noted earlier, is often emphasized by the presence of white linear or streaky surface-aggregates of froth, constituted of countless small bubbles of liberated oxygen. Any such vista as this augurs an early ensuing population explosion of zooplankton, consumers of the phytoplanktonic algae.

Red hemoglobins color the blood-cells of vertebrates, and appear also in the circulating fluids of many invertebrate animals, notably in annelid worms and a few echinoderm and molluscan species, serving always as oxygen carriers.

Muscle hemoglobins, called myoglobins, are manifest also in the more active, voluntary muscles of both vertebrate and invertebrate animals, notably, within the latter, in some polychaete worms, in the foot and adductor muscle, as well as in gills and brain, of the Pacific pismo clam *Tivela stultorum*; and in the posterior adductor muscle and heart of the so-called "pile-worm", a wood-boring bivalve, *Bankia setacea*. Hemoglobins are of scattered incidence through a number of other invertebrate phyla, including Platyhelminthes (flatworms and nemerteans), Nemathelminthes (threadworms) and holothurian echinoderms. These pigments are found even in a few crustaceans, such as the small freshwater, pelagic "water-flea" *Daphnia magna* and the brine shrimp *Artemia salina*, which, like the freshwater pond-snail *Planorbis corneus*, exhibit varying shades of redness reflecting corresponding changes in hemoglobin content according to environmental conditions (reviewed by Fox, 1953).

Many invertebrate species, whether lacking or supplied with but degenerate vascular systems, may carry their respiratory proteins in corpuscles suspended in the coelomic fluid. Examples include polychaete genera, e.g., *Polycirrus*, *Capitella* and *Glycera* and the echiurid *Urechis*, all characterized by the presence of hemoglobins, and the gephyrean or sipunculid worms *Phascolosoma*, *Sipunculus* and *Dendrostoma*, which employ hemerythrin instead. Some polychaetes such as *Terebella lapidaria*, *Travisia forbesii* and *Thoracophelia* (= *Euzonus*) *mucronata* contain their hemoglobin acellularly, as a suspended colloid in the vascular blood, while the former two species carry erythrocytes in the coelomic fluid. The "blood-worm" *Thoracophelia* (=*Euzonus*) *mucronata*, inhabiting moist sand-beds of the Pacific coast of North America, owes its bright red and purplish colors to the presence of (non-cellular) hemoglobin in rich supplies.

The late H. M. Fox (1947) examined the blood pigments of a score of species, in 15 genera of serpulid and sabellid worm families, and reported chlorocruorin, a green iron-hemin protein in some species of *Spirorbis*, hemoglobin in others, while some were found to carry both biochromes, e.g., *Serpula* which carries more hemoglobin when young but a preponderance of chlorocruorin when older. Both the red and the green iron-hemoproteins possess the power to combine reversibly with oxygen. The affinity of most hemoglobins for oxygen is, however, higher than that of chlorocruorins, which involve a different porphyrin group from that of hemoglobins. Hemerythrin, reddish when oxygenated and colorless when not, involves combined iron, as do hemoglobin and chlorocruorin, but unlike them, contains no porphyrin moiety.

A marine gephyrean worm, *Bonellia viridis*, bears an integumentary pigment (bonellin) which is green in dilute aqueous systems, exhibiting dichroism when more concentrated, and manifests a red fluorescence. Its chemical and spectral properties are listed elsewhere (Fox, 1953), and likened it rather closely to another porphyrin, mesopyrrochlorin. Another sipunculid, *Thalassema lankasteri*, exhibits a bluish-green pigment in its skin and mucus, which manifests chemical and spectral characteristics closely resembling those of bonellin. No physiological function has yet been recognized for these unusual porphyrins. Protoporphyrin IX (Fig. 9) has been detected as solid, intracellular granules within the red-orange freckles over the siphon of *Bankia setacea*. The same pigment has been recognized also in red sea-pens, e.g., *Pennatula*, resulting in high photosensitivity in the soft parts of this coelenterate.

Certain porphyrins, combined in all probability as complex calcareous salts, contribute red or pink colors, showing fluorescence in ultraviolet light, to many skeletal structures, including not only those of some land mammals, but the eggshells of numerous birds, and very notably the shells of molluscs. In the latter, uroporphyrin I (carrying 8 carboxyl groups, Fig. 9) and to a lesser degree coproporphyrin (equipped with but 4 such radicals, Fig. 9) give rise to the arresting pink, red, or darker ruddy colors, and the UV induced red fluorescence of shells belonging to many gastropod and bivalve species, particularly the inner shell surfaces. A New Zealand brachiopod, *Terebratella rubicunda*, also stores uroporphyrin in its pink shells.

Despite wide, genetically guided variation in depth or shade of coloration and color-patterns of molluscan shells, uroporphyrin I is the prominent cyclic tetrapyrrolic or porphyrin pigment encountered in shells of

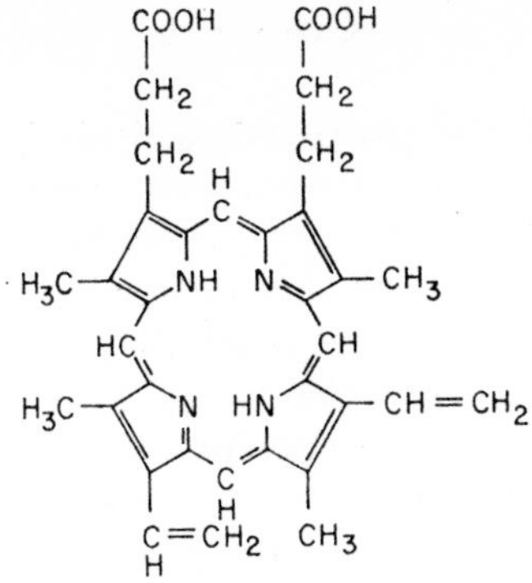

Protoporphyrin IX (= 1,3,5,8-Tetramethyl-2,4-divinylporphin-6,7-dipropionic acid)

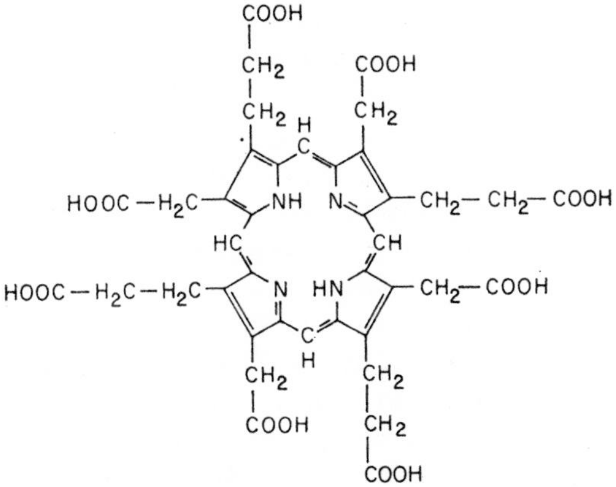

Uroporphyrin I (= Porphin-1,3,5,7- tetra-acetic-2,4,6,8- tetrapropionic acid)

(Uroporphyrin III is Porphin-1,3,5,8-tetra-acetic-2,4,6,7- tetrapropionic acid)

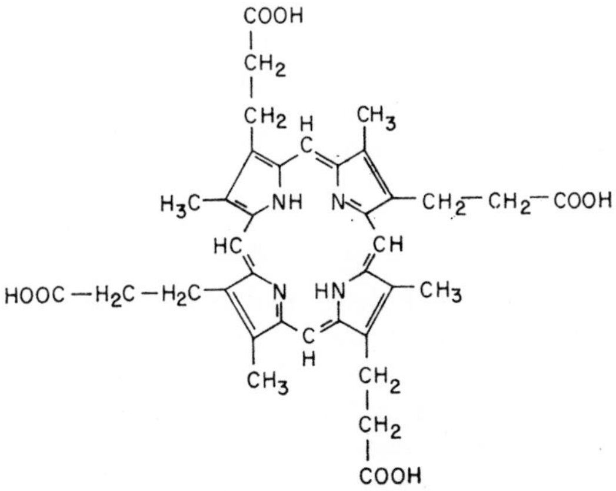

Coproporphyrin I (=1,3,5,7- Tetramethylporphin-2,4,6,8 -tetrapropionic acid)

(Coproporphyrin III is 1,3,5,8 - Tetramethylporphin-2,4,6,7-tetrapropionic acid)

FIG. 9. Some common porphyrin types.

204   DENIS L. FOX

50 or more living genera of gastropods and bivalves, and of some dozen fossil species (Comfort, 1951). Porphyrins seem not to occur in shells of land snails, nor in freshwater gastropods save in certain Neritidae. Uroporphyrin I seems to occur only in marine species shells wherein it may amount to as much as 1% by weight, e.g., in the shell of the Persian lingah oyster *Pinctada vulgaris* (Comfort, 1951).

Some bivalves, including *Venus fasciata*, *Placuna*, and several species of *Pteria* conjugate uroporphyrin I with the calcareous material of their shells (Fox, 1972a, and citations therein). Some pearls owe certain red or green hues to the presence of porphyrins, whether in combination with lead, zinc, or non-metallic salts. Both the green type, with relatively more metallic, and the red, involving greater proportions of non-metallically conjugated porphyrin, manifest red fluorescence under UV light (cited by Fox, 1972a).

Reports of porphyrins in the echinoderm phylum are very few, but interesting. Many years ago Crescitelli (1945) recorded the occurrence of a red porphyrin from the erythrocytes of the holothurian *Cucumaria miniata*, and another unusual oxygen-combining biochrome from the red blood-cells of another sea-cucumber, *Molpadia intermedia*. Goodwin (1969) cites the more recent finding of respiratory hemoglobins in some holothurians, including *Cucumaria miniata Caudina* and *Thyonella* spp, by Prosser and Brown in 1961, and Manwell and Baker in 1963. Kennedy and Vevers, in 1953–54, reported the presence of chlorocruorporphyrin and protoporphyrin in the sea-stars *Astropecten irregularis* and *Luidia ciliaris*, and protoporphyrin also in another asteroid *Asterias rubens* (cited by Goodwin, 1969).

There is ample ground for the view that certain porphyrins may play a role in activating pituitary hormones in some animals, and may initiate functions of some biocatalysts concerned with estrous periods (e.g., Klüver, 1944a, b). Moreover, free integumentary porphyrins, notably in some molluscs and coelenterates, are believed to be photosensitive receptors of biologically effective light fractions.

B. Bilichromes

These so-called linear, better designated as open-chained or non-cyclic, tetrapyrroles, once popularly known as bile pigments, reflecting the site of their earliest recognized source, appear in many soft and hard parts of animals. Passing over the examples of jaundice or yellowing of skin and eye-sclera through the excessive accumulation of bilirubin, and the manifestation of green or blue-green biliverdin compounds in young green locusts, in the shells of some land-birds' eggs as well as pterobilin (= mesobilirubin) in the wings of a number of pierid butterflies, we shall adhere to orientation in this discussion toward organisms of the marine environment.

Oöcyan, the blue pigment now believed on good evidence to be biliverdin is familiar not only in the shells of numerous land-birds' eggs, but occurs in varying degrees also in the eggshells of certain gulls.

Biliverdin is incident in skin, body-fluids and some glands of many animals, and has been recognized in the green or blue-green bones of belonid, cottid, and allied fishes, e.g., in skeletons of *Belone, Cottus, Strongylura, Zoarces* and others, as well as in the bones of a genetically recessive phase of the ocean skipjack *Katsuwonus pelamis* (Fox and Millot, 1954a).

Various bilichromes (Fig. 10) may be derived by invertebrate animals in several alternative, but not necessarily mutually exclusive ways, e.g., whether metabolically formed irrespective of their presence in food, or rendered accessible in the food as native bilichrome, phycobilin, hemoglobin or chlorophylls (With, 1968).

A red bilichrome, rufin, is encountered in the integument of the land-slug *Arion rufus*, and a similar orange compound is recovered from the shell of the red abalone *Haliotis rufescens*. Indeed, Chapman and Rüdiger have conducted searching analyses of shells from a dozen *Haliotis* species, and have found in every instance this red bilirubinoid haliotisrubin (= rufescine), which carries keto groups in place of the terminal —OH radicals present in bilirubin. The investigators report that this conspicuous bilichrome is derived by the large marine snails' consumption of red algae, which synthesize the allied red compound phycoerythrin (personal communication, also Fox, 1972a).

Indeed, Dr D. L. Leighton donated to me three small whole shells of the abalone *H. rufescens*, wherein is recorded evidence of the animals' diet over a year's time. One specimen, fed on red algae continuously for eleven months developed an overall pink to red-surfaced outer shell, while two others which had been fed instead upon brown and upon red algae through alternating months, display six white and six pink-to-red bands across the length of the shell's upper surface. The red algal erythrobilin or its metabolic derivative haliotisrubin more likely, had been deposited as a complex calcareous salt into the shell material.

Rüdiger and Chapman encountered another bilichrome which they called haliotisviolin, likewise endowed with keto radicals in the same respective terminal-ring positions as in haliotisrubin. This violet compound which turned up in nine of the twelve species investigated (as a peptide in one of these, the black abalone *H. cracherodii*), apparently is yet unknown in algae (cited by Fox, 1972a).

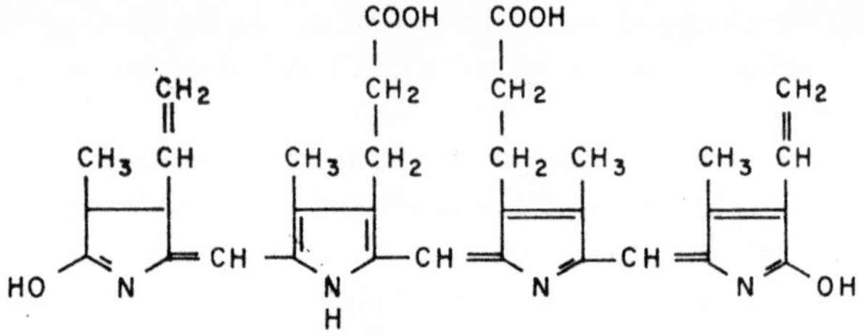

Biliverdin

$C_{33}H_{34}O_6N_4$

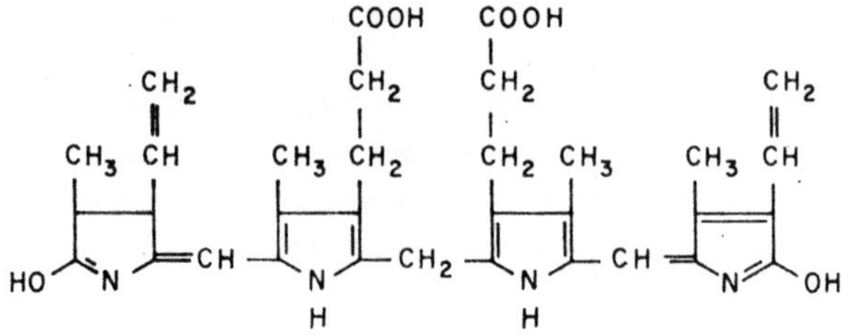

Bilirubin

$C_{33}H_{36}O_6 N_4$

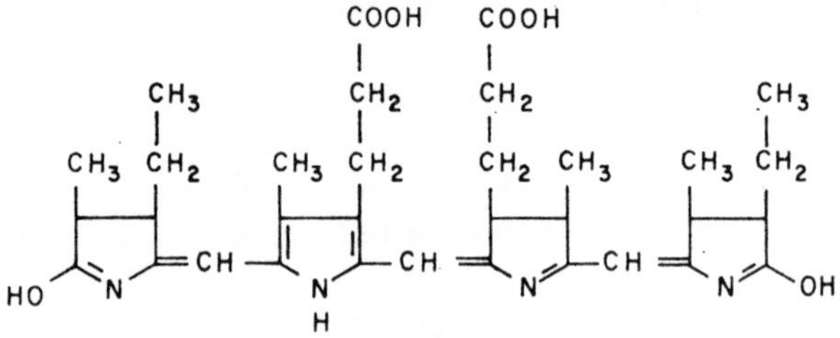

Glaucobilin

$C_{33}H_{38}O_6N_4$

FIG. 10. Some common bilichrome compounds.

Some naked marine gastropods secrete bilichromic compounds into their ink-glands. An example is the sea-slug (sea-hare) *Aplysia californica*, which derives the deep violet-colored bilichrome aplysioviolin from the erythrobilin in the red algae of its seaweed diet. Slugs which had been de-inked on a diet of brown algae failed to develop new ink until returned to red-algal food.

Although *Aplysia* ink possesses one or more chemicals rendering it malodorous and distasteful to predators, it is not believed that the purple aplysioviolin component is at all responsible as a defensive material. More probably it is merely a waste product (Chapman and Fox, 1969).

The arresting calcareous-skeletal color of the blue Alcyonarian coral *Heliopora caerulea*, native to Australian and West Indo-Pacific waters, challenged naturalists for many years, until it received more careful chemical study by Tixier-Durivault (1942, 1943) and Tixier (1945). They recognized the pigment as a member of the bilichrome class of compounds and called it helioporobilin. However, Rüdiger *et al.* (1968) isolated the colored compound and identified it as biliverdin-IXα.

VII. Hemocyanins

There remain a number of miscellaneous pigments, many of unknown chemical structure, observable in various marine animal species, as we saw among some of the gorgonian corals and other coelenterates. We shall not undertake to discuss these save for one exception, the hemocyanins. These important, complex, copper-containing chromoproteins, colorless when chemically reduced or de-oxygenated and blue in the oxygenated state, are not conspicuous in intact organisms, and are limited to certain classes of but two phyla, i.e., occurring widely in collidal solution in the circulatory fluids among only the arthropods and molluscs.

Within the arthropod phylum, the chief incidence of hemocyanins is in the crustacean class, but they have been found also in the blood of some arachnids, including certain spiders and scorpions and notably in the marine xiphosuran arachnid *Limulus polyphemus*, or so-called "horse-shoe crab" (Goodwin, 1971).

Redmond (1971) has reviewed some of the older and much of the recent investigations of arthropod hemocyanins, and Ghiretti and Ghiretti-Magaldi (1972) have similarly surveyed the hemocyanin literature concerning the molluscan phylum, wherein the chromoprotein is encountered in the amphineurans, cephalopods, and many species of the gastropod class. Tabulated data remind us that the amino acid content of arthropod and molluscan hemocyanins seem to be fairly similar *inter se*, varying somewhat in the relative proportions of the 18 different amino acids and ammonia linked together in constituting the large protein molecule. The dicarboxylic acids, aspartic and glutamic, show high proportions, amounting to nearly one-fourth of the total identified components. The hemocyanins of different species exhibit also considerable differences in their oxygen-binding capacity, hence in the slope of the respective sigmoid oxygen-equilibrium curves (Redmond, 1971, pp. 132-135).

Native hemocyanin has been classified as a globular protein with two Cu-atoms situated at each site whereon an oxygen molecule may be bonded. The average copper content of arthropod hemocyanin comes to ca 0.17%, and that of molluscan hemocyanin around 0.25%. Hemocyanin molecules of most arthropods are characterized by lower molecular weights (units of 10^5) than those of most molluscs, which in many instances are in units of 10^6. There appear to be in hemocyanins no characteristic prosthetic groups linking the metal atoms to the protein, as occurs in hemoglobin. The copper seemingly is bonded directly to the protein, whether or not through sulfhydryl or through alkaline N-groups remains uncertain. Nor has the valence state of the copper been clearly established. It would appear to be all cuprous (Cu^+) in the colorless de-oxygenated state; and, while the conspicuous blue color of oxygenated hemocyanin strongly suggests the cupric (Cu^{2+}) condition, evidence that all of the copper may change to this oxidized form seems not to be final. Oxygenated hemocyanins exhibit very broad absorption maxima, centering from about 565 to 580 nm, depending on the species bled.

Little seems to have been learned of the manner or site in which hemocyanins are synthesized, although there is evidence for concentration of copper in the hepatopancreas of crustaceans, and that the levels of copper so stored may vary inversely with the hemocyanin concentration in the blood. In the spider-crab *Maja squinado*, blood-hemocyanin and hepatopancreatic copper have been observed to decline in concentration following a molt, and to rise steadily as the new skeleton is forming and hardening. Similar observations have been made after experimental bleeding of certain crabs (Zuckerkandl, cited by Redmond, 1971).

Hemocyanins serve in many species not only as transporters of oxygen for respiration, but as a kind of oxygen-storage bank, notably in species which remain for periods in unfavorable environments, burying themselves in oxygen-poor loci exemplified by watery mud in the floor of their burrows.

Other functions which have been assigned to hemocyanins, which often may represent the chief blood protein, include a role as a principal factor in maintaining a balanced blood-colloidal osmotic pressure, perhaps aiding against water losses otherwise induced by blood hydrostatic pressures (Prosser and Brown, cited by Redmond, 1971). As blood-proteins, hemocyanins likely serve importantly as pH-buffers, as do hemoglobins. It is assumed that the copper-protein serves to combine with carbon dioxide, transporting it to sites of discharge, although the precise manner of combination and transport remains unclear.

Some interesting discoveries were made by Pilson (1965) during part of his doctoral researches carried out in this laboratory, wherein he was concerned with the ranges of hemocyanin concentration in plasma of abalones of the *Haliotis* genus. While the non-protein nitrogen in plasma from four *Haliotis* species showed but minor variations in concentration, there were wide differences in the content of hemocyanin, HCy, as follows:

	H. fulgens	*H. cracherodii*	*H. corrugata*
Non-HCy Protein			
Mean value%	0.20	0.14	0.24
HCy Protein			
Median %	0.54	0.38	0.15
Low	0.03	0.21	0.0017
High	1.89	2.03	1.53
Range	63-fold	10-fold	900-fold

H. rufescens also exhibited wide variability in HCy concentrations in its plasma. Pilson could detect no relationships between blood-HCy concentration and an animal's weight, sex, reproductive condition (by gonadal index), nutritional status, depth of habitation, or annual season. He concluded very logically, "This enormous range in concentration of hemocyanin in the blood appears to be incompatible with any physiological function which has so far been suggested".

That hemocyanin, with its reversible oxygen-combining power, is of direct and indispensable usefulness to many, perhaps to most of the species in which it is present, is beyond reasonable doubt. But its function as a direct respiratory role in *Haliotis* is open to puzzling questions.

VIII. References

Atz, J. W. (1939). *Bull. N.Y. Zool. Soc.* **42**, 128.

Beatty, R. A. (1941). *J. exp. Biol.* **18**, 144–152.

Beatty, R. A. (1949). *J. exp. Biol.* **26**, 125–136.

Blackburn, M. (1950). *Aust. J. mar. and Freshwat. Res.* **1**, 155–198.

Chaffee, R. R. and Mazia, D. (1963). *Devl. Biol.* **7**, 507–512.

Chapman, D. J. and Fox, D. L. (1969). *J. exp. Biol. Ecol.* **4**, 71–78.

Cheesman, D. F. (1958). *Proc. R. Soc. Ser. B* **149**, 571–587.

Cheesman, D. F. (1967). *Biol. Rev.* **42**, 132–160.

Comfort, A. (1951). *Biol. Rev.* **26**, 285–301.

Crescetelli, F. (1945). *Biol. Bull.* **88**, 30–36.

Davies, B. H., Hsu, W. J. and Chichester, C. O. (1965). *Biochem. J.* **94**, 26P.

Davies, B. H., Hsu, W. J. and Chichester, C. O. (1970). *Comp. Biochem. Physiol.* **33**, 601–615.

Denton, E. J. and Nicol, J. A. C. (1964). *J. mar. biol. Ass. U.K.* **44**, 219–258.

Dimelow, E. J. (1958). *Nature, Lond.* **182**, 512.

Drumm, P. J. and O'Connor, W. F. (1940). *Nature, Lond.* **145**, 425.

Drumm, P. J., O'Connor, W. F. and Renouf, L. P. (1945). *Biochem. J.* **39**, 208–210.

Fox, D. L. (1937). *Proc. natn. Acad. Sci.* **23**, 295–301.

Fox, D. L. (1944). *Sci. mon.* **59**, 394–396.

Fox, D. L. (1953). "Animal Biochromes and Structural Colours". Cambridge University Press. (Revised, supplemented edition, University of California Press, 1974).

Fox, D. L. (1957). The Pigments of Fishes: Chapter 7, *In*: "The Physiology of Fishes", Vol. 2, (Brown, M. E., Ed.). Academic Press, New York and London.

Fox, D. L. (1966). *In* "Physiology of Molluscs", Pigmentation of Molluscs: Chapter 8 (Wilbur, K. M. and Yonge, C. M., Eds.), Academic Press, New York and London.

Fox, D. L. (1972*a*). *Am. Scient.* **60**, 436–447.

Fox, D. L. (1972*b*). *Comp. Biochem. Physiol.* **43B**, 919–929.

Fox, D. L. (1973). *Comp. Biochem. Physiol.* **44B**, 953–962.

Fox, D. L. and Anderson, L. J. (1941). *Proc. natn. Acad. Sci. U.S.A.* **27**, 333–337.

Fox, D. L. and Crane, S. C. (1942). *Biol. Bull.* **82**, 284–291.

Fox, D. L., Crane, S. C. and McConnaughey, B. H. (1948). *J mar. Res.* **7**, 567–585.

Fox, D. L., Crozier, G. F., and Smith, V. E. (1967*a*). *Comp. Biochem. Physiol.* **22**, 177–188.

Fox, D. L. and Haxo, F. T. (1958). Sect. III. No. 3. *In Proc. XV Intern. Congr. Zool.* 280–282.

Fox, D. L. and Hopkins, T. S. (1966*a*). *Comp. Biochem. Physiol.* **17**, 841–856.

Fox, D. L. and Hopkins, T. S. (1966*b*). *In* "Physiology of Echinodermata". (Boolootian, R., Ed.), The Comparative Biochemistry of Pigments; Chapter 12. Interscience Publishers, John Wiley and Sons.

Fox, D. L. and Kuchnow, K. P. (1965). *Science* **150**, 612–614.

Fox, D. L. and Millott, N. (1954*a*). *Experientia* **10**, 185–187.

Fox, D. L. and Millott, N. (1954*b*). *Proc. R. Soc. Ser. B* **142**, 392–408.

Fox, D. L. and Pantin, C. F. A. (1941). *Phil. Trans. R. Soc. Ser. B* **230**, 415–450.

Fox, D. L. and Pantin, C. F. A. (1944). *Biol. Rev.* **19**, 121–134.

Fox, D. L. and Scheer, B. T. (1941). *Biol. Bull.* **80**, 441–455.

Fox, D. L., Smith, V. E., Grigg, R. W. and Macleod, W. D. (1969*a*). *Comp. Biochem. Physiol.* **28**, 1103–1114.

Fox, D. L., Smith, V. E. and Wolfson, A. A. (1967*b*). *Experientia* **23**, 965–966.

Fox, D. L., Updegraff, D. M. and Novelli, G. D. (1944). *Arch. Biochem.* **5**, 1–23.

Fox, D. L. and Wilkie, D. W. (1970). *Comp. Biochem. Physiol.* **36**, 49–60.

Fox, D. L., Wolfson, A. A. and McBeth, J. W. (1966*b*). *Comp. Biochem. Physiol.* **29**, 1223–1239.

Fox, H. M. (1947). *Nature, Lond.* **160**, 825.

Ghiretti, F. and Ghiretti-Magaldi, A. (1972). *In* "Chemical Zoology," Vol. VII Mollusca, 210–217. Academic Press, New York and London.

Gilbert, P. W. (1963). *In* "Sharks and Survival, Chapter 9. The Visual Apparatus of Sharks." D. C. Heath and Co., Boston.

Goodwin, T. W. (1969). *In* "Chemical Zoology", Vol. III. 135–147. (Florkin, M. and Scheer, B. T., Eds.). Academic Press, New York and London.

Goodwin, T. W. (1971). *In* "Chemical Zoology VI, Arthropoda Part B" pp. 290–306, (Florkin, M. and Scheer, B. T., Eds.). Academic Press, New York and London.

Gordon, M. (1948). *N.Y. Acad. Sci., Spec. Publ. IV.*

Karrer, P. and Solmssen, U. (1935). *Helv. chim. Acta* **18**, 915–921.
Klüver, H. (1944a). *Science* **99**, 482–484.
Klüver, H. (1944b). *J. Psychol.* **17**, 209–227.
Lederer, E. (1938). *Bull. Soc. Chim. biol.* **20**, 567–610.
Mason, H. S., Kahler, H., MacCardle, R. C. and Dalton, A. J. (1947). *Proc. Soc. exp. Biol., N.Y.* **66**, 421–431.
McBeth, J. W. (1972a). *Comp. Biochem. Physiol.* **41B**, 55–68.
McBeth, J. W. (1972b). *Comp. Biochem. Physiol.* **41B**, 69–77.
Millott, N. and Jacobson, F. W. (1952). *J. Invest. Derm.* **18**, 91–95.
Mukai, T. (1958). *Mem. Fac. Sic. Kyushu Univ., Ser. C. Chem.* **3**, 29–33.
Mukai, T. (1960). *Bull. Chem. Soc. Japan* **33**, 453–456.
Nicol, J. A. C. (1961). *J. mar. biol. Ass. U.K.* **41**, 271–277.
Nicol, J. A. C. (1964). *J. Fish. Res. Bd. Can.* **21**, 1089–1100.
Pilson, M. E. Q. (1965). *Biol. Bull.* **128**, 459–472.
Powell, V. H. and Sutherland, M. D. (1967). *Aust. J. Chem.* **20**, 541–553.
Powell, V. H., Sutherland, M. C. and Wells, J. W. (1967). *Aust. J. Chem.* **20**, 535–540.
Redmond, J. R. (1971). In "Chemical Zoology". Vol. VI. Arthropoda, Part B, Blood Respiratory Pigments—Arthropoda, pp. 119–144. Academic Press, New York and London.
Rüdiger, W., Klose, W., Tursch, B., Houvenaghel-Crevecoeur, N. and Budzikiewicz, H. (1968). *Justus Leibigs Annln Chem.* **713**, 209–211.
Schmidt, P. J. and Baker, E. G. (1969). *J. Fish. Res. Bd. Can.* **26**, 357–360.
Smith, V. E. (1968). *Doctoral Dissertation, S.I.O., Univ. of Calif., San Diego.*
Sumner, F. B. (1940). *Biol. Rev.* **15**, 351–375.
Sumner, F. B. and Fox, D. L. (1935). *Proc. natn. Acad. Sci. U.S.A.* **21**, 330–340.
Sutherland, M. D. and Wells, J. W. (1959). *Chem. and Ind. (Rev.)* **78**, 291–292.
Sutherland, M. D. and Wells, J. W. (1967). *Austr. J. Chem.* **20**, 515–553.
Thomson, R. H. (1957). "Naturally Occurring Quinones". Butterworths Science Publications, London.
Tixier, R. (1945). *Monaco: Ann. Inst. Oceanogr.* **22**, 243–297.
Tixier, R. and Tixier-Durivault, A. (1942). *Bull. Soc. Chim. biol.* **24**, 276–279.
Tixier, R. and Tixier-Durivault, A. (1943). *Bull. Soc. Chim. biol.* **25**, 98–102.
Vallentyne, J. R. (1956). *Limnol. Oceanogr.* **1**, 252–262.
With, T. K. (1968). "Bile Pigments. Chemical, Biological and Clinical Aspects". Academic Press, New York and London.
Wood, R. W. (1934). "Physical Optics". Macmillan, New York.
Yadon, V. L. (1968). Personal communication.
Yamaguchi, M. (1957). *Bull. chem. Soc. Japan* **30**, 111–114.
Yamaguchi, M. (1958). *Bull. chem. Soc. Japan* **31**, 51–55.
Yamaguchi, M. (1959). *Bull. chem. Soc. Japan* **32**, 1171–1173.
Yamaguchi, M. (1960). *Bull. chem. Soc. Japan* **33**, 1560–1562.

Biochemical Genetic Studies of Fishes: Potentialities and Limitations

FRED M. UTTER, HAROLD O. HODGINS AND FRED W. ALLENDORF

Northwest Fisheries Center,
National Marine Fisheries Service,
National Oceanic and Atmospheric Administration,
2725 Montlake Boulevard East,
Seattle, Washington 98112

I. Introduction

A new dimension in the understanding of protein variation had its basis in two events of the 1950's. The model of the structure of the DNA molecule proposed by Watson and Crick (1953) led to an understanding of the direct relationship between genes and proteins. Starch gel electrophoresis (Smithies, 1955) enhanced by the application of histochemical staining methods (Hunter and Markert, 1957) added simplicity and sensitivity to the study of protein variation. These events led to an explosion of information concerning protein variation and its significance which started in the 1960's and has continued in this decade. Studies of genetic variation at the protein level have made major contributions to a diverse array of biologically oriented disciplines. Many fishes have been studied electrophoretically (reviewed by de Ligny, 1969, 1972) and our co-workers and we have been involved in some of these investigations.

This is not a review paper in a comprehensive sense, although considerable literature will be cited to illustrate various points. Rather, we attempt to evaluate the present status of biochemical genetic research of fishes from the particular perspective of our experience, giving a background for this perspective and an indication of what we regard as fruitful areas for future research. At the same time we urge readers of this review to also read a complementary review of a similar topic by Robertson (1972). For the reader who is not studying genetic variation at the protein level, we hope to clarify what can and cannot be accomplished using electrophoretic-histochemical methods. The less experienced reader who is involved to some degree in the type of research reviewed, will, we hope, through the sharing of our experiences, have enlarged his perspective. Finally, the equally or more experienced reader will find that some of our thoughts will at least complement his own experience, and we hope that some will be found provocative.

II. Background of Serological and Biochemical studies

Our present research program had its roots in the mid-1950's when George Ridgway of the U.S. Bureau of Commercial Fisheries (now the National Marine Fisheries Service) proposed that Pacific salmon, *Oncorhynchus* spp., caught on the high seas might be identified as to continent of origin by serologically detected differences reflecting, presumably, genetic differences. The problem was approached through studies of serum proteins and erythrocyte antigens.

Serum protein variants were detected in sockeye salmon, *O. nerka*, using immunodiffusion and immunoelectrophoresis and with antisera developed mainly in rabbits. One variant (called the A or SM antigen) was found in maturing females of all salmonid species tested and has proven useful for maturity rather than for racial studies. A pair of serum antigens (called antigens I and II), which appeared to vary markedly in their distribution between fish of Asian and American origin, initially seemed highly promising for assigning the continental origin of fish taken on the high seas. However, difficulties in producing adequate quantities of specific and potent antisera, coupled with strong indications that these variations reflected artifacts of differential preservation rather than valid genetically based differences, led to discontinuance of study of these antigens.

Erythrocyte antigens were examined with an array of agglutinating substances, including normal sera from a variety of animals, phytohemagglutinins, heteroimmune (xenogeneic) sera and isoimmune (allogeneic) sera; only the isoimmune sera gave any real promise for reliable identification of genetic differences in Pacific salmon species. These reagents, produced in rainbow trout, *Salmo gairdneri*, were used to identify differences between populations of sockeye; chum, *O. keta*; and pink salmon, *O. gorbuscha*; and to indicate inbreeding in a population of cut-throat trout, *S. clarki*. However, these antisera were difficult to produce and purify in adequate quantities; their patterns of reactivity were very complex and there was a loss of potency detectable after frozen storage from one year to the next.

In a detailed review of the above studies, Hodgins (1972) concluded: "Significant advances have been made in concepts of the nature of fish populations as a result of extensive serological and immunochemical studies. The magnitude of two types of problems militates against further extensive studies of this nature in our laboratory at present: (1) technical problems, such as difficulty in producing vast quantities of specific high-titered antisera and the critical nature of sample preservation; and (2) theoretical problems, such as relating blood groups to genes. These procedures can be used, however, to differentiate stocks under certain conditions."

An alternate approach was needed after a decade of limited success and considerable frustration through the use of serological methods in attempting to identify genetic differences in fish populations. Thus we turned to starch gel electrophoresis. It soon became apparent that this was the method of choice over any previous methods that we had used for the identification of intra-species genetic variations of fish species that we were studying.

The major advantage of biochemical genetic data obtained through gel electrophoresis compared to immunological data is that valid genetic interpretations can be made directly from raw data. Codominant expression of most variant alleles occurs on starch gels. That is to say that, in an individual containing different alleles of a given locus, each of the alleles is expressed as a single, distinct protein. Such a situation commonly permits designation of the genotypes of individual samples based on staining patterns on gels. The frequency with which a given gene occurs in a population of individuals can be directly determined and the distribution of phenotypes can be tested for deviations from expected values based on simple genetic models, an important one being the Hardy-Weinberg model. This states that in a randomly mating population in the absence of a variety of disturbing forces (e.g. selection, mutation, etc.) the expected distribution of genotypes is determined by the random combination of alleles. In the case of two alleles (A — B) the expected genotypic frequency, therefore, is

$$q^2(AA) : 2q(1 - q)(AB) : (1 - q)^2(BB)$$

where q is the portion of A alleles in the population. Inferences can also be made directly from raw data regarding the subunit structure of proteins based on the number of bands observed for a particular system in heterozygous individuals. This is discussed in detail in a later section. Two other significant practical advantages of the starch gel method are that antisera are not required and that sample preparation and preservation have proven to be much simpler. The overall advantages of the starch gel method have resulted in an enormous differential in the amount of genetic information concerning intraspecies variation per unit of effort obtainable by comparing blood grouping and starch gel electrophoretic methods. As a hypothetical example (but based on actual experience), one worker examining a previously unstudied species could obtain more valid genetic information in a day using starch gel electrophoretic methods than a team of workers could obtain in 6 months using blood grouping methods.

By 1968, the major portion of our program's genetic efforts had shifted from blood grouping to electrophoretic and histochemical staining methods. The early studies in this phase were exhilarating because we were not yet accustomed to finding genetic variations with the ease made possible by our change in methods. A summary of our research through 1970 (Utter *et al.*, 1972) reported genetic variants at one or more loci in 16 species of marine and anadromous fish.

III. Methodology

The methodology we use is similar to that employed in other laboratories using horizontal starch gel electrophoresis to study a wide variety of organisms. Some modifications which we routinely employ, which may result in saving time and money, may not be apparent from the literature. Although we continually look for ways to simplify our methods while maintaining or improving the quality of the protein resolution, the basic electrophoretic methodology described below is that taught to us by C. N. Stormont and Y. Suzuki of the University of California at Davis.

Starch gel preparation generally follows that described by Kristjansson (1963). The gel is poured into a frame composed of a 7 x 10-1/2 inch glass plate bounded by removable 1/4-inch thick Plexiglass strips which are held in place with paper clamps. We have observed a strong preference among our colleagues for this arrangement rather than complete preformed electrophoretic units. After pouring, the gel is cooled to room temperature, and covered with plastic wrap. The cooling process may be accelerated in the refrigerator and the gel made ready for electrophoresis within 15 min after pouring. Freshly poured gels are preferred, although we have obtained satisfactory results from gels that have been poured as long as 48 h previously. We have found that placing a glass plate on top of the gel immediately after pouring is not desirable.

Prior to application of samples, the gel is cut 3 cm from the end along the long side and the smaller section of the gel is pulled back about 1 cm. Samples are drawn onto filter paper inserts as small as 4 × 6 mm and placed side by side along the gel separated by 1 mm; in this manner, more than 40 samples can be tested on a single gel. For electrophoresis, the smaller gel section is placed firmly against the larger section and inserts. Plastic wrap is folded back to expose about 1 cm at each end of the gel.

Buffer trays used are plastic dishes containing 150-200 ml of buffer. The tray buffer is transmitted to the gel with disposable utility cloths which are re-used indefinitely. Initial current for electrophoresis varies according to the buffer systems; we use as much as 300V but never exceed 100 ma. Sample inserts are removed and discarded following 10 min preliminary electrophoresis. Gel sections are again placed firmly together. Ice packs composed of gelled refrigerant repackaged to fit on a gel-sized glass plate are placed on top of the gel, and electrophoresis is continued until dye marker or boundary has migrated the appropriate distance. Following electrophoresis, the 1/4-inch plastic strips are removed from the gel frame. The gel is sliced into four layers, with nylon thread guided by 1/16th-inch plastic strips placed sequentially on top of one another; each layer is placed in a separate staining tray and stained differentially as desired.

It is desirable to use a variety of buffer systems in the initial phases of an electrophoretic investigation. No single buffer system can resolve all of the proteins that can be detected by starch gel electrophoresis. We have generally found, however, that a protein that is clearly resolved by a particular buffer system in one species is usually resolved in other species by the same system. We routinely use four different buffer systems. The components of these systems and the proteins that they best resolve are given in Table I. Additional buffer systems have been described elsewhere which may provide better resolution for certain proteins (Shaw and Prasad, 1970).

Staining methods we routinely use have followed directly or were modified from procedures of Shaw and Prasad (1970) in most instances. We have saved a considerable amount of money and time by reusing many solutions requiring tetrazolium salts. These salts plus substrates, cofactors

TABLE I: Different buffer systems for starch gel electrophoresis.

Buffer System	Comments
1. After Ridgway *et al.*, 1970. *Electrode Buffer*—LiOH 0·06M, H_3BO_4 0·3M, pH 8·3 *Gel Buffer*—Tris 0·03M, citric acid acid. 0·005M, pH 8·0 Use 99% gel buffer and 1% tray buffer for making gels. Use electrode buffer undiluted in electrode compartments.	Best system for resolving LDH,* PGM, Tfn, SDH, CrK. Gives only fair to poor resolution for some proteins that are clearly resolved by other systems.
2. After Markert and Faulhaber, 1965. *Stock Solution*—Tris 0·9M, H_3BO_4 0·5M, NaEDTA 0·02M, pH 8·6 Dilute stock solution 1 : 20 for making gel, 1 : 4 for electrode compartments	Best system for AGPD. Adequate resolution for most proteins.
3. After Wolf *et al.*, 1970. *Stock Solution*—Na_2HPO_4 0·1M with pH adjusted to 6·5 with 0·1M NaH_2PO_4. Dilute stock solution 1 : 10 for gel, use undiluted in electrode compartments.	The only system of those listed to adequately resolve IDH and all MDH isozymes.
4. After Utter and Hodgins, 1972. *Stock Solution*—one part 0·3M H_3BO_4 to four parts 0·3M tris, pH 9·0 Dilute stock solution 1 : 10 for making gel, use undiluted for electrode compartments.	Gives adequate resolution for most proteins. Particularly useful for resolution of non-specific muscle proteins migrating towards the anode.

*See Table II for full names of abbreviated proteins.

and coupling enzymes can be quite expensive. We have found that some solutions are stable for periods exceeding 1 week when stored refrigerated in the dark. Appropriate controls are of course required to ensure that negative results are caused by no enzyme activity rather than by deterioration of a component of the staining solution.

We have found that many samples can be prepared for electrophoresis with a minimum of effort. Skeletal muscle that has been frozen more than a few days is more readily extractable and can be placed in a suitable container with an equal amount of extracting fluid and tested without further treatment. Liver samples can often be similarly extracted.

Stability of proteins under frozen storage varies according to the particular protein itself and the tissue in which it is expressed. Muscle proteins have been quite stable with some enzymes, e.g., PGM and LDH (see Table II for full names of abbreviated enzymes), retaining good activity well in excess of a year. Proteins expressed only in the liver have proven considerably more labile, some enzymes—e.g., ADH—losing activity after only a few weeks.

Reliable blood samples for transferrin testing are obtained up to 48 hours after the fish was killed, provided the fish was promptly placed on ice but not frozen. The flesh is pierced with a sharp scalpel and the blood removed from the pericardial cavity with a Pasteur pipette; the blood is then expressed into a suitable container holding an anticoagulant solution. Dilution of the blood as much as 5:1 still results in reliable transferrin phenotypes. This method has proven useful in small fish where only minimal amounts of blood can be collected. Although freshly collected serum is preferable, transferrin phenotypes can also be found in eye fluids (Utter, 1969). The investigator should always run tests initially with both serum and eye fluids to ensure that parallel phenotypes are being expressed. We found that transferrins in eye fluids are sometimes detectable only after concentration.

It may be noted here that we have not included polyacrylamide gel electrophoresis in this discussion largely because of our more limited experience with it. We can use a somewhat smaller volume of sample with acrylamide but starch is faster and more flexible, results are more repeatable and we see little significant decreased resolution of fish protein bands with starch rather than acrylamide.

We may also mention that electrophoresis can be combined with immunodiffusion to give immunoelectrophoresis (IEP, see Williams and Chase, 1971). This test has great resolving power for mixtures of proteins. In human serum, for example, about 6 components are detected by electrophoresis in agar alone, while 30 or more components are detected by IEP. IEP in agar gels has been used in this laboratory to characterize serum proteins in sockeye salmon (Krauel and Ridgway, 1963) and in other species for population studies. Methods routinely used are described in the paper by Krauel and Ridgway in which they reported that as many as 25 components were detected in sockeye salmon serum. We have also extensively applied this procedure to identifying and characterizing immunoglobulin in rainbow trout serum (Hodgins et al., 1965; Hodgins et al., 1967).

Other matrices than agar (such as cellulose acetate, agarose, and polyacrylamide) are used in IEP. We have used a combination of an acrylamide disc gel imbedded in agar in several experiments. This entails prior electrophoresis of the sample in a disc gel followed by immediate embedding of the entire unfixed gel in agar on a 2 × 3inch glass slide and then proceeding with usual IEP procedures of cutting troughs and applying antiserum. With specific antisera and this technique, we have characterized female-specific antigens in salmonid fish sera and eggs (Gronlund *et al.*, 1973).

IV. Criteria for Mendelian variation

In spite of the many advantages that we have found for using biochemical data for studying genetic variation within species, it is necessary to impose some rather stringent restrictions on all biochemical variation observed on starch gels before assuming that a particular pattern of variation is actually a reflection of genetic variation. The strongest data are those obtained from progeny of parents having known biochemical differences; a genetic basis is regarded as confirmed if these data conform to models of simple Mendelian inheritance. A genetic basis is assumed for variants that are similar, and presumably homologous, to variants observed in closely related species for which breeding data have been obtained.

In the absence of breeding data, several requirements are imposed. The bands resolved in starch must fit interpretable patterns based on simple genetic hypotheses. These patterns must be repeatable with sub-sampling of the same tissue of an individual. It is useful, though not essential, if a pattern of variation is expressed over a broad range of development of individuals in a species. Any significant deviations of phenotypic distribution from predictions of a genetic model must be explained.

We have observed four basic types of variation that may be interpreted as reflecting simple codominant expression of allelic genes (Fig. 1). Patterns

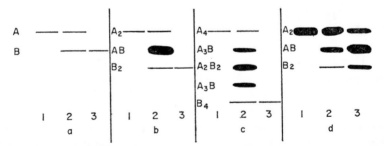

FIG. 1. Patterns of biochemical genetic variation observed in fishes reflecting codominant expression of two alleles. Subunit compositions are given to left of each band. 1. fast homozygote, 2. heterozygote, 3. slow homozygote. a—monomeric pattern, b—simple dimeric pattern, c—tetrameric pattern, d—dimeric pattern with two loci—one fixed and one polymorphic. Similar patterns with more phenotypes are observed when three or more alleles are found.

a, b, and c are typical of allelic expression for their respective subunit combinations in a wide variety of organisms. The enzymes yielding patterns a, b, and c are proteins consisting of one, two, and four subunits respectively. In each case two genetically distinct enzymes are involved and both are present in the heterozygote. That is, we are dealing in the examples shown with two codominant alleles. The patterns assume that the individual subunits of the enzymes in question combine in a random manner. For the interested reader, Shaw (1964) has discussed in detail the patterns expected for allelic proteins having from one to four polypeptide chains in the active molecule. In Fig. 1d we show a more complex pattern for a dimeric protein with two genetic loci—one fixed and one polymorphic. Pattern d has been reported for MDH and IDH in salmonid fishes (Bailey et al., 1970; Wolf et al., 1970; Allendorf and Utter, 1973). An hypothesis of two loci, one fixed and one polymorphic, has been verified for both loci through actual breeding data (Bailey et al., 1970; Allendorf and Utter, 1973). The need for breeding data must be emphasized for proper interpretation of this kind of variation; it had been previously proposed only on the basis of observation that IDH variation reflected tetrasomic inheritance at a single locus (Wolf et al., 1970).

Initial observations of patterns a, b, and c in Fig. 1 have generally, but not invariably, proven to be valid reflections of simple codominant allelic variation. Exceptions have been with some non-specific muscle proteins and esterases in various tissues where pattern a-type variations have not been repeatable on subsampling of the same individual. We have also encountered difficulty in obtaining liver esterase patterns in progeny of raibow trout based on phenotypes of parents, although subsampling data from individual fish are consistent. The nature of this variation remains questionable and may reflect ontogenetic changes in gene expression superimposed on typical Mendelian variation.

With a few clearly noted exceptions, the remaining sections of this review are devoted to a consideration of the applications of codominantly expressed genetic variants that have met our requirements for allelism. Such variations are an intrinsic property of the individual fish, and their expression is not directly affected by the environment (although there may be an indirect effect through processes of natural selection).

There may be considerable variation expressed on starch gel other than that reflecting (or appearing to reflect) codominant allelic variation. Although this variation may be partially or totally under genetic control, it cannot be analyzed with the precision that is possible for data reflecting codominant gene expression. Such variation includes the presence or absence of single or multiple bands, or variations in the intensity of such bands. This can often reflect artifacts arising from storage and preservation of samples (see Eppenberger *et al.*, 1971). It may also arise from different environmental or pathological conditions (see Hochachka, 1971) or may reflect gene modifications that do not alter the electrophoretic mobility of the protein in question (see Yoshida, 1973). Null (i.e., inactive) alleles may also account for some of this kind of variation; such variation can be analyzed similarly to codominant variants (i.e., a genotype can be assigned to each phenotype) if the active alleles of this locus interact with gene products of another locus, as for a null allele described for an LDH locus in rainbow trout (see Wright and Atherton, 1970). Some variation may have a codominant basis but cannot be clearly resolved electrophoretically; this observation emphasizes the need for using a variety of buffer systems in the early phases of an electrophoretic study. Regardless of cause, non-codominant variation is generally excluded from detailed analysis in our investigations.

V. Genetic variations of proteins within species

A. Relative Frequency of Variations

We have observed that some protein systems tend to vary more than others. A survey of protein variation observed in as many as 20 species that we have studied (Table II) indicates that over half of the species tested were polymorphic for PGM while none were polymorphic for 6PGD, ADH, or CrK. This observation is not unusual. It is well documented from amino acid sequencing data comparing similar proteins of diverse organisms (Dickerson, 1972) that proteins evolve at different rates; the less variable proteins observed here may represent more conservative evolutionary lines. It is interesting that G. B. Johnson (1971) predicted that ADH should be among the more variable proteins generally, on the basis of observed variations in *Drosophila* spp. and the equilibrium constants of ADH. It is possible of course that ADH of teleost fishes and *Drosophila* have different evolutionary dynamics. In a more practical vein, the data summarized in Table II offer a guide to other workers studying polymorphisms in fishes, whose methods do not yet include staining techniques for some of the proteins listed, by establishing priorities on the basis of the more variant proteins.

B. Different Patterns of Variations

The diversity of biochemical genetic variation within species that has been observed in our studies is typical not only of that observed by other workers in fishes (reviewed by de Ligny, 1969, 1972), but also of a continually broadening sampling of animal and plant species in general (see

TABLE II: Average frequencies of polymorphic proteins in fish species.

Protein	Number of species tested	Number of species polymorphic	Proportion of species polymorphic
Phosphoglucomutase (PGM)	19	11	0·58
Alpha-glycerophosphate dehydrogenase (AGPD)	19	9	0·47
Transferrin (Tfn)	7	3	0·43
Lactate dehydrogenase (LDH)	20	8	0·40
Malate dehydrogenase (MDH)	20	7	0·35
Esterase (Est)*	20	5	0·20
Sorbitol dehydrogenase (SDH)	6	1	0·17
Aspartate aminotransferase (AAT)	20	3	0·15
Isocitrate dehydrogenase (IDH)	20	3	0·15
Non-specific muscle protein	20	3	0·15
Diaphorase (Dia)	10	1	0·10
Tetrazolium oxidase (TO)	20	2	0·10
Peptidase (Pep)	20	1	0·05
6-phosphogluconate dehydrogenase (6PGD)	20	0	0
Alcohol dehydrogenase (ADH)	10	0	0
Creatine kinase (CrK)	10	0	0

*As noted earlier (p. 221), we found artifacts in analyses of esterases and urge investigators to be particularly stringent in their criteria for allelism in this class of enzymes.

Gottlieb, 1971). Intraspecies patterns of geographic variation that we have observed have been quite different, even among closely related species. It is clear from these observed patterns of genetic variation that the potential of biochemical genetic variation for use in our original purpose of stock identification differs considerably among species. We have interpreted significant differences at one or more loci between two groups of fish of a particular species as evidence for different breeding groups. Conversely, however, non-significant differences between two groups of fishes are regarded only as indicating that these groups are not necessarily different, but not as evidence that they are the same. An analogous situation at the individual level is a paternity test where the putative parent is exonerated when his genotype conflicts at one or more loci with that of the offspring in question, but is not incriminated in the absence of conflicting genetic data. This interpretation has generally proven to be accurate when other kinds of evidence, such as data from tagging experiments and growth studies, are also available for different groups of a species that we are investigating. It is fallible, however, in instances where subgroups of a single breeding population reflect differential selective pressures through different allelic frequencies, and that topic is discussed separately. In

Table II are presented some of these different patterns of polymorphism that have been interpreted as described above.

The simplest pattern of variation was found in the Pacific hake, *Merluccius productus* (Utter, 1969; Utter and Hodgins, 1969, 1971; Utter *et al.*, 1970). Two major population units were identified one in Puget Sound, Washington, and the other in the northeastern Pacific Ocean extending from British Columbia southward to Mexico. Highly significant differences of allelic distribution were found in each of four polymorphic systems; LDH, muscle protein, transferrin and an esterase system expressin in vitreous fluids. Very large fish occasionally caught in Puget Sound were originally suspected of being immigrants from the Pacific Ocean but were determined to be indigenous to Puget Sound by their gene frequencies.

A clinal distribution, where the frequency of polymorphism gradually changes with the geographical area, has been observed for protein variation in some species. The distribution of variant forms of two proteins—PGM and LDH—are clinal in sockeye salmon; much higher frequencies of variant forms are found in the more westward range of the species and gradual decreases occur as sampling progresses eastward (Hodgins *et al.*, 1969; Utter and Hodgins, 1970; Hodgins and Utter, 1971). It can therefore be assumed that samples of sockeye salmon taken on the high seas, lacking LDH variants and having low frequencies of PGM variants, originated from the southeastern range of the species. A clinal distribution has also been observed for a muscle protein variation in the greenstriped rockfish, *Sebastes elongatus*, in samples collected between Queen Charlotte Sound, B. C. and Puget Sound (Johnson, 1972).

An abrupt reversal of gene frequencies has been found for a three allele system of the serum protein transferrin in samples of coho salmon, *O. kisutch*, in different areas of Washington State (Utter *et al.*, 1970; Utter *et al.*, 1973. The A allele has a frequency greater than 90% in samples taken from streams entering the Columbia River. The C allele is most frequent in samples taken from streams entering Puget Sound and the Pacific Ocean, including those entering the ocean adjacent to the Columbia River. A third allele (B) is found at fairly high frequencies in fish from coastal and Puget Sound streams, but has not been detected in Columbia River fish. The pattern of transferrin variation in coho salmon from streams south of the Columbia River and north of Washington State is presently unknown, and knowledge of whether or not similar discontinuities occur in other areas must await further sampling. This system appears to have considerable potential for determining whether fish caught in the ocean in the vicinity of the Columbia River are destined to spawn in Columbia River tributaries or other areas.

Genetic variation of TO in chinook salmon, *O. tshawytscha*, may be correlated to the time of their return to freshwater (i.e., their spawning migration). A low frequency of a variant TO allele (F) was observed in four populations of fish returning to rivers of origin in the spring in populations sampled from southeastern Alaska and the Columbia River. However, the average frequency of the variant TO allele in five populations of fish from the Columbia River and Puget Sound that returned to the rivers in the fall was much higher (Utter *et al.*, 1973).

The rainbow trout is of particular interest with regard to the use of biochemical polymorphisms for racial studies because of the relatively large number of variant loci that have been reported in this species. Fourteen loci identified by electrophoretic variants are listed in Table III.

TABLE III: Polymorphic loci of rainbow trout.

Locus	Number of alleles	Reference
LDH E	2	Wright and Atherton, 1970.
LDH B^1	2	Wright and Atherton, 1970.
LDH B^2	2	Williscroft and Tsuyuki, 1970; Utter and Hodgins, 1972.
MDH A	2	Allendorf, unpublished data.
MDH B	4	Bailey *et al.*, 1970; Allendorf, unpublished data.
Tfn	3	Utter and Hodgins, 1972; Utter *et al.*, 1973.
PGM	2	Roberts *et al.*, 1969.
AGPD	2	Engel *et al.*, 1971; Utter and Hodgins, 1972.
TO	3	Utter, 1971; Utter and Hodgins, 1972.
Est I	2	Kingsbury and Masters, 1972.
Est II*	2	Allendorf, unpublished data.
IDH	4	Wolf *et al.*, 1970; Allendorf and Utter, 1973.
SDH	2	Engel *et al.*, 1970.
Hemoglobin	2	Tsuyuki and Ronald, 1970.

*Some question exists concerning the genetic nature of the variation. See discussion under "Criteria for Mendelian Variation".

Some of this variation has proven useful for distinguishing groups of rainbow trout, particularly when data from more than one locus are included in comparisons (Utter and Hodgins, 1972; Utter *et al.*, 1973). A comparison of gene frequencies in anadromous rainbow trout (steelhead) taken from hatcheries of tributary streams of the Columbia River (Table IV) indicates considerable heterogeneity in a group of fish that is managed in some instances as a single population. Of particular interest is a comparison of allelic frequencies of LDH and TO in

TABLE IV: Gene frequencies of biochemical genetic variants at seven loci in Columbia River steelhead trout populations.

Population	Frequency of most commonly observed allele						
	AGPD	Est II	IDH	LDH	MDH-B	PGM	TO
Upstream							
Clearwater	1·00	—	—	0·29	0·97	0·99	1·00
Pahsimeroi	0·99	0·72	0·45	0·40	0·99	1·00	0·94
Hell's Canyon	1·00	0·45	0·53	0·46	1·00	1·00	0·94
Deschutes	1·00	0·50	0·28	0·40	0·98	1·00	0·96
Downstream							
Big Creek	1·00	0·74	0·40	0·99	0·79	1·00	0·55
Skamania	0·92	0·81	0·10	0·80	0·88	0·99	0·60
Cowlitz	0·98	0·60	0·22	0·83	0·83	1·00	0·70

—no data available.

upstream and downstream populations. Data from these loci alone may be useful in predicting the extent of migration in steelhead entering the Columbia River.

Two species that have considerable biochemical polymorphism but which have failed to reveal any significant variations among populations— in spite of fairly extensive sampling—are the Pacific herring, *Clupea harengus pallasi*, and the Pacific saury, *Cololabis saira*, (see Utter *et al.*, 1972, and Utter, 1972, for preliminary reports of these variants). The variations of these species are summarized in Table V. Herring have been

TABLE V: Variant loci of Pacific herring and Pacific saury.

Species	Average frequency of most common allele							
	LDH-A	LDH-B	Est	AAT	MDH	PGM	AGPD	IDH
Pacific herring	>0·99	>0·99	0·85	0·98	0·96	0·74	1·00	1·00
Pacific saury	1·00	1·00	0·68	1·00	0·63	1·00	0·70	0·52

sampled from off the coast of Oregon northward to Kodiak Island and Saury were sampled off the coast of Japan and from southern California northward to British Columbia. The only differences in gene frequencies were for the LDH variants of herring; variants at two loci occured at low frequencies in samples collected in Puget Sound and in the North Pacific Ocean near Vancouver Island, but so far these variants have not been observed in samples collected from Alaskan waters. It is premature to speculate in any detail concerning possible reasons for the stability of allelic frequencies over broad geographic areas in these two species. It is interesting, however, that both species serve as forage fish for larger carnivorous species (the herring in inshore areas and the saury on the high seas) and thus occur in large numbers throughout their respective ranges. It may be that their gene frequencies are less subjected to random factors because of large breeding groups and that gene frequencies are stabilized among these groups, in the absence of strong selection, by some degree of genetic interchange.

To summarize the potential of biochemical genetic data in racial studies of fish, we believe that data based on reliable Mendelian variants should be collected in the initial phases of any study examining the population structure of a fish species. If the geographic range of sample collection is sufficiently broad, a situation similar to that observed in the Pacific hake should soon become apparent and biochemical genetic data, alone, may be sufficient to define adequately the populations in question. In instances such as the one exemplified by the sockeye salmon, biochemical genetic data may be used for broad definition of stocks taken at sea, and alternate methods, such as scale characteristics, may then be employed for more precise identification. Conversely, in the case of Columbia River fish vs. populations of coho salmon from coastal streams, biochemical genetic data may be used for more precise definition, but may be less useful alone for identification over a broader geographic range. Similarly, differences in gene frequencies of LDH and PGM variants between populations of sockeye salmon within major river systems have been observed in Alaska and may be useful for characterization of these stocks (Hodgins and Utter, 1971; Utter, unpublished observations). Finally, in instances such as that of Pacific saury, biochemical genetic data are of value in supporting hypotheses of homogeneity established by other criteria, but cannot be used alone for either broad or restricted definitions of populations.

The unique value of biochemical genetic studies is their direct reflection of genetic differences coupled with their ease of application. As indicated above, we believe that biochemical genetic methods should generally complement rather than replace other methods of examining population differences.

C. Gene-Environment Interactions

Considerable uncertainty presently exists concerning the role of natural selection in the maintenance of biochemical polymorphisms in populations. The large amount of biochemical polymorphism observed in most organisms has been explained, both on the basis of selection (Prakash *et al.*, 1969) and random factors (King and Jukes, 1969). The potential usefulness of biochemical polymorphisms could be extended considerably with a better understanding of the degree to which allelic forms of different proteins interact with components of the environment. Variants could then have greater value in examining patterns of dispersal and questions of systematics.

In fishes, as in other groups of organisms that have been studied, very little is presently known concerning the extent to which genotype-environment interactions of biochemical genetic variants can account for the amount of variation that has been described. Koehn and his associates have related allelic esterases of certain fresh water teleosts to temperature on the basis of geographic distribution and biochemical activities (Koehn, 1969; Koehn et al., 1971). M. S. Johnson (1971) observationally and experimentally related allelic forms of LDH in the high cockscomb, Anoplarchus purpurescens, to environmental temperature differences.

In our own studies (Johnson et al., 1971; Johnson, 1972), significant excesses of heterozygous individuals were observed for PGM and AGPD polymorphisms in deep water collections of Pacific ocean perch, Sebastes alutus, that were not observed in collections taken from shallower waters. There were also significant non-random associations of phenotypes observed for these two polymorphic systems at greater depths which were not found in shallower collections. These observations suggested that selective forces were acting upon these two loci at greater depths and pointed to the need for including other gene products of the individual, as well as external influences, when considering components of natural selection at a particular locus. This is one instance where selective forces may be acting differently on different segments of the same breeding population (see also Williams et al., 1973).

Our current studies include a detailed examination of the kinetics of the allelic forms of LDH in sockeye salmon described above. These studies were carried out by L. J. Guilbert during a 1-year post-doctoral appointment with our research program. The data—to be published in detail elsewhere—have indicated firstly, that differences in Michaelis constants (Km's) exist in different phenotypes of liver LDH in sockeye salmon at physiological temperatures*; and secondly, that forms of the enzyme appearing identical in electrophoresis have different Km's. The latter results suggest that "silent" alleles (see Boyer, 1972) reflected in substitutions of similarly charged amino acids may be present, altering functional properties of the enzymes without changing their electrophoretic mobilities. These investigations have proven to us the value of a

---

*A similar phenomenon has been described by Merritt (1972) for allelic LDH variants of the fat head minnow and related to a north-south cline observed in this species.

team approach to studies of complex biological problems, where complementary training and skills are emphasized in different members of the team.

Gaining a better understanding of the relationships between biochemical genetic variation and environmental factors is a major, although difficult, direction for intensive investigation during the present decade. The vast amount of descriptive data collected during the 1960's and early 1970's, that illustrate considerable biochemical genetic variation in diverse organisms, have provided an empirical basis for a change in thinking regarding the amount of genetic variation that can be tolerated in natural populations and the role of this variation in evolutionary processes. The far more complex task of demonstrating the biological significance of this variation lies ahead.

D. Fish Culture

The relating of biochemical genetic variants at single loci to production characteristics such as growth rate has not been notably successful to date (Robertson, 1972). This may be explained, in part, by the highly multigenic nature of most production characteristics of economic importance and very likely by the complexities of gene-gene interactions where the product of a single gene may interact with a variety of mechanisms relating to production characteristics. The limited success of the use of biochemical genetic variation in breed improvement appears to be related to the limited present knowledge concerning the degree of interaction of this variation with components of the external or internal environment of the organism. As more knowledge accumulates concerning the nature of gene-environment and gene-gene interactions, it seems likely that this knowledge can be usefully applied to animal breeding programs, perhaps from the simultaneous consideration of data from many loci.

Biochemical genetic variation may find useful application in many areas of fish culture other than direct relationships to production characteristics, however. For example, one problem of concern to fish culturists is inbreeding. Relative degrees of inbreeding could be directly estimated by comparing average heterozygosities observed in different brood stocks of a given species. A routine registry of gene frequencies in artificially maintained stocks of widely cultured fish species, such as rainbow trout, is a feasible means of facilitating such extension of biochemical genetic methods to fish culture.

Degrees of genetic differences can be directly estimated from biochemical genetic data. Methods for comparing genetic similarity at the species level (discussed in a later section) can be applied to intraspecies variation as well. Knowledge of the nature of genetic differences among stocks can be very useful when it is desirable to introduce additional genetic variation into a particular stock. As a rational basis for maximum extension of the gene pool, introductions could be made from stocks showing the greatest biochemical-genetic differences. Similarly, two stocks showing large biochemical-genetic differences could be reared separately and the F_1 progeny between them used as production fish in attempts to maximize heterotic effects.

Biochemical identification of the hatchery of origin, based on a limited number of samples, is a possibility that presently exists for rainbow trout (Utter and Hodgins, 1972). It is possible to breed towards unique homozygous combinations involving multiple loci where no strong environmental interactions have been detected. A unique genetic brand on all fish of a given hatchery could thereby be achieved in a few generations, and could be used in many situations where it is desirable to identify fish of that hatchery.

E. Hybrids

Some of the advantages offered by biochemical genetic methods for studies of intraspecific variation can be extended through studies of species hybrids because of the greater amount of genetic variation that exists between any two species than that which exists within either of them. More loci are available for studies of linkages if sufficient fertility exists between the F_1 generation and either of the parent species (Wheat et al., 1973) (see also the discussion in the following section). Evidence can be obtained for identification of parent species in natural and artificial hybrids (Abramoff et al., 1968; Aspinwall and Tsuyuki, 1968). If hybridization coupled with subsequent backcrossing and intercrossing has been part of a breeding program, the exotic portion of the genome can be estimated by biochemical-genetic data. Intraspecific differences in activation and control of allelic genes can be observed in hybrid individuals (Hitzeroth et al., 1968; Whitt et al., 1972). Evidence for heterosis has been described in hybrid sunfish (Whitt et al., 1973). On several occasions it has been possible to determine whether or not certain individual salmonids suspected of being hybrids are indeed hybrids. We have also used hybrid data to examine questions of allelism and sub-unit composition of proteins that are monomorphic in individual species but which differ among them (Utter et al., 1973).

F. Cytogenetic and Linkage Studies

Biochemical genetic data, in conjunction with classical cytogenetic techniques, have played a major role in the determination of both the existence and mechanisms of gene duplication. The role of gene duplication in providing the material for vertebrate evolution has been discussed in detail by Ohno (1970*b*) who has postulated that gene duplication by means of autotetraploidy occurred in an ancestral salmonid species on the basis of DNA content, chromosomal characteristics, and the duplication of LDH loci (Ohno *et al.*, 1969; Ohno, 1970*a*). Subsequent reports of numerous duplicated biochemical genetic loci in salmonids have supported this hypothesis (Bailey *et al.*, 1970; Wilkins, 1972). A recent paper, however, using karyotypes and an LDH locus has reported the presence of non-deleterious trisomy in brook trout, *Salvelinus fontinalis*, and has suggested recurrent trisomy as a possible alternate means of gene duplication in salmonids (Davisson *et al.*, 1972).

Our own studies do not indicate as much gene duplication in salmonids as is generally accepted (Ohno, 1970*a*, Bailey *et al.*, 1970). We recently reported polymorphisms of six loci in rainbow trout of which only two demonstrated evidence of gene duplication (Utter and Hodgins, 1972).

Single locus variants represent a basic tool of the geneticist. The genome of an organism such as *Drosophila* has been extensively mapped through the use of external morphological traits controlled by single loci, and has provided information which is extremely useful when utilizing this species as an object of genetic research. This type of variation is largely absent in fish species and because of this there have been very few reports of linkage relationships in fish. Biochemical genetics, however, provides a number of single locus markers which makes the study of linkages possible. We have examined the linkage relationships of five biochemical loci—LDH, MDH, TO, AGPD, and IDH (Allendorf, unpublished data)—as a part of our investigations of rainbow trout. Each two-way segregation was examined with the exception of LDH-AGPD where a double-heterozygote parental fish was not available. In no instance did the observed two-way segregation differ significantly from that expected with independent assortment. Only in a single case (AGPD-IDH) did recombination approach a non-random rate. However, Wheat *et al.* (1973) examined possible linkage relationships among six loci which varied in hybrid sunfish and found linkage between 6PGD and AGPD loci (recombination frequencies 15%–22%).

Because of the large number of potential linkage groups in rainbow trout—2N = 60 (Wright, 1955)—the chances of finding linkage between any two randomly chosen loci is small. Potentially more promising areas for detecting unusual two-way segregation are duplicated loci. Linkage between duplicated loci is more likely than between two randomly chosen loci because of the nature of gene duplication mechanisms. If duplication occurs through some form of tandem duplication these loci will lie on the same chromosome. If duplication occurs through tetraploidy or tetrasomy, the expected fusion of homologous chromosomes (Ohno et al., 1969) would result in linkage between duplicated genes. Wheat et al. (1973) found no significant deviation from random segregation between duplicated MDH loci in sunfish hybrids. Morrison (1970) however, did discover non-random segregation of two LDH loci in brook trout x lake trout, *Salvelinus namaycush*, hybrids. There are additional instances of gene duplication in salmonids where linkages between duplicated loci could be determined (e.g., MDH-A and MDH-B loci). One of our future objectives is to discover variants at these loci so that the linkage relationships can be explored.

VI. Variation at the species level

An extension of studies of allelic protein variation within a species is a comparison of the total amount of such variation that exists in a species relative to the total number of loci that can be reliably detected. If it is assumed that data from a limited number of loci coding for proteins can be extrapolated to the entire genome (Lewontin and Hubby, 1966), such data can be used to estimate the proportion of polymorphic loci and average individual heterozygosity in a species. These estimates can provide a useful basis for comparison among species.

We have estimated the proportion of polymorphic loci and average individual heterozygosity in three scorpaenid and six salmonid species (Johnson et al., 1973; Utter et al., 1973). In all species but rainbow trout, our estimates have been considerably lower than the average values reported in a wide variety of plant and animal species (reviewed by Gottlieb, 1971). The higher values observed in rainbow trout may reflect the greater habitat diversity of this species relative to any of the closely related species of Pacific salmon. If extrapolation to the remainder of the genome is valid, the greater genetic variation of rainbow trout suggests that heritabilities for a given characteristic should be higher in rainbow trout than in Pacific salmon, a possibility of potential interest to fish culturalists.

Biochemical genetic methods have proven useful in examining relationships among related animal species ranging from *Drosophila* (Ayala et al., 1970) to mammals (Johnson and Selander, 1971). This is not surprising because of the direct relationship between the protein and the gene. We have used criteria for allelism that are similar to those imposed for intraspecies variations for comparing protein differences between closely related species, although it is more difficult to obtain direct evidence for allelism by breeding tests. Homology for a single locus has been assumed for proteins differing among closely related species that are expressed as single bands. Differences of multiple banded protein patterns can be analyzed on the basis of two fixed loci if the banding patterns conform to those typically observed for intra-species variants (e.g., invariant 5-banded LDH patterns or 3-banded MDH patterns). There are occasional instances where multiple-banded protein patterns do not conform to known models of molecular interaction but the repeatability of such patterns strongly suggests that the variation observed is genetically based. In these instances two fixed loci have also been assumed and comparisons between species have been made by ratios obtained by dividing common bands by total bands (Utter et al., 1973).

We have examined biochemical genetic variation among related species for two somewhat different objectives, firstly, to identify species from biochemical characteristics of unknown samples and secondly, to examine systematic relationships based on biochemical data. Rockfishes of the northeastern Pacific Ocean have been examined primarily for the first objective, although a considerably more detailed study of this group towards its second objective is in progress.

In a study of muscle proteins from 27 Pacific Ocean species of *Sebastes* (Johnson et al., 1972), 14 different biochemical groups were identified, based on non-specific muscle protein patterns and five enzyme systems. No group had more than 4 species, and 10 species had unique biochemical profiles. These biochemical criteria have considerable practical potential in these species because of the economic importance of rockfishes and the fact that some species in different groups are morphologically indistinguishable to the non-expert. Accidental or intentional mislabeling can be minimized through routine testing of muscle samples obtained from commercial sources.

We have examined the phylogenetic relationships of the six species of Pacific salmon and two related trout species, based on biochemical genetic data. Interspecies comparison of muscle protein patterns (including enzymes) were made, based on related proteins of eight loci and more complex protein patterns assumed to reflect four additional loci (Utter et al., 1973). A dendrogram (Fig. 2), constructed from indices of similarity which reflected pairwise interspecies protein differences, separated the species into two major groups; one group contained the two trout species which were paired closely and, more distantly, the masu salmon, *O. masu*, while the other group contained the remaining five species of salmon. The data generally support classifications of this group of fishes, based on other criteria (reviewed by Behnke, 1965 and 1968), and indicate that the masu salmon is phylogenetically closer to the two trout species than to the other species of Pacific salmon.

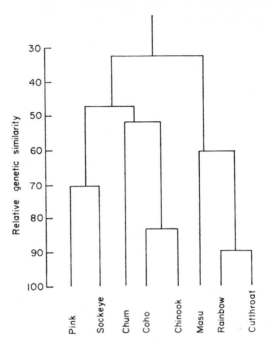

FIG. 2. Genetic relationships of eight salmonid species based on biochemical genetic data from 12 loci (from Utter *et al.*, 1973).

It is of interest to examine the data from which Fig. 2 was drawn from the point of view of species identification as a practical exercise because of the considerable potential for commercial mislabeling of salmonid flesh. Figure 3 is a biochemical key for identifying these species from samples of muscle tissue.

The genetic potential of electrophoretic criteria becomes less and less applicable for systematic comparisons between organisms as the taxonomic differences rise beyond the species level. Even electrophoretic identity within a species does not necessarily mean genetic identity, but merely that there is no detectable difference. (The Km data for sockeye salmon LDH certainly suggests a genetic difference for electrophoretically identical isozymes). It has been estimated that only 40% of the amino acid substitutions can be electrophoretically detected on the basis of charge differences (Nei, 1971). It seems, because of the time elapsed between common ancestry, that many electrophoretic identities of analogous proteins beyond the family level may be merely fortuitous. On the other hand, immunological methods—which have been of very limited use to us

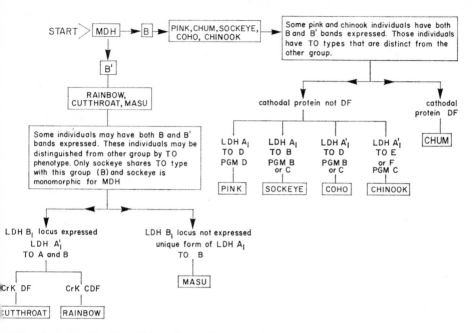

FIG. 3. A key for identifying six species of Pacific salmon and two related trout species based on genetic variations of skeletal muscle proteins. See Utter *et al.*, 1973 for allelic designations.

for studies of genetic variations within species—appear to have greater potential for comparative studies as the taxonomic distance increases (Gorman *et al.*, 1971).

VII. Summary

We have reviewed potentialities and limitations of biochemical genetic studies of fishes using electrophoretic methods. These methods can be usefully applied in numerous areas of fisheries research by virtue of their direct reflection of genetic differences and their ease of application.

The principle conclusions presented were:

(1) The potential of biochemical genetic data for use in identifying fish populations varied considerably among species. This ranged from clear separation of populations based on significant differences at many loci to no detectable differences over a broad geographic range in spite of the existence of numerous polymorphic loci.

(2) Studies of gene-environment interactions appear to be a difficult but rewarding field for the application of biochemical genetic methods.

Variations in catalytic activities of genetic variants of certain enzymes have been observed. These have occurred in some forms of enzymes that are electrophoretically identical as well as in electrophoretically distinct forms.

(3) Biochemical genetic data may presently be applicable to aspects of fish culture that include detection of inbreeding, brood stock identification and registration, and estimations of genetic differences among stocks. However, useful associations between genetically related proteins and production characteristics have not yet been found.

(4) Hybrids between species are readily detected by biochemical genetic methods.

(5) Proteins that vary either among or within species are useful for studying genetic linkages and ancestral gene duplication.

(6) Protein differences among species can be used for estimating relationships and for identification of species of tissues of questionable identity.

VIII. References

Abramoff, P., Darnell, R. and Balsano, J. (1968). *Am. Nat.* **102**, 555–558.
Allendorf, F. W. and Utter, F. M. (1973) *Genetics*, **74**, 647–654.
Aspinwall, N. and Tsuyuki, H. (1968). *J. Fish. Res. Bd. Can.* **25**, 1317–1322.
Ayala, F. J., Mourão, C. A., Pérez-Salas, S., Richmond, R. and Dobzhansky, T. (1970). *Proc. Natn. Acad. Sci. U.S.A.* **67**, 225–232.
Bailey, G. S., Wilson, A. C., Halver, J. E. and Johnson, C. L. (1970). *J. biol. Chem.* **245**, 5927–5940.
Behnke, R. J. (1965). *Ph.D. thesis, Univ. Calif. Berkeley.* 273 p. (1968). *Mitt. hamb. zool. Mus. Inst.* **66**, 1–15.
Boyer, S. H. (1972). *Nature, Lond.* **239**, 453–454.
Davisson, M. T., Wright, J. E. and Atherton, L. M. (1972). *Science* **178**, 992–994.
Dickerson, R. E. (1972). *Sci. Amer.* **226**(4), 58–72.
Engel, W., Op't Hop, J. and Wolf, U. (1970). *Humangenetik* **9**, 157–163.
Engel, W., Schmidtke, J. and Wolf, U. (1971). *Experientia* **27**, 1489–1491.
Eppenberger, H. M., Scholl, A. and Ursprung, H. (1971). *FEBS Letters* **14**, 317–319.
Gorman, G. C., Wilson, A. C. and Nakanishi, M. (1971). *Syst. Zool.* **20**, 167–185.
Gottlieb, L. D. (1971). *Biol. Sci.* **21**, 939–944.
Gronlund, W. D., Hodgins, H. O. and Blood, E. (1973). *Int. N. Pac. Fish. Comm. Anna. Rep.* 1971, 101–107.
Hitzeroth, H., Klose, J., Ohno, S. and Wolf, U. (1968). *Biochem. Genet.* **1**, 289–300.
Hochachka, P. W. (1971). *Am. Zool.* **11**, 81–82.

Hodgins, H. O. (1972). *In* "The Stock Concept in Pacific Salmon" (Simon, R. C. and Larkin, P. A., Eds.) pp. 199–208. H. R. MacMillan Lectures in Fisheries, University of British Columbia, Vancouver, B.C.

Hodgins, H. O., Ames, W. E. and Utter, F. M. (1969). *J. Fish. Res. Bd Can.* **26**, 15–19.

Hodgins, H. O., Ridgway, G. J. and Utter, F. M. (1965). *Nature, Lond.* **208**, 1106–1107.

Hodgins, H. O. and Utter, F. M. (1971). *Int. Counc. Explor. Sea, Rapp. Proc.-Verb Re'un.* **161**, 100–101.

Hodgins, H. O., Weiser, R. S. and Ridgway, G. J. (1967). *J. Immun.* **99**, 534–544.

Hunter, R. L. and Markert, C. L. (1957). *Science* **125**, 1294–1295.

Johnson, A. G. (1972). *Ph.D. Thesis, University of Washington*, 56 pp.

Johnson, A. G., Utter, F. M. and Hodgins, H. O. (1971). *Comp. Biochem. Physiol.* **39B**, 285–290.

Johnson, A. G., Utter, F. M. and Hodgins, H. O. (1972). *Fishery Bull. Fish Wldl. Serv. U.S.* **70**, 403–413.

Johnson, A. G., Utter, F. M. and Hodgins, H. O. (1973). *Comp. Biochem. Physiol.* **44B**, 397–406.

Johnson, G. B. (1971). *Nature, Lond.* **232**, 347–349.

Johnson, M. S. (1971). *Heredity* **27**, 205–226.

Johnson, W. E. and Selander, R. K. (1971). *Syst. Zool.* **20**, 377–405.

King, J. L. and Jukes, T. H. (1969). *Science* **164**, 788–798.

Kingsbury, N. and Masters, C. J. (1972). *Biochem. biophys. Acta* **248**, 455–465.

Koehn, R. K. (1969). *Science* **163**, 943–944.

Koehn, R. K., Perez, J. E. and Merritt, R. B. (1971). *Am. Nat.* **105**, 51–69.

Krauel, K. K. and Ridgway, G. J. (1963). *Int. Archs Allergy appl. Immum.* **23**, 246–253.

Kristjansson, F. K. (1963). *Genetics* **48**, 1059–1063.

Lewontin, R. C. and Hubby, J. L. (1966). *Genetics* **54**, 595–609.

Ligny, W. de (1969). *Oceanogr. mar. Biol. Ann. Rev.* **7**, 411–513.

Ligny, W. de (1972). *In* "XIIth European Conference on Animal Blood Groups and Biochemical Polymorphism," pp. 55–65. Dr. W. Junk N.V., The Hague.

Markert, C. L. and Faulhaber, I. (1965). *J. exp. Zool.* **159**, 319–332.

Merritt, R. B. (1972). *Am. Nat.* **106**, 173–184.

Morrison, W. J. (1970). *Trans. Am. Fish. Soc.* **99**, 193–206.

Nei, M. (1971). *Am. Nat.* **105**, 385–398.

Ohno, S. (1970*a*). *Trans. Am. Fish. Soc.* **99**, 120–130.

Ohno, S. (1970*b*). "Evolution by gene duplication", 160 pp. Springer-Verlag, New York.

Ohno, S., Muramoto, J., Kline, J. and Atkin, N. B. (1969). *In* "Chromosomes Today." (Darlington, C. D. and Lewis, K. R., Eds.), Vol. 2, pp. 139–147. Oliver and Boyd, Edinburgh.

Prakash, S., Lewontin, R. C. and Hubby, J. L. (1969). *Genetics* **61**, 841–858.

Ridgway, G. J., Sherburn, S. W. and Lewis, R. D. (1970). *Trans. Am. Fish. Soc.* **99**, 147–151.

Roberts, F. L., Wohnus, J. F. and Ohno, S. (1969). *Experientia* **25**, 1109–1110.

Robertson, F. W. (1972). *In* "XIIth European Conference on Animal Blood Groups and Biochemical Polymorphism." (Kovacs, G. and Papp, M., Eds.) pp. 41–54. Dr. W. Junk N.V., The Hague.

Shaw, C. R. (1964). *In* "Subunit Structure of Proteins, Biochemical and Genetic Aspects." pp. 117–130. Brookhaven Symp. Biol. No. 17. Brookhaven Nat. Lab., New York.

Shaw, C. R. and Prasad, R. (1970). *Biochem. Genet.* **4**, 297–320.

Smithies, O. (1955). *Biochem. J.* **61**, 629–641.

Tsuyuki, H. and Ronald, A. P. (1970). *J. Fish. Res. Bd Can.* **27**, 1325–1328.

Utter, F. M. (1969). *J. Fish. Res. Bd Can.* **26**, 3268–3271.

Utter, F. M. (1971). *Comp. Biochem. Physiol.* **39B**, 891–895.

Utter, F. M. (1972). *In* "The Stock Concept in Pacific Salmon." (Simon, R. C. and Larkin, P. A., Eds.) pp. 191–197. H. R. MacMillan Lectures in Fisheries, Vancouver, B.C.

Utter, F. M., Allendorf, F. W. and Hodgins, H. O. (1973). *Syst. Zool.* **22**, 257–270.

Utter, F. M., Ames, W. E. and Hodgins, H. O. (1970). *J. Fish. Res. Bd. Can.* **72**, 2371–2373.

Utter, F. M. and Hodgins, H. O. (1969). *J. exp. Zool.* **172**, 59–67.

Utter, F. M. and Hodgins, H. O. (1970). *Comp. Biochem. Physiol.* **36**, 195–199.

Utter, F. M. and Hodgins, H. O. (1971). *Cons. Perma. Int. Explor. Mer., Rapp. Proc.-Verb. Re'un* **161**, 87–89.

Utter, F. M. and Hodgins, H. O. (1972). *Trans. Am. Fish. Soc.* **101**, 494–502.

Utter, F. M., Hodgins, H. O., Allendorf, F. W., Johnson, A. G. and Mighell, J. (1973). *In* "Genetics and Mutagenesis of Fish" (Schröder, J. H. Ed.), pp. 329–339. Springer-Verlag, Berlin.

Utter, F. M., Hodgins, H. O. and Johnson, A. G. (1972). *Int. N. Pac. Fish. Comm. Anna. Rep.* **1970**, 98–101.

Utter, F. M., Stormont, C. N. and Hodgins, H. O. (1970). *Anim. Blood Grps. Biochem. Genet.* **1**, 69–82.

Watson, J. E. and Crick, F. H. C. (1953). *Nature, Lond.* **171**, 964–967.

Wheat, T. E., Whitt, G. S. and Childers, W. F. (1973). *Genetics* **74**, 343–350.

Whitt, G. S., Childers, W. F., Tranquilli, J. and Champion, M. (1973). *Biochem. Genet.* **8**, 55–72.

Whitt, G. S., Cho, P. L. and Childers, W. F. (1972). *J. exp. Zool.* **179**, 271–282.

Williams, C. A. and Chase, M. W. (1971). "Methods in immunology and immunochemistry." Vol. III, Academic Press, New York.

Williams, G. C., Koehn, R. K. and Mitton, J. B. (1973). *Evolution* **27**, 192–204.

Williscroft, S. N. and Tsuyuki, H. (1970). *J. Fish. Res. Bd Can.* **27**, 1563–1567.

Wilkins, N. P. (1972). *J. Fish. Biol.* **4**, 487–504.

Wolf, U., Engel, W. and Faust, J. (1970). *Humangenetik* **9**, 150–156.

Wright, J. E. (1955). *Prog. Fish. Cult.* **17**, 172.

Wright, J. E. and Atherton, L. M. (1970). *Trans. Am. Fish. Soc.* **99**, 179–192.

Yoshida, A. (1973). *Science* **179**, 532–537.

Biochemical Aspects of Fish Swimming

E. BILINSKI

Fisheries Research Board of Canada
Vancouver Laboratory
Vancouver, B.C.
Canada

I. Introduction

Although the fundamental biochemical processes taking place during muscular work are essentially similar in fish and in higher vertebrates, some of the metabolic aspects of fish locomotion present distinct features, which make fish swimming an important area of concern to students of comparative biochemistry. It is apparent that the swimming ability and biochemical organization of fish show a great degree of adaptation to various modes of life. Many fish are particularly well suited to furnish a short term, vigorous muscular effort essential for catching prey or escaping predators. On the other hand, some pelagic fish are known to swim continuously during their life time and the reproduction of certain species requires an extreme and prolonged swimming effort during the spawning migration. Because several of the main constituents of the animal body are denser than water, fish have developed different devices to overcome sinking, which are energetically more or less efficient under various modes of life in the aquatic environment. In this chapter selected topics related to the biochemical aspects of fish locomotion are reviewed with emphasis placed on the differences existing between fish and mammals in the energy supplying metabolism.

II. Energy Metabolism During Swimming
A. Pathways Generating Energy for Muscular Work

Before considering energy metabolism in fish, it might be in order to outline our present understanding of the energy supplying processes during the muscular effort of animals in general. There is a large body of evidence indicating that the hydrolysis of adenosine triphosphate (ATP) to adenosine diphosphate (ADP) and inorganic phosphate (Pi) serves as the immediate source of energy for muscular contraction (see Fig. 1). Muscular contraction itself is an anaerobic process not requiring O_2 within the tissue, but the generation of ATP by capture of part of the energy released by a stepwise degradation of organic compounds may take place under aerobic or anaerobic conditions. On a long-term basis the formation of ATP by aerobic processes is quantitatively much more important than that by anaerobic reactions. Molecular oxygen does not participate directly in many of the enzyme reactions involved in biological oxidations but it serves as a final acceptor in the electron transport system. This chain of reactions (respiratory chain) provides a means for reoxidation of the reduced forms of nicotinamide adenine dinucleotide (NADH) and nicotinamide adenine dinucleotide phosphate (NADPH) needed for the continued operation of catabolic pathways. Glycolysis, consisting of the conversion of glycogen and other sugar derivatives to lactic acid, is quantitatively the most important mechanism by which ATP is produced for muscular work under anaerobic conditions. The reduction of pyruvate to lactate in the final step of glycolysis leads to the oxidation of NADH to NAD required in the earlier stages of glycolysis. This reaction makes the operation of

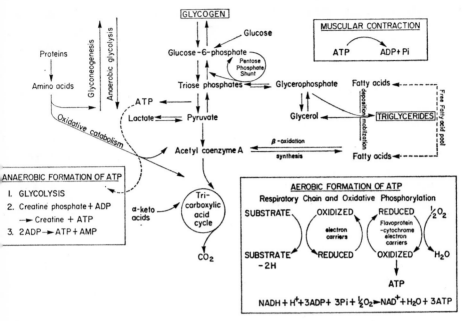

FIG. 1. Schematic representation of the major metabolic pathways related to the supply of energy for muscular contraction.

glycolysis independent of the supply of O_2. Under anaerobic conditions in muscle, ATP is also formed from creatine phosphate, which is the main form of storage of directly available energy in vertebrate muscle, and by a reaction between two molecules of ADP mediated by the enzyme myokinase. Aerobic metabolism (β-oxidation of fatty acids, oxidative decarboxylation of pyruvate, the tricarboxylic acid cycle, etc.), operating in conjunction with the terminal respiratory chain, provides a means for generating ATP by complete oxidation of organic molecules to CO_2 and H_2O. The amount of ATP produced by aerobic pathways is considerably greater than by anaerobic metabolism, due to the high yields of ATP during oxidative phosphorylation occuring at the final reactions of the respiratory chain. For example, when glycogen is converted to lactic acid there is a net synthesis of 3 molecules of ATP for each 6-carbon unit of glycogen metabolized. A complete oxidation of a 6-carbon unit of glycogen generates 39 molecules of ATP, 34 of which are formed by oxidative phosphorylation. Comparable yields of ATP may be obtained when the degradation of glycogen involves the participation of the pentose phosphate shunt. On the other hand, a complete oxidation of palmitic acid, a 16-carbon fatty acid, yields 129 molecules of ATP.

Although the generation of ATP by anaerobic glycolysis appears as a comparatively inefficient process, the obvious physiological implication is that this mechanism allows the animal to furnish an additional muscular effort exceeding the capacity of the muscle for the production of energy by aerobic pathways. When a moderate muscular effort is initiated, the energy is first supplied to the muscle by anaerobic metabolism until the organism has increased its O_2 supply to meet the needs of aerobic metabolism. Subsequently a steady state level is reached in which an equilibrium is established between the work performed and the energy produced by aerobic metabolism. Under these circumstances the amount of O_2 consumed is about proportional to the amount of work accomplished. In vigorous effort exceeding the capacity of aerobic metabolism, ATP is generated by anaerobic glycolysis—the depletion of muscle glycogen is intensified but its degradation does not go beyond the formation of pyruvate which is reduced to lactate. Lactic acid accumulates in the muscle and diffuses into the bloodstream. The muscular effort sustained by anaerobic metabolism can continue for a considerable time until the glycogen supply is exhausted, or until the activity is decreased due to the harmful effects of an excessive accumulation of lactic acid. Following a muscular effort involving the generation of energy by anaerobic pathways, the animal pays the oxygen debt—despite the fact that muscular work is decreased or terminated, the O_2 consumption remains for some time at higher levels. This oxygen is consumed in processes restoring the metabolic conditions of the rested organism, which comprise the conversion of lactate to other metabolites and replenishment of muscle stores of ATP, creatine phosphate and glycogen.

When aerobic conditions prevail in the muscle, glycogen and long chain fatty acids are the two main sources of energy for muscular work. The ability of the animal to store carbohydrates in the form of liver and muscle glycogen is quite restrained when compared to the almost unlimited capacity to deposit triglycerides. The earlier concept that the energy for muscular contraction can be supplied only through the intermediacy of carbohydrates has had to be revised in light of the evidence accumulated during the past two decades showing that long chain fatty acids also play a major role as a source of energy. Indications are that the relative importance of these two substrates in supplying energy to the muscle through aerobic metabolism depends on the extent of exercise and the nutritional state of the animal. It has been shown in studies with mammals that fatty acids are the prevalent source of energy during prolonged periods of moderate excercise, whereas carbohydrates are used more extensively in the earlier stages of the effort and following absorption of food.

There is strong evidence indicating that fish utilize the same metabolic pathways as mammals to derive energy for muscular work from the degradation of organic molecules. The differences in the energy metabolism of fish and mammals appear to reflect adaptive changes in metabolic organization. Several recent reviews are available concerning intermediary metabolism in fish and its control (Tarr, 1969, 1972; Hochachka, 1969; Fry and Hochachka, 1970; Lee and Sinnhuber, 1972). The biochemical properties of fish muscle have been reviewed by Hamoir (1955), Dyer and Dingle (1961), Tomlinson and Geiger (1962), Jones (1962), Buttkus and Tomlinson (1966), Bilinski (1969), Connell (1970), Love (1970), Wittenberger (1972b), and Tsuyuki (1973). The energy requirements of fish have been reviewed by Phillips (1969) and Halver (1972). With reference to mammals, a collection of papers presented at a symposium on "Muscle Metabolism During Exercise" (Pernow and Saltin, 1971) gives recent developments in this field.

B. Utilization of Energy Reserves in Severe and Moderate Swimming Effort

In many investigations on biochemical changes in fish during swimming a distinction has been drawn between severe and exhaustive short-term swimming effort (for instance, chasing of fish for 5-10 min), and swimming at moderate speeds reflecting more the sustained cruising activity. Investigations on the swimming performance of fish showed that the speed is a function of the total body length (Brainbridge, 1958, 1962). Consequently the comparison between the swimming performance of various fish is best made in terms of body length (B.L.) per unit time. Fish seem capable of burst speeds equivalent to 10 B.L./s, salmonids and scombroids being among the fastest swimmers; 2 to 4 B.L./s can be maintained for long periods by most species (Blaxter, 1969).

The earlier studies on utilization of carbohydrates during swimming of fish were the subjects of several reviews (Drummond and Black, 1960; Black et al., 1961; Dean and Goodnight, 1964; Beamish, 1966; Drummond, 1967, 1971; Dando, 1969). Essentially, it has been established that the changes related to glycogen utilization during a strenuous muscular effort are basically similar in fish and mammals. There are, however, significant quantitative differences in the pattern of recovery, namely, in fish the metabolism of lactate and the replenishment of glycogen reserves takes place at a much slower rate than in mammals. Black found that a severe exercise of salmonids results in a very rapid depletion of muscle glycogen with a concurrent increase of lactate in the muscle and bloodstream. Pyruvate levels follow the same course as lactate but liver glycogen and blood glucose do not change significantly. During recovery after a severe swimming effort, muscle lactate begins to decline upon cessation of the activity but it does not return to the pre-exercise levels until the 6th to 8th hour of recovery. Blood lactate continues to rise for 2 to 4 h following cessation of the exercise and does not approach the pre-exercise level for 8 to 12 h (see Fig. 2) whereas muscle glycogen may remain at

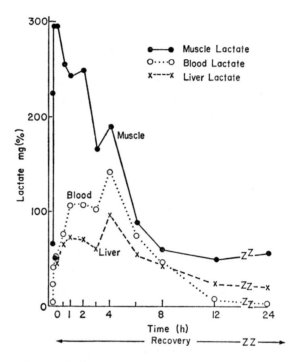

FIG. 2. Changes in muscle, blood and liver levels of lactate in $1\frac{1}{2}$ year old rainbow trout during 15 min of strenuous exercise and 24 h recovery. (Reproduced from Black *et al.*, 1962 with the permission of the *J. Fish. Res. Bd Can.*).

a decreased level for up to 24 h. In mammals the blood lactate may continue to rise for a short time following a severe muscular effort but it drops very rapidly thereafter. For instance, in man subjected to maximal bicycle exercise, muscle and blood lactate concentrations return to normal levels within 30 min of rest (Karlsson, 1971).

A rapid depletion of muscle glycogen during vigorous swimming and a slow rate of removal of lactate have been observed in a large number of species of fish. For example in haddock, *Melanogrammus aeglefinus*, subjected to a strenuous muscular exertion during capture by otter trawl, blood lactate reaches peak levels 1.5 to 4.5 h after capture and subsides to resting levels within 12 h (Beamish, 1966). In cod, *Gadus morhua*, blood lactate continues to increase for one hour after strenuous swimming and decreases in 16 h to levels similar to those of unexercised fish (Beamish, 1968). In yellowfin tuna, *Thunnus albacares*, and skipjack, *Katsuwonus pelamis*, which have a higher body temperature than the other species, blood lactate returns to normal levels already after 2 h of recovery (Barrett and Robertson-Connor, 1964). The latter observation supports the suggestion made by Black *et al.* (1961) that the slow rate of diffusion at low body temperature may be one of the main reasons for prolonged elevation of blood lactate levels during recovery of fish. Hochachka (1961) and Hammond and Hickman (1966) found that physical conditioning of rainbow trout, *Salmo gairdneri*, enables them to produce higher concentrations of lactate during extended periods of exercise and to remove it more rapidly during recovery. In a study conducted with several freshwater species Burton and Spehar (1971) provided conclusive evidence that short chain free fatty acids are not an end product of metabolism in fish exposed to hypoxia but, as in higher vertebrates, lactic acid is formed.

Hammond and Hickman (1966) observed that the exercise of trout resulted in oscillating fluctuations in inorganic phosphate levels in the muscle and blood plasma, and they suggested that this may reflect the operation of a feedback mechanism prone to overshoot. The rate of generation of ATP, depleted during muscular effort, is related to the concentrations of the two breakdown products, i.e. ADP and inorganic phosphate. The energy yielding processes show a tendency to overcompensate for the loss of ATP, resulting in a pronounced decrease in ADP and inorganic phosphate levels. The synthesis of ATP is resumed only when ADP and inorganic phosphate have again reached the critical level.

As far as the utilization of carbohydrates by fish during sustained swimming is concerned, the picture is less clear than in the case of a vigorous swimming effort. Black *et al.* (1961, 1962), working with trout, *S. gairdneri*, obtained inconclusive results. They found that at a moderate speed the depletion of muscle glycogen was small or negligible but blood lactate levels showed a detectable increase. Beamish (1968) found that in Atlantic cod, *G. morhua*, swimming at low speeds, muscle glycogen was notably reduced although blood lactate did not change appreciably from the unexercised level. At higher speeds lactate accumulated in the muscle and blood-stream and the depletion of muscle glycogen was more pronounced. These findings appear to suggest that, in cod, muscle glycogen is utilized at any swimming velocity but at moderate speeds it is presumably being subjected to a complete oxidation. Connor *et al.* (1964) examined the changes in glycogen and lactate levels in migrating salmonids during ascents of 1:16 and 1:8 gradients in experimental fishways. The fish tested were chinook salmon, *Oncorhynchus tshawytscha*, sockeye salmon, *O. nerka*, and steelhead trout, *S. gairdneri*, derived from their upstream migration in the Columbia River. During short ascents of the 1:16 fishway the muscle glycogen levels of all species tested were not significantly decreased, but in fish ascending the steeper (1:8) fishway the muscle glycogen content did show a decrease. When muscle glycogen was utilized during prolonged ascents the expenditure appeared to be progressive and to depend on the total amount of muscular exercise performed. The magnitude of the blood lactate increase was very similar to that observed by Black *et al.* (1961, 1962) in moderately exercised trout. In view of the lack of clear evidence for utilization of muscle glycogen in the mild form of exercise, Connor *et al.* (1964) have considered the possibility that the energy for this type of activity could be supplied from metabolism of protein and/or fat or other carbohydrate substrates.

In some studies the biochemical transformations during swimming of fish were measured separately in the predominant white muscle as well as in the red muscle. Fraser *et al.* (1966) found that in cod, *G. morhua*, forced to struggle until near exhaustion, the glycogen of the white muscle was significantly depleted but there was no appreciable change in the red muscle; lactate accumulated in red and white muscle suggesting partial activation of glycolysis in both tissues. Bone (1966), working with the dogfish, *Scyliorhinus canicula*, observed that after a period of vigorous movement the glycogen levels in the white muscle fell markedly while those in the red muscle did not change. After long periods of sustained slow swimming, the fat levels in the red muscle fell and there was also a slight decrease in glycogen levels in these tissues. Pritchard *et al.* (1971) studied the biochemical changes in jack mackerel, *Trachurus symmetricus*, after forced swimming at different speeds. Glycogen, lactate and total fat were determined separately in the red and white muscle; changes in liver glycogen were also measured. Mackerels were subjected to swimming at a speed at which 50% of the fish would fatigue during 6 h (sustained speed threshold, 7.7 B.L./s), as well as at subthreshold (6.7 B.L./s) and superthreshold speeds (9.5 B.L./s). In fish exercised for 8 min at superthreshold speed, glycogen decreased in the white muscle only, but lactate levels were considerably above the controls in both the white and red muscle. Fish that swam continuously for 6 h at threshold and subthreshold speeds showed a decrease in liver and red muscle glycogen, but there was no change in the glycogen content of the white muscle. There was a reduction in the fat content of the red muscle only when fish swam at subthreshold speed. Pritchard *et al.* (1971) have concluded that at sustained cruising speeds and at higher velocities, the energy needed for the swimming of mackerel is derived primarily from glycogen and that white muscle is the principal locomotor organ; at lower speeds the utilization of lipids is apparent in the red muscle which probably also has a locomotor role.

Johnston and Goldspink (1973*a*) measured glycogen and lactate levels in the red and white skeletal muscle and liver of coalfish, *G. virens*, following swimming at various velocities. The sustained swimming speed (50% fatigue level after 6 h) amounted to 4 B.L./s. At speeds in excess of 2 B.L./sec a significant increase in lactate concentration occurred in the white muscle (see Table 1). Glycogen levels of the white muscle showed a decrease only at higher speeds (3.1 B.L./s and above). In the red muscle glycogen was also depleted at lower speeds. In liver depletion of glycogen and elevation in lactate levels took place at higher swimming speeds. Johnston and Goldspink (1973*a*) have concluded that the red muscle alone is used by the coalfish at speeds below 2 B.L./s; at higher swimming speeds the extra work imposed on the fish requires the deployment of an increasing number of white fibres.

Similar observations were made by Johnston and Goldspink (1973*b*) in a study of the swimming performance of the Crucian carp, *Carassius carassius*. They found that during the recovery of carp after sustained swimming, muscle lactate and glycogen return much faster to the pre-exercise level in the red than in the white muscle. It might be pointed out that these differences are consistent with the greater ability of the red muscle to oxidize lactate and synthetize glycogen (see Section III.C.)

Krueger *et al.* (1968) found that 16:0 and 18:1 fatty acids were most extensively utilized in juvenile coho salmon, *O. kisutch*, forced to swim for 24 h at 52 cm/s. When fish were swimming at water velocity greater than that which could be sustained (59 cm/sec), 20:5 and 20:6 fatty acids were depleted. Saddler and Cardwell (1971) observed that tagging of juvenile pink salmon, *O. gorbuscha*, resulted in extensive mobilization of specific fatty acids from muscle to liver. These fatty acids were primarily monosaturated and saturated types.

C. Respiratory Changes

The energy metabolism of an animal during muscular work is reflected, among other things, by the evolution of the respiratory gases. Muscular activity is manifested by an increased rate of O_2 consumption and CO_2 production. As mentioned above (see Section II.A.), in muscular effort at

TABLE 1: The concentrations (mg/100g tissue) of glycogen and lactate in the red and white myotomal muscles and the liver of non-exercised and exercised coalfish (*Gadus virens*)

Forced Swimming speed (Bodylengths/s)	No. of fish	Red muscle (*musculus lateralis superficialis*) (Mean ±S.E.)		White muscle (*m. lateralis profundis*) (Mean ±S.E.)		Liver (Mean ±S.E.)	
		Glycogen	Lactate	Glycogen	Lactate	Glycogen	Lactate
0	9	1276±92	97·6±10·8	275±47	86·2±12·8	3609±384	35·6±3·6
1·58	10	1102±202	129·2±13·0	246±45	78·0±16·0	3094±429	37·0±6·2
2·37	8	831±173	114·0±11·4	249±48	194·0±17·8	2586±302	48·8±15·8
3·14	6	272±50	146·0±39·4	145±39	177·6±9·6	1850±296	73·6±11·2
4·62	10	164±23	75·0±7·0	131±24	239·8±45·2	1204±219	113·2±10·0

(Reproduced with permission of Johnston and Goldspink, 1973*a*).

a steady state level the volume of O_2 consumed is proportional to the amount of work performed. The rate of O_2 consumption taken as a general measure of energy expenditure of an animal is called the metabolic rate. In fish, swimming may elevate the metabolic rate by as much as 15 times (Brett, 1970). The metabolic rate of fish has been studied extensively at various degrees of swimming effort in relation to different factors, such as temperature, O_2 and CO_2 tension, salinity, size of fish, etc. These physiological aspects of the energetics of swimming of fish are the subject of several reviews (Fry, 1957; Fry and Hochachka, 1970; Brett, 1962, 1970, 1972; Beamish and Dickie, 1967; Randall, 1970; Doudoroff and Shumway, 1970). Brett (1972) compared the metabolic demand for O_2 in fish and other vertebrates. The maximum rate of O_2 consumption among fish varies at least five-fold between different species with a probable ceiling of about 700 cm³ O_2/kg/h. Small mammals and birds have a capacity for continuous supply of O_2 which is 10 to 100 times that of fish, amphibians and reptiles. Brett (1972) points out that, to compensate for the comparatively limited capacity for O_2 consumption, fish have developed an ability to tolerate a greater oxygen debt. The rate of recovery following the anaerobic effort is, however, remarkably slow in fish. For instance, Heath and Pritchard (1962) found that, immediately following severe muscular activity, bluegill sunfish, *Lepomis macrochirus*, showed a 65% increase of O_2 consumption over the resting rate, followed after one hour by a gradual decline, the pre-exercise level being reached between 10 and 24 h. This pattern of recovery contrasts markedly with that in mammals where severe exercise may be followed by a ten-fold or more increase in O_2 consumption which then falls rapidly, reaching the pre-exercise level in 15-30 min. Schmidt-Nielsen (1972) used metabolic rates to calculate the energy cost of locomotion of different forms of animals and

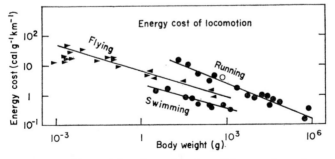

FIG. 3. Energy cost of locomotion for swimming, flying, and running animals in relation to body size (Reproduced with permission of Schmidt-Nielsen (1972), *Science* 177. 222. Copyright 1972 by the American Association for the Advancement of Science).

found that fish swimming require comparatively less energy expenditure than birds and insects flying, or terrestrial animals running (see Fig. 3).

With reference to the scope of the present review, the study of Smit *et al.* (1971) on O_2 consumption and efficiency of swimming in the goldfish, *Carassius auratus*, is of particular interest since it provides an indication of the extent of aerobic and anaerobic metabolism. Smit *et al.* have determined the efficiency of swimming by calculating the ratio of the power generated by swimming over the O_2 uptake. Essentially, a high efficiency value means that, for the same amount of swimming effort the fish have consumed lower amounts of O_2. In agreement with the concept that at

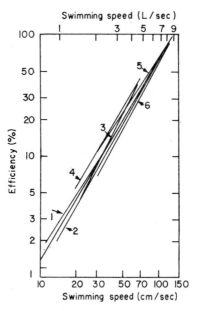

FIG. 4. Relation between speed and efficiency of swimming goldfish. The heavy line is the mean of the six thin lines and represents 618 measurements of oxygen consumption rate. (Reproduced after Smit *et al.*, 1971 with permission of Microfilm International Marketing Corporation Exclusive Copyright Licensee of Pergamon Press Journal Back Files).

higher swimming speeds fish draw more energy from anaerobic metabolism, Smit *et al.* found that the efficiency increases with swimming speed. Furthermore, they observed a straight line relationship between efficiency and swimming velocity, suggesting a very progressive change from aerobic to more anaerobic metabolism in the muscular tissues (see Fig. 4).

Indications on the types of substrate utilized by the animal during muscular effort may be gained by measuring concurrently the evolution of O_2 and CO_2. The respiratory quotient (R.Q.), representing the ratio of the volume of CO_2 expired to the volume of O_2 consumed, depends on the gross chemical composition of the material being metabolized. When glycogen or glucose are converted to CO_2 the R.Q. amounts to 1.0; triglycerides, being more highly reduced compounds than carbohydrates, have an R.Q. of approximately 0.7. This picture is, however, distorted when the muscular effort reaches a significant level of anaerobic glycolysis since the liberation of CO_2 increases due to the action of lactic acid upon sodium bicarbonate from body fluids. When, subsequently, the effort decreases, some of the consumed oxygen is used to repay the oxygen debt and part of the CO_2 formed during the metabolism serves to replenish the bicarbonate stores of the body. Kutty (1968) made a study of the R.Q. in goldfish, *C. auratus*, and rainbow trout, *S. gairdneri*, and he reviewed the earlier investigations conducted in this field. He indicates that the limitations of the techniques presently available for determination of CO_2 output of aquatic organisms restrict valid observations on the R.Q. in fish to great changes such as those occurring under anaerobiosis. Kutty found, for instance, that the R.Q. remained at 1.02 in goldfish spontaneously active in ambient O_2 above 50% air saturation, and rose to 1.94 below 25% air saturation. Rainbow trout under similar conditions displayed an R.Q. of 0.96 and 1.40. The latter R.Q. could be sustained for only a short time. Using a modified technique for measuring CO_2 evolution, Kutty (1972) showed that, in *Tilapia mossambica* exercised for 6 h at a swimming speed of 2 B.L./s, the R.Q. remained at a high value of 1.2, implying a continuous utilization of energy through anaerobic means. Kutty (1972) also found that the ammonia excretion showed a progressive increase with duration of exercise and he suggested that coupling of an increased protein metabolism and anaerobic energy utilization may be of advantage in preventing acidosis and conserving sodium ions in fish. According to Nagai and Ikeda (1972) proteins are the primary source of energy for the carp, *Cyprinus carpio*, which appears to show a pattern of energy utilization comparable to that found in diabetic mammals. This conclusion is based on studies of the effects of diet on glucose and glutamate metabolism in the liver of carp. Mercy Bai (1970) also expressed the view that proteins are more important as an energy source than either carbohydrates or fat in the fish, *Ophiocephalus punctatus*, exercised for 1 to 6 h. In mammals the breakdown of proteins is of little significance in supplying the energy for muscular work performed under normal nutritional conditions.

D. Fuel for Spawning Migration

During the spawning migration of fish, the energy demands may be so great that, in addition to carbohydrates and triglycerides, tissue proteins are utilized to a great extent. This is very apparent in the migration of salmon. The spawning migration of salmonids covers the upstream journey from the mouth of the river to the spawning grounds and it may amount to as much as 1000 miles. Since salmon do not feed during this period they must rely entirely on body constituents to provide the energy for swimming and to supply the material for the formation of gonads. The representatives of the five species of Pacific salmon (*Oncorhynchus tshawytscha*, *O. nerka*, *O. kisutch*, *O. gorbuscha*, and *O. keta*) die after the first spawning, whereas the Atlantic salmon, *Salmo salar*, and the sea-runs of trout, *S. gairdneri*, return to the sea and resume feeding.

The pronounced depletion of fat reserves and muscle protein during spawning migration of salmon has long been well established. The review of Greene (Greene, 1926) covers the earlier investigations in this field whereas Idler and Tsuyuki (1958) and Drummond and Black (1960) have reviewed the more recent studies. Greene (1926) measured the chemical changes in king salmon, *O. tshawytscha*, migrating in the basins of the Sacramento and Columbia Rivers and found a 30% loss of muscle proteins in spawned fish. The neutral fat content of the muscle was reduced from 20% at the beginning of migration to 1–2% at spawning. It is noteworthy that Greene made farsighted observations by pointing out that this picture is incompatible with the theory then generally accepted that the energy for muscular work can be supplied only through degradation of carbohydrates. Idler and associates (Idler and Tsuyuki, 1958; Idler and Bitners, 1958, 1959, 1960; Idler and Clemens, 1959) made extensive studies of energy expenditure and biochemical changes during spawning migration of sockeye salmon, *O. nerka*, in the basin of the Fraser River. They showed that during the first 250 miles of migration fish withdrew 30–40% more fat than protein from the muscle, but during the remaining 465 miles they used about twice as much muscle protein as fat. The overall effect of spawning migration was that on a weight basis fish consumed about similar amounts of fat and protein from the muscle, but significant amounts of fat were also depleted from other parts of the body whereas the muscle was the only main source of protein. The total consumption of females was about 30% greater than that of males due to greater demands for gonad formation.

Despite the fact that salmon has quite limited reserves of glycogen, carbohydrate metabolism remains as one of the major pathways supplying energy during the spawning migration. Fontaine and Hatey (1952) found that in Atlantic salmon, *S. salar*, the blood glucose shows similar levels in fish on the spawning grounds as in fish starting migration, and they concluded that salmon are very well adapted to regulate the formation of carbohydrates from other body constituents. Jonas and MacLeod (1960) observed only a slight decrease in blood glucose levels following a 715 mile spawning migration of sockeye salmon, *O. nerka*, in the Fraser River. Chang and Idler (1960) found that the liver glycogen levels decreased during the earlier phase of spawning migration of sockeye salmon, *O. nerka*, but increased during the later stage so that the levels at the spawning grounds were approximately twice those at the mouth of the river. The rise in liver glycogen was coincident with a change from predominantly fat metabolism to increased protein utilization at the later stages of migration. Although it is known that proteins serve to replenish liver glycogen (glyconeogenesis) in fasted mammals, the glycogen concentration is usually far below that encountered under normal physiological conditions. Chang and Idler suggested that the significant rise in liver glycogen in fish approaching the spawning grounds might be related to the appearance of elevated cortical steroid levels which would stimulate glyconeogenesis. Butler (1968) showed that in the North American eel, *Anguilla rostrata*, plasma hydrocortisone is involved in regulating glyconeogenesis. It might be pointed out that, in addition to the conversion of amino acids to glycogen, other pathways might be involved in the utilization of muscle proteins. For instance, the complete oxidation of carbon skeletons such as pyruvate, oxalacetate and α-ketoglutarate, as well as acetyl coenzyme A, formed during the degradation of some amino acids, provides a direct route for the generation of energy from proteins but the extent to which such pathways might be operating in migrating salmon is not known. It has been occasionally suggested that fat reserves may serve in fish to replenish the supply of carbohydrates. It must be stressed, however, that this is an unlikely possibility, as it would involve the glyoxalate pathway which has not so far been found in fish. Glycerol is the only moiety of triglycerides which can readily participate in the biosynthesis of carbohydrates and there is strong evidence that animals generally have no ability to form glycogen from long chain fatty acids (Legge, 1971).

Several recent investigations are relevant to the utilization of fat reserves by salmon during spawning migration. Members of the R/V "Alpha Helix" expedition of the University of California (Lobanov-Rostovsky, 1968) made comparative studies on pink salmon, *O. gorbuscha*, caught along the Pacific coast of Canada before starting migration and on fish collected on spawning grounds after a 60 mile freshwater migration. Robinson and Mead (1970) showed that the hump developed by the male pink salmon is largely composed of lipids which are progressively depleted and replaced by water during the spawning migration. The initial high proportion of triglycerides in hump lipids decreases during migration, and there is a relative increase of cholesterol esters and free fatty acids. Patton *et al.* (1970) found that the migration of pink salmon resulted in an approximately 60% decrease in the concentration of total lipids in blood serum, due largely to depressed levels of phospholipids and triglycerides. Phleger (1971) observed that both hepatic lipogenesis and cholesterol synthesis were decreased in pink salmon caught on the spawning grounds. Trams (1969) found that, at the end of spawning migration, pink salmon showed a decreased ability to oxidize free fatty acids *in vivo*. Since the migration of pink salmon covered a comparatively short distance, it seems likely that the abnormalities in lipid metabolism observed in the above studies were due more to physiological changes during the spawning period than to an exhaustion of body reserves.

Nakai *et al.* (1970a, b, c) studied changes in the activity levels of enzymes involved in energy metabolism in the muscle and liver of kokanee salmon, *O. nerka*, before and during spawning, and found that glycolytic metabolism was largely superceded by fat metabolism during the spawning period. In a study on the fatty acid composition of the muscle lipids of chinook salmon, *O. tshawytscha*, and coho salmon, *O. kisutch*, Iverson (1972) obtained indications that the long chain monounsaturated fatty acids are preferentially utilized by fish during spawning migration.

III. Functional Differentiation of the Skeletal Muscle

A. Contractile Properties

The dissimilar functions performed in the animal by various muscle types are reflected in the differentiation of metabolic and contractile properties of muscle fibers: the fibers which are able to give sustained contractions are well adapted to meet energy demands through aerobic processes, whereas those capable of providing intense but short-term contractions rely more on anaerobic glycolysis for the supply of energy. The red (slow) and white (fast) fibers of the skeletal muscle are representatives of these two types. Heart tissues and other types of red muscle show the general properties of the first category. The existence of intermediary types of fibers, is however, well established in the skeletal muscle and other tissues. In animals where the red and white fibers of the skeletal musculature are well confined to separate tissues (for instance, rabbit, some birds and some fish, frogs, etc.), the red muscle has been implicated in slow and sustained motion of the animal and the white in rapid movements.

The structural and contractile properties of red and white fibers from fish skeletal muscle show many similarities to the slow and fast fibers from mammals (Boddeke *et al.*, 1959; Barets, 1961; Andersen *et al.*, 1963; Franzini-Armstrong and Porter, 1964; Bone, 1966; Bishop and Odense, 1967; Nishihara, 1967; Nakajima, 1969; Waterman, 1969; Nag, 1972*a*, *b*). The organization of the red and white fibers within the musculature of fish shows a certain degree of diversity which has been related to the mode of life of fish (Boddeke *et al.*, 1959): the red muscle appears to occur in larger amounts in species subjected to more continuous swimming ("stayers"), whereas a large mass of white muscle common to most species of fish is a characteristic of "sprinters". In fish the red fibers are often confined to the musculature along the lateral line but in some species a deep-seated red muscle is also encountered. In the following discussion of the enzyme activity in the red and white muscle of fish, the terms "red" (or dark) and "white" (or light) are used to describe tissues composed mainly of one of the two types of fiber, but they do not imply that the preparations used were homogenous from a histological point of view.

There is at present a good understanding of the relationship existing between the shortening velocity of the muscle and the enzyme activity of the contractile proteins (Barany, 1967; Katz, 1967, 1970). If the rate of conversion of ATP to ADP by muscle preparations *in vitro* reflects the maximum rate of energy liberation by the intact muscle, one would expect a correlation between the ATPase activity of the contractile proteins of a given muscle and the maximum shortening velocity of the muscle. Such a relationship has indeed been established; in mammals the maximum shortening velocity of the white muscle is about three times greater than that of the red skeletal muscle and this ratio is similar to the ratio of myosin ATPase activities in the two types of muscle. Katz (1970) pointed out the physiological significance of the low ATPase activity which limits the contractile velocity of the cardiac and red skeletal muscles. Because the contractile activity of these muscles is generally carried out without prior recovery, it is essential that the rate of ATP utilization be equal to the rate of ATP production. This attenuation of one aspect of functional capacity is the price paid for the ability for sustained activity.

Differences in myofibrillar ATPase activity are also known to exist in the red and white muscle of fish. Syrovy *et al.* (1970) found that in carp, *Cyprinus carpio*, the Ca^{2+} independent ATPase activity of the myosin from the red skeletal muscle was about half that of the white muscle myosin. The activity of Ca^{2+} dependent ATPase was slightly higher in the myosin isolated from the red muscle but Syrovy *et al.* suggested that this was presumably due to the greater instability of the myosin from the white muscle. Johnston *et al.* (1972) measured the ATPase activities in fresh myofibril preparations from the red and white muscles of coalfish, *Gadus virens*, and cod, *Gadus morhua*. They found that in both species the activity of Ca^{2+} dependent ATPase of the red muscle amounted to about a quarter of that found in the white muscle. The activity of Ca^{2+} independent ATPase was also considerably lower in myofibril preparations from the red muscle. Using preparations from rainbow trout, *Salmo gairdneri*, Nag (1972a) found that the actomyosin of the white muscle fibers has an average of three times the ATPase activity of the red muscle fibers. These results, showing strong similarities between fish and mammals in the pattern of ATPase activity in the two types of myofibrils, provide an additional support to the view that the red and white muscle of fish correspond to the fast and slow fibers of mammals.

Direct evidence for the involvement of the red and white muscle of fish in different patterns of swimming was obtained in various types of experiments conducted with live animals. Bone (1966) found that during slow swimming of dogfish, *Scyliorhinus canicula*, only red muscle fibers generated an electrical potential. During vigorous swimming the white fibers were active but it was not possible to establish if the red fibers were also contracting. Rayner and Keenam (1967), working with skipjack tuna, *Katsuwonus pelamis*, detected an electrical current in the white muscle at high speeds only and they found that the red muscle was active at any speed. Greer Walker (1971) measured the hypertrophy of the red and white muscle fibers of coalfish, *Gadus virens*, to determine their involvement in swimming at various velocities. In fish swimming for 42 days at a sustained speed of 0.93 B.L./s, the red fibers hypertrophied by 55% compared with the controls, whereas the white fibers showed no significant change. Greer Walker indicates that this result does not imply that the white fibers were not used but only that they were used no more than in the control fish. At the speed of 2.0 B.L./s the hypertrophy of the red fibers increased to 61% and that of the white fibers to 32%. At 3.0 B.L./s the red fibers hypertrophied by 68% and the white by 60% (Greer Walker and Pull, 1973). These authors concluded that the white fibers are increasingly active at higher sustained speeds whereas the red fibers become progressively less active. Smit *et al.* (1971), using O_2 consumption to determine the swimming efficiency of goldfish, *Carassius auratus* (see Section II.B.), have reached the conclusion that the red and white muscle fibers are simultaneously active at all swimming velocities, but at low speeds a high percentage of active muscle fibers are red whereas at high speeds they form only a small percentage of the total number of active fibers. Studies on biochemical changes in the red and white muscle of fish during swimming provide further support for the concept that the effort produced at higher speeds involves a greater participation of the white fibers (see Section II.B.).

B. Terminal Oxidation

Terminal oxidation, involving the operation of the tricarboxylic acid cycle and the respiratory chain, represents pathways common to carbohydrates, fats and amino acids. At this stage of catabolism the intermediate products of degradation, consisting of acetyl coenzyme A, oxalacetate or α-keto-glutarate, enter the tricarboxylic acid cycle to be converted to CO_2 (see Fig. 1). The reduced coenzymes formed in the operation of this cycle and in other pathways are oxidized by the respiratory chain, leading to formation of ATP by oxidative phosphorylation. The red muscle, showing a high capacity for carrying on these processes, is able to furnish a continuous effort by working at a steady state level, i.e. the utilization of ATP is compensated by its formation though oxidative phosphorylation. The high capacity of the red muscle for aerobic metabolism results from a combined effect of several factors, namely, (1) an abundant supply to the muscle fibers of blood carrying O_2, (2) a high concentration of myoglobin which functions as an O_2 reserve and may compensate for short periods of insufficient O_2 supply, (3) an abundance of mitochondria which are the site of aerobic metabolism within the cell. Lawrie (1953a,b) demonstrated a close correlation between the activity of succinic dehydrogenase, cytochrome oxidase and amounts of myoglobin in different types of horse muscle, and he observed that the myoglobin concentration in the muscle is directly related to the capacity of the muscle to synthesize ATP aerobically. It is also known from studies with mammals that the number of capillaries around muscle fibers is directly proportional to the oxidative metabolism of the fiber (Romanul, 1971). There is a large body of evidence to suggest that the red muscle of fish is similar in these respects to that from mammals. The investigations conducted in this field have been the subject of several reviews (Hamoir, 1955; Braekkan, 1959; Bone, 1966; Bilinski, 1969; Love, 1970; Wittenberger, 1972b).

In recent studies Malessa (1969), working with eel, *Anguilla vulgaris*, found that the activities of cytochrome oxidase and succinate dehydrogenase were, respectively, 6 to 10 and 2 to 3 times higher in the red than in the white muscle. In the musculature posterior to the anus the red muscle contains over 60% of the total cytochrome oxidase activity even though it makes up only 11 to 15% of the total muscle. Malessa suggested that an increase in aerobic metabolism in the red muscle is of importance during winter migration of eels. He found that, following cold acclimation of eels, the aerobic metabolism was greatly enhanced in the red muscle as indicated by an extremly high cytochrome oxidase activity concurrent with a relative increase in the amount of the red muscle.

Boström and Johansson (1972) have studied the activities of enzymes in various metabolic pathways in the white and red muscles of eel, *A. anguilla* (see Table II). They measured the enzyme levels at three developmental stages of the eel associated with different manners of living: glass eel (pelagic, actively swimming), yellow eel (bottom dwelling, sluggish), and silver eel (migratory, actively swimming). The citric acid cycle enzymes (fumarase and malate dehydrogenase) and cytochrome oxidase showed much higher levels of activity in the red than in the white muscle. Both types of muscle in silver eel had twice the activity of the corresponding muscle in the yellow eel, indicating that a high aerobic capacity of the muscle is related to a more active mode of life. These findings are in agreement with those of Fukuda (1958) who found that pelagic fish show a greater succinate dehydrogenase activity in the muscle than non-pelagic species. Boström and Johansson (1972) point out that the level of activity of oxidative enzymes in red and white muscle of silver eel, compared with those of the yellow eel, resemble the differences between a trained and untrained muscle. It is known that, in mammals, training results in an increased content and oxidative activity of muscle mitochondria (Hoch, 1971).

Gordon (1972a, b) compared the oxygen uptake rates by red and white muscle preparations from ten species of shallow water, marine fish. The ratio between oxygen consumption by the red and by the white muscles generally remained within the range 2 to 7, showing differences between species and variations within the same species. Certain ecological and physiological factors contribute significantly to a higher oxygen uptake (cool environmental temperature, epipelagic environment, high activity, etc.). Gordon found that the oxygen uptake rates (at 15–25°C) by the red muscle from a number of epipelagic species are either comparable or substantially higher than those measured at 35°C for similar preparations of the thigh muscle of the white rat, and he suggested that this feature may be largely responsible for the great capacity for sustained high speed swimming shown by most epipelagic fish.

C. Carbohydrate Metabolism

There are several reports indicating that, in fish, glycogen is present in larger amounts in the red than in the white muscle (Buttkus, 1963; Bone, 1966; Wittenberger, 1968, 1972a; Pritchard *et al.*, 1971; Johnston and Goldspink, 1973a, b). Bokdawala and George (1967a), using histochemical techniques, found that in carp, *Cirrhina mrigala*, uridine diphosphoglucose glycogen synthetase occurs at higher concentrations in the red than in the white muscle. Ingram (1970) made a study of uridine diphosphoglucose glycogen synthetase in rainbow trout, *Salmo gairdneri*, and showed that the soluble enzyme was about 15 times more active in the red

TABLE II. Activities of enzymes in various metabolic pathways in body muscle of the glass eel and white and red muscle of the yellow and silver eel.

Enzymes	Glass eel	Yellow eel		Silver eel	
	Body muscle	White muscle	Red muscle	White muscle	Red muscle
Hexokinase	4·93±0·19	40·5±2·0	34·2±5·1	26·2±0·7	44·3±0·7
Glucose 6-phosphate dehydrogenase	4·26±0·32	1·05±0·13	4·72±0·77	1·86±0·28	8·69±0·28
6-Phosphogluconate dehydrogenase	6·17±0·22	2·38±0·003	8·90±0·65	4·90±0·31	16·9±2·20
Glycerolphospate dehydrogenase	471±22	423±6	423±34	721±114	218±30
Pyruvate kinase	3260±232	3260±310	3900±570	3000±73	1060±500
Lactate dehydrogenase	5920±180	16,800±1840	14,700±600	16,480±950	8260±1000
Lactate dehydrogenase isoenzyme ratio	0·52±0·01	0·59±0·02	0·80±0·03	0·68±0·01	1·00±0·01
Succinate dehydrogenase		31±4		58±5	
Fumarase	14·7±0·7	6·2±0·5	66·2±5·1	17·8±1·3	128±8·0
Malate dehydrogenase	2680±60	1500±70	11,900±1200	2550±150	18,500±900
Cytochrome oxidase	8210±838	372±131	11,700±886	10,400±99	49,400±1750

Enzyme activities are expressed as mU/mg protein. The results are given as means ±S.E.M. of ten animals. The SDH activity was measured in the mitochondrial fraction. (Reproduced with permission of Bostrom and Johansson, 1972).

muscle than in the white. The activity of the liver was more comparable to the red muscle than to the white. The particulate enzyme, representing less than 20% of the total activity, was equally active in both red and white muscle but was about 30 times more active than in the liver preparations. Wittenberger (1972a) injected [^{14}C]glucose into the red muscle of mackerel, *Scomber scombrus*, and found that in this tissue the radioactivity was incorporated into glycogen at a rate about 40 times higher than in the white muscle or in the liver. Yamamoto (1968) purified glycogen phosphorylase from the muscle of rainbow trout, *S. gairdneri*, and studied its kinetic properties. He found that the total activity of phosphorylase (a + b) was more than twice as high in the white as in the red muscle.

Using tissue homogenates from dogfish, *Scyliorhinus canicula*, silver eel, *Anguilla anguilla* and trout, *S. gairdneri*, Crabtree and Newsholme (1972) showed that, as in mammals, glycogen phosphorylase has a higher level of activity in the white than in the red muscle of fish. They found also that in fish, as in mammals, hexokinase has a greater activity in the red muscle, suggesting that free glucose might be a more important source of energy in the red than in the white muscle. The study of Crabtree and Newsholme (1972) on the activities of cytoplasmic and mitochondrial glycerol 3 – phosphate dehydrogenases provides no support for the proposition that the glycerol 3 – phosphate cycle is a quantitatively important pathway in the energy metabolism of fish muscle. This cycle is of major significance in insect flight muscle, where it facilitates the reoxidation of cytoplasmic NADH formed during glycolysis. It would appear that in the muscle of fish and higher vertebrates the reoxidation of NADH is achieved mainly by the conversion of pyruvate to lactate.

Hamoir *et al.* (1972) studied the occurrence of glycolytic enzymes in the sarcoplasmic proteins from the white muscle and from different types of red muscle of the mirror carp, *Cyprinus carpio*. The proteins were separated by starch-gel electrophoresis. Glycolytic enzymes were apparent in preparations from the white as well as the red muscles but some of the bands were more intensively stained in the former (phosphoglucose isomerase, aldolase and glycerylaldehyde phosphate dehydrogenase). The sarcoplasmic protein pattern of the cardiac muscle differed to some extent from that of the red skeletal muscle by having a higher content of myoglobin, and by lacking some low molecular weight proteins devoid of glycolytic activity. In a study on eels referred to earlier (see Section III.B.), Boström and Johansson (1972) found that pyruvate kinase, used as the indicator of glycolytic activity, showed somewhat higher levels in the red than in the white muscle of yellow eel, but in silver eel the activity in the red muscle amounted to only one-third of that in white muscle. (See Table II, p. 259).

The differentiation in carbohydrate metabolism between the red and white muscle is well apparent at the final stage of anaerobic glycolysis involving lactate-pyruvate interconversions. Lactate dehydrogenase (LDH), mediating this transformation, occurs in vertebrates in different isoenzyme forms which appear to be specifically adapted to the anaerobic or aerobic type of metabolism by showing a different degree of inhibition by pyruvate (Dawson *et al.*, 1964). In the white muscle there is a need for sporadic release of large amounts of energy in the relative absence of oxygen. This energy is supplied by glycolysis which produces large amounts of pyruvate and requires its reduction to lactate. The molecular form of LDH found in the white muscle maintains activity at relatively high pyruvate concentrations. On the other hand, the LDH of the aerobic (heart) type is strongly inhibited by an excess of pyruvate. In this way the pyruvate formed in heart by anaerobic glycolysis, rather than being converted to lactate, may be efficiently utilized as a source of energy through complete oxidation in mitochondria. LDH has been studied extensively in the muscle and other tissues of fish (see reference cited by Hamoir *et al.*, 1972). As in higher vertebrates, the red skeletal muscle of fish exhibits an LDH isoenzyme pattern comparable to heart muscle, although minor differences appear to exist between these two aerobic muscles (Hamoir *et al.*, 1972).

Bilinski and Jonas (1972) made a comparative study on *in vitro* oxidation of lactate to CO_2 by various tissues of rainbow trout, *S. gairdneri*. Lactate, labelled specifically in C-1 or C-3, was used to distinguish between two different stages in the degradation of lactate (see Fig. 5). The

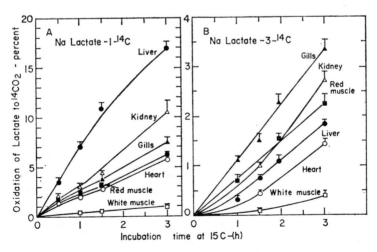

FIG. 5. Oxidation of lactate (5mM) to CO_2 by rainbow trout tissues. (Reproduced after Bilinski and Jonas, 1972 with permission of the *J. Fish. Res. Bd Can.*).

liberation of $^{14}CO_2$ from [1-^{14}C]lactate indicates the conversion of lactate to pyruvate followed by the oxidative decarboxylation of the latter. After being converted to acetyl coenzyme A, the remaining two carbons of lactate may undergo a complete oxidation by the action of the enzymes of the tricarboxylic acid cycle. This oxidative process is reflected by the liberation of $^{14}CO_2$ from [3-^{14}C]lactate. In agreement with the evidence indicating that the red muscle has a high capacity for aerobic metabolism, the rate of oxidation of both substances was 5 to 6 times higher in this tissue than in the white muscle. The level of activity of the red muscle was comparable to that found in the heart. With respect to the oxidation of [1-^{14}C]lactate, liver had a much higher activity than the other tissues, implying that, in trout, this organ has an important role in the catabolism of lactate. When [3-^{14}C]lactate was used as substrate, gills were the most active tissue, suggesting that the complete oxidation of lactate may be of significance in supplying energy for exchange reactions in gills. It may be noted that chloride cells isolated from the gill tissue of salt water adapted *Anguilla anguilla* have a relatively high capacity to oxidise (U-^{14}C) lactic acid (Sargent and Bornancin, unpublished data, 1972). Such findings underline the importance of the acid-base balance in relation to lactic acid metabolism because gills are major excretory organs of H^+, NH_4^+ and HCO_3^- in marine teleosts. It is intriguing to note that the mammalian kidney is a powerfully gluconeogenic organ. Whether the same is true for the gills of marine teleosts remains to be established, although electron miscroscopy has revealed that chloride cells contain substantial amounts of glycogen (Philpott and Copeland, 1963).

Bokdawala and George (1967a) provided histochemical evidence that the pentose phosphate shunt enzymes, glucose 6-phosphate dehydrogenase and 6-phosphogluconate dehydrogenase, occur at higher concentrations in the red than in the white muscle of carp, *Cirrhina mrigala*. They have suggested that this alternative pathway of glucose catabolism used in the red muscle for the generation of NADPH is essential for fatty acid biosynthesis. This picture is consistent with the high lipid content of the red muscle of fish. Boström and Johansson (1972) also found that these two dehydrogenases show a considerably higher level of activity in the red than in the white muscle of eel, *A. anguilla* (see Table II, p. 259).

Wittenberger and Diaciuc (1965) studied the interrelation between carbohydrate metabolism in the white and red muscles of fish. They found that an electrical stimulation *in situ* of the lateral musculature of carp resulted in quite different effects on the two types of muscular tissue. For example, following a moderate stimulation of the musculature, glycogen was significantly depleted in the white muscle only, but the corresponding excess of lactic acid and an increase in O_2 consumption appeared in the red muscle. After a prolonged electrical stimulation leading to the exhaustion of a number of carp, the white muscle showed a pronounced decrease in glycogen and an increase in lactate; in the red muscle a slight decrease in glycogen was accompanied by a rise in lactate much above the level observed in the white muscle. These results and other observations (Wittenberger, 1967, 1968, 1972a, b; Wittenberger et al., 1969) have led Wittenberger to support the hypothesis that in fish the red muscle has a liver-like function (see Section III.E.).

Studies on the energy-supplying metabolism in animal tissues have revealed the existence of enzyme groups of constant proportion within the main metabolic systems, i.e. glycolysis, β-oxidation of fatty acids, citric acid cycle, respiratory chain, glycogenesis, etc. Within each of these groups the relative enzyme activities were often found to be constant although the absolute enzyme activities of each group may vary by orders of magnitude when different tissues or organs are compared (Bass et al., 1969; Pette, 1971; Staudte and Pette, 1972). Nakai et al. (1970a, b, c), working with kokanee salmon, Oncorhynchus nerka, found that the levels of activity of various enzymes are closely related to the physiological condition of the fish, and the principle of constant proportion groups of glycolytic enzymes in the muscle was observed only prior to the spawning period. Boström and Johansson (1972) showed that in the red and white muscle of eel, A. anguilla, there is no constant ratio between the enzymes of the citric acid cycle and the respiratory chain. In the case of other pathways the picture was more comparable to that found in higher vertebrates; for instance, the ratio between glycolysis and the citric acid cycle was much higher in the white than in the red muscle of eel.

D. Lipid Metabolism

The energy supplied to the animal by the breakdown of lipid reserves comes primarily from the oxidation of fatty acids. In contrast to the liver where the oxidation of fatty acids may stop at the stage of formation of ketone bodies, in skeletal muscle and other extrahepatic tissues fatty acids are subjected to complete oxidation to CO_2 and H_2O and there is very strong evidence to suggest that this process takes place entirely within mitochondria. Bilinski (1963) found that, in vitro, the oxidation of long chain saturated fatty acids to CO_2 occurs at a much higher rate in the red than in the white muscle of rainbow trout, Salmo gairdneri, and he related these differences to the biochemical organization of the two types of muscle. Jonas and Bilinski (1964) showed that acetate is also oxidized at a faster rate by the red muscle of fish. Studies with mitochondria isolated from the dark (red) lateral line muscle of salmonids suggest that the conversion of long chain fatty acids to CO_2 takes place through β-oxidation and the tricarboxylic acid cycle according to the mechanism demonstrated in mitochondria from other animal sources (Bilinski and Jonas, 1964). The mitochondria from the red muscle of fish appear, however, to differ from those studied by Brown and Tappel (1959) in carp liver. By showing an increase in oxidative activity after addition of coenzyme A, Bilinski and Jonas (1970) studied the effects of coenzyme A and carnitine on the oxidation of palmitate and oleate by mitochondria from the red lateral line muscle, white muscle, heart and other tissues of rainbow trout. With all the mitochondrial preparations the oxidation of fatty acids was considerably increased by the addition of carnitine and, in the presence of added carnitine, coenzyme A gave a further increase in oxidative activity. Coenzyme A alone stimulated the oxidation of fatty acids by mitochondria from the red muscle and heart. Mitochondria from the red muscle and heart also showed a similar range of maximal activity which was considerably above that found with mitochondria from other tissues. The results showing similarities between the mitochondria from trout heart and the red lateral muscle provide further support to the concept that fatty acids play a prominent role as sources of energy for the red skeletal muscle of fish. It is known that in mammals fatty acids represent the most important fuel for heart in the postabsorptive state (Bing, 1965). Although it was found in studies with trout that the activity in the white muscle was invariably much below that of the red muscle, it is noteworthy that under optimum assay conditions *in vitro*, (i.e. using a medium fortified with carnitine and other cofactors) the white muscle preparations exhibited a significant level of oxidative activity. The stimulation of fatty acid oxidation by carnitine reflects the role played by this compound in facilitating the transfer of extramitochondrial fatty acids to their site of oxidation in mitochondria (Bremer, 1968; Fritz, 1968). The carnitine-dependent oxidation of long chain fatty acids is known to occur widely in the animal kingdom as indicated by the distribution of the enzyme carnitine palmityl transferase, which was also shown to be present in cod (Norum and Bremer, 1966). The observations showing that the red muscle is better provided than the white with the enzymes involved in fatty acid oxidation were confirmed in several studies conducted with fish (George and Bokdawala, 1964; Dean, 1969) and with other vertebrates (Bass *et al.*, 1969; Pette, 1971; Pande and Blanchaer, 1971).

Studies of George and his associates emphasize the importance of fatty acids as sources of energy for the red muscle of fish. Besides showing the importance of aerobic metabolism in the red muscle of fish, these investigators observed that lipase activity is greater in this tissue than in the white muscle (George, 1962; George and Bokdawala, 1964; Bokdawala and George, 1967*a*, *b*; Bokdawala, 1967). Bilinski and Gardner (1968) found that, in trout, free fatty acids occur at 3 to 4 times higher concentrations in the red than in the white muscle.

In addition to a high ability for utilization of fatty acids, the red muscle of fish has a greater capacity than the white to store triglycerides. It has been shown in a large number of studies conducted with various species of fish that the red muscle has a higher total lipid content than the white muscle (see references cited by Love, 1970). These differences are particularly apparent in so-called "fatty fish" such as mackerel, herring or salmon which store triglycerides within the muscular tissue. Bosund and Ganrot (1969) studied the distribution of various classes of lipids in the red and white muscle of Baltic herring, *Clupea harengus*. They found that the differences between the red and white muscle in the total lipid content are mainly due to the triglyceride fraction, which is also responsible for the seasonal variations in fat content of the muscle. The fatty acid composition of triglycerides was the same in the red as in the white muscle of herring but certain differences were found in the pattern of phospholipids. Robinson and Mead (1973) found that the dark muscle of rainbow trout, *Salmo gairdneri*, is considerably more active than the white in depositing muscle lipids. When [1-^{14}C] palmitic acid was force-fed to trout, the incorporation rate of ^{14}C into the triglycerides of the white muscle was only 5% of that found in the dark muscle. Robertson and Mead point out, however, that in trout the white muscle is a quantitatively more important site of deposition of fat due to a much larger proportion of this type of muscle in the musculature.

E. Functions of the Red and White Muscle of Fish

The existing evidence on the structure and biochemical properties of the white and red skeletal muscle of fish strongly suggests that these two types of tissue play a role comparable to white (fast) and red (slow) voluntary muscle of mammals. As far as the energy metabolism is concerned, the red muscle of fish is better adapted than the white for carrying on aerobic pathways, whereas the latter shows a greater capacity for anaerobic glycolysis. In the white muscle glycolysis leads to a more extensive conversion of pyruvate to lactate, while in the red muscle pyruvate is readily subjected to complete oxidation. The red muscle is better adapted than the white to synthetize both glycogen and triglycerides. Indications are that the magnitude of aerobic metabolism in fish muscle is subjected to variations with changing physiological and environmental conditions.

The differences in contractile properties and biochemical organization of the red and white muscle has led to the concept that, for sustained swimming, fish utilize fat chiefly through aerobic processes in the red muscle, and for short bursts of activity they draw energy mainly from anaerobic degradation of glycogen in the white muscle (Boddeke et al., 1959; George, 1962). Bone (1966) reviewed thoroughly the earlier studies in this field and he concluded from his work with dogfish, *Scyliorhinus canicula* (see Section IIB and IIIA), that the red and white muscle fibers represent two separate motor systems which operate independently and utilize different metabolites. According to Bone this picture also holds for other kinds of fish, but such an arrangement is not economical for terrestrial animals which cannot afford to carry around a large mass of normally inactive white muscle. It is the essential weightlessness of fish which makes the possession of large amounts of normally inactive white muscle an economic proposition energetically. The concept that the white muscle provides a mechanism for rapid acceleration of fish in chase and escape reactions is supported by various observations made with live fish (see Section IIIA).

It has been postulated by some investigators that, in fish, the red muscle fulfils a metabolic role different from that carried out by this tissue in higher vertebrates. Braekkan (1956, 1959) and Mori et al. (1956) suggested that the main role of the red muscle of fish is not contractile activity but a function similar to that of mammalian liver with respect to carbohydrate and lipid metabolism. This concept is based on a similarity in biochemical properties of the red muscle and liver of fish (level of oxidative enzymes, vitamins and fat content, etc.) and on an apparent inability of the liver to perform the same function as in mammals due to a poor circulatory system in fish. According to this hypothesis the red muscle carries out synthetic processes and stores nutrients which are utilized for the muscular activity in the white muscle. Wittenberger (see Section IIIC) considers that the red muscle ensures essential metabolic conditions for the muscular effort of the white muscle: catabolites of anaerobic glycolysis (lactate, pyruvate) formed in the white muscle may be oxidized completely in the red muscle which in turn stores energy for the white muscle in the form of glycogen. The assumption that, in fish, there is metabolic interdependence of red and white muscle has been supported by Pritchard et al. (1971) who found a parallel pattern in glycogen depletion in the red muscle and liver of jack mackerel, *Trachurus symmetricus*, swimming near and above the threshold for sustained speed (see Section IIB). They suggested that at these speeds the red muscle, like liver, provides the energy reserves to the white muscle which serves as a primary locomotor agent; at low speeds both systems might be active. Robinson and Mead (1973) point out that in salmonids the red muscle, representing only a small portion of the total body musculature, has only a limited role as a kinetically responsible, contractile body but its function might be more related to the metabolic processes.

It is apparent that at the present state of knowledge it is difficult to draw firm conclusions about some of the distinct functions of the red muscle in fish. In our opinion, further experimental evidence is needed to establish whether the red muscle in fish plays a metabolic role which differs radically from that performed by this tissue in higher vertebrates. With reference to the liver-like function of the red muscle of fish, it might be pointed out that the metabolic reactions cited in support of this hypothesis (glycogen synthesis, oxidative activity, etc.) are also known to be predominant in the red muscle fibers of mammals. Because the red and white muscle fibers in fish are generally more confined to separate tissues than in mammals, certain basic physiological processes common to fish and mammals may appear more evident in fish (Bilinski, 1969). The possibility that, compared with mammals, the red and white muscle of fish represent two separate locomotor systems which utilize different metabolites seems unlikely from a biochemical point of view. Although pronounced quantitative differences in enzyme activities exist in fish between the red and white skeletal muscle, there is no evidence for a strict compartmentation between the two types with respect to the aerobic and anaerobic pathways or with respect to the utilization of fatty acids and carbohydrates.

IV. Mobilization of Lipid Reserves

A. Mechanism of Lipid Mobilization

Our present understanding of the processes involved in the mobilization of lipid reserves of fish is still fragmentary but indications are that significant differences exist between fish and higher vertebrates. The recent developments in this field are reviewed in this section. A brief outline of the main features of lipids mobilization, based on studies with mammals, follows.

Because of their limited solubility in water, triglycerides and other lipids combine with natural emulsifying agents to be able to move from one part of the organisms to another. Most of the lipids circulating in blood serum and in the lymph surrounding the cells are present in water soluble combinations with proteins in the form of lipoproteins or albumin-bound free fatty acids (FFA). The lipoprotein complexes are composed of proteins, triglycerides, cholesterol, cholesterol esters and phospholipids, staying in loose associations involving few covalent bonds. In mammals lipoproteins play a major role in the transport of fat derived from the diet and in the transfer of hepatic lipids. In oviparous animals, including fish (Ho and Vanstone, 1961; Takashima et al., 1971, 1972; Plack et al., 1971;

Plack and Fraser, 1971), they are implicated in the mobilization of lipids required for gonad formation. There is strong evidence that in mammals lipoproteins are not involved in the transfer of lipids from adipose tissue to the sites of utilization, this function being performed by albumin-bound FFA. Energy demands, due to muscular effort or starvation, lead in mammals to an increase in FFA levels in the bloodstream. As far as the processes of deposition and mobilization of tissue triglycerides are concerned, at least two general types of lipolytic activities occur in mammals – lipoprotein lipases, which facilitate the transfer of lipoprotein-bound triglycerides into the tissue, and hormone-sensitive lipases which are responsible for hydrolysis of adipose tissue triglycerides during the mobilization of lipid reserves. The mobilization of FFA from mammalian adipose tissue depends upon the rate of triglyceride biosynthesis and the level of activity of the hormone-sensitive lipase, and a close interrelation exists between carbohydrate and fatty acid utilization (Masoro, 1968). Generally speaking, factors enhancing the utilization of one type of substrate depress the mobilization of the other. This effect can be traced to the interreaction between carbohydrate and fatty acid metabolism, and indications are that the availability of L-α-glycerophosphate in the metabolic pool plays a key role in this mechanism (see Fig. 1, p. 241). Since L-α-glycerophosphate serves as substrate for the esterification of FFA, high levels of this intermediate promote the biosynthesis if triglycerides and other lipids. The L-α-glycerophosphate arises as a product of carbohydrate degradation. When tissues utilize large amounts of carbohydrates, the rate of L-α-glycerophosphate generation is high, leading to the formation of triglycerides from FFA. Besides providing L-α-glycerophosphate for FFA esterification, glucose triggers insulin secretion. Insulin causes a decrease in serum FFA by increasing glucose oxidation and by inhibiting the activity of the hormone-sensitive lipase. The depression of carbohydrate catabolism by a high rate of fatty acid utilization has been ascribed to multiple mechanisms involving specific inhibitions of glycolytic enzymes. These effects are largely related to increased levels of acetyl coenzyme A arising from fatty acid oxidation.

B. Lipid Reserves

The distribution of lipid deposits within tissues differs significantly among various species of fish (Vague and Fenasse, 1965). A well-developed subcutaneous adipose tissue typical for warm-blooded animals is not found in fish but extra- and intra-abdominal adipose deposits occur to some extent. In many fish the liver or the skeletal muscle appear as the main sites of lipid storage. By contrast to warm-blooded animals where fatty liver is pathological, the deposition of large amounts of lipids is normal for many teleost (cod, carp, halibut, etc.) and elasmobranch fish (dogfish, shark). It has been suggested that the liver serves as a storage site of lipid reserves in sluggish bottom-dwelling fish, whereas skeletal muscle might play this role in more active species (Tashima and Cahill, 1965). In some fish, in addition to triglycerides, the neutral lipid deposits contain significant amounts of hydrocarbons, diacyl glyceryl ethers and wax esters (Malins and Wekell, 1970). Little is known about the involvement of these lipids in energy-supplying metabolism, and their role in fish has been related to the control of buoyancy (see Section V.B.). The triglycerides, which appear as a main energy reserve of fish, are composed of long chain fatty acids having carbon chain lengths that generally range from 12 to 24 carbons and they are characterized by the presence of relatively high proportions of polyunsaturated fatty acids typical for marine lipids (Ackman, 1964; Stansby, 1967; Malins and Wekell, 1970; Lee and Sinnhuber, 1972).

C. Free Fatty Acids

In the blood plasma of fish FFA are usually found at concentrations comparable to mammals (i.e. 0.2–1.0 μEq/ml), and there is strong evidence suggesting that they play a role in lipid transport. The interrelation between carbohydrate and fatty acid utilization established in mammalian systems appears to exist also in fish. It was found in studies conducted with various species that the injection of glucose into live fish leads to a decrease in serum FFA levels (Farkas, 1967a, b, 1969; Minick and Chavin, 1972a; Mazeaud, 1973). This effect of glucose was also observed in vitro with adipose tissue preparations (Farkas, 1969). The stressing of fish induces a decrease in serum FFA (Farkas, 1969; Mazeaud, 1969b; Minick and Chavin, 1972a), presumably due to a rise in blood glucose levels resulting from the secretion of adrenalin (Nakano and Tomlinson, 1967). The administration of insulin to lamprey, Lampetra fluviatilis, and to various species of teleosts, was shown to decrease serum FFA (Leibson et al., 1968; Plisetskaya and Mazina, 1969). Minick and Chavin (1972a) found that both serum FFA and total serum phospholipid levels were lowered after administration of insulin to goldfish, Carassius auratus. Hypoglycemia occurred only 2 h after injection of insulin whereas serum FFA was decreased in 30 min. This observation led Minick and Chavin to conclude that the control of lipid metabolism may be the primary function of insulin in goldfish. The administration of mammalian diabetogenic agents to goldfish was found to cause an initial drop followed by an elevation in serum FFA levels (Minick and Chavin, 1972b).

The possible involvement of adrenalin, noradernalin and other hormones in the mobilization of fatty acids in fish has been the object of several investigations but variable responses were observed in different studies. Leibson *et al.* (1968) showed that the injection of adrenalin to lamprey, *Lampetra fluviatilis*, and scorpion fish, *Scorpaena corpus*, increased plasma FFA concentrations. Plisetskaya and Mazina (1969) provided further evidence for the lipolytic action of adrenalin in lamprey. According to Larsson (1973), eels, *Anguilla anguilla*, increase their level of plasma FFA in response to adrenalin treatment. Farkas (1967*a*, *b*, 1969), using carp, *Cyprinus carpio*, pike perch, *Lucioperca lucioperca*, and bream, *Abramis brama*, found that adrenalin, noradrenalin and other hormones known to be lipolytic in mammals had a depressing effect on serum FFA and lipolytic activity in fish. Farkas suggested that, by contrast to mammals, a cyclic AMP-dependent, hormone-sensitive lipase is absent in fish, and cyclic AMP formed under the influence of hormones acts only on carbohydrate metabolism, leading to a rise in L-α-glycerophosphate concentration, a condition promoting FFA re-esterification. Minick and Chavin (1970) studied the effects of pituitary hormones upon serum FFA in goldfish, *C. auratus*, and found that porcine adrenocorticotrophic hormone was significantly lipolytic. Growth hormone and ovine prolactin were also shown to promote rises in FFA levels. Larsson and Lewander (1972) found that glucagon does not appear to have any effect on plasma FFA in eels. Perrier *et al.* (1972) observed no change in plasma FFA levels after injection of adrenalin into rainbow trout, *Salmo gairdneri*. Mazeaud (1971) found that hypophysectomy in carp, *Cyprinus carpio*, altered the regulation of plasma FFA. During rest the level of plasma FFA of hypophysectomized fish was lower than that of controls. Adrenalin caused a greater decrease in FFA in hypophysectomized fish than in controls. Hypophysectomized carp were not able to regulate their glucose levels after adrenalin injection (Mazeuad, 1969*a*). In carp in a state of physiological shock induced by asphyxia or muscular stress, the levels of plasma FFA decreased while blood sugar increased (Mazeaud, 1969*b*). Using several species of freshwater fish, Mazeaud (1973) observed that the administration of adrenalin or noradrenalin may cause a decrease or a rise in plasma FFA levels. He also found that in goldfish, *C. auratus*, maintained at 20°, the half-life of palmitate was 4.6 min whereas that of glucose amounted to 21.4 min. In carp, *C. carpio*, the half-life of palmitate was 5.3 min. Comparable figures were obtained by Herodek (1969) who used carp in his study. Bilinski and Lau (1969) found that, *in vitro*, adrenalin did not stimulate lipolysis in the red muscle of rainbow trout, *S. gairdneri*, but it showed some effect by partially preventing loss of activity during preincubation. Lipshaw *et al.* (1972)

showed that adrenalin or noradrenalin decreases the *in vitro* incorporation of [^{14}C]acetate into the hepatic lipids of the nurse shark, *Ginglimostoma cirratum*, and they have suggested that these hormones have a lipolytic role in elasmobranch fishes.

In mammals starvation leads to a rapid and marked increase in serum FFA but changes of comparable magnitude were not observed in fish. Tashima and Cahill (1965) found a decrease in serum FFA of the toadfish, *Opsanus tau*, following 30–90 days of starvation. Bilinski and Gardner (1968) observed only a moderate increase in the level of serum FFA measured at seven intervals during a 70-day starvation of rainbow trout, *S. gairdneri*, and they have suggested that this might be partially due to a lower metabolic rate in fish under prolonged starvation. A similar picture was obtained by Robinson and Mead (1973) in rainbow trout subjected to a 35-day starvation. Mazeaud (1973) found no evidence for an increase in the level of serum FFA in several species of freshwater fish after 100 days of starvation. Larsson and Lewander (1973) found that during 95 days of starvation, eels, *A. anguilla*, tended to utilize liver triglycerides but there was no increase in plasma FFA levels. Between 95 and 145 days of starvation the eels utilized great amounts of both liver and muscle triglycerides and this state of prolonged starvation led to a marked elevation of plasma FFA levels. Mayerle and Butler (1971) observed a significant rise in plasma FFA level in North American eels, *A. rostrata*, following 150–180 days of starvation. The results of these studies appear to indicate that, by contrast to mammals, only a very prolonged starvation leads to a marked increase in serum FFA in fish.

D. Other Aspects of Lipid Mobilization

Mills and Taylaur (1971) found that there are significant differences between mammals, birds and fish in the distribution and composition of serum lipoproteins, but the members of each class of vertebrates show a degree of similarity. Lee and Puppione (1972) made a study on serum lipoproteins in the Pacific sardine, *Sardinops caerulea*. They indicate that the absence of lipoproteins with pre-β and β-mobility in sardines suggests major differences between fat transport in mammalian and fish systems. Lee and Puppione (1972) have not detected significant amounts of chylomicrons in sardines; likewise Sargent *et al.* (unpublished observation, 1972) found no chylomicrons in dogfish, *Squalus acanthias*.

Lauter *et al.* (1968) showed that in dogfish, *S. acanthias*, the FFA content of the lipoprotein fraction is significantly higher than in comparable fractions from mammals. Taking into account the anatomy of the dogfish, *S. acanthias*, Malins and Wekell (1970) suggested that in these fish the ingested fat is likely to be transported by the vascular system. Results implying differences between fish and mammals in the mechanism of fat transport were obtained by Robinson and Mead (1973). They force-fed [^{14}C] palmitic acid to rainbow trout, *S. gairdneri*, and found that, after the first 2 h, free fatty acids contained almost all the radioactivity present in the blood lipids. This picture contrasts with that observed in mammals where long chain fatty acids, whether fed in unesterified form or bound in triglycerides, appear in the bloodstream almost quantitatively as regenerated triglycerides. Phleger (1971) postulated that in spawning pink salmon, *Oncorhynchus gorbuscha*, cholesterol is used to mobilize the fatty acids of triglycerides in the form of cholesterol esters. He observed a significant incorporation of ^{14}C from [^{14}C]acetate into liver cholesterol esters, despite the fact that both the hepatic synthesis of triglycerides and cholesterol were depressed in spawning salmon.

Lech (1970) studied glycerol utilization in rainbow trout, *S. gairdneri*, and found that the liver contains a substantial activity of the enzyme glycerol kinase which mediates the conversion of glycerol to L-α-glycerophosphate. Being highly water soluble, the glycerol formed by the complete hydrolysis of triglycerides diffuses readily from the site of storage of fat into the bloodstream. The phosphorylation of glycerol to L-α-glycerophosphate appears essential for its re-entry into the metabolic pool. Fried *et al.* (1969) found that the teleost liver has a comparatively low L-α-glycerophosphate dehydrogenase activity implying a greater dependence of fish on L-α-glycerophosphate production through the glycerol kinase reaction. They observed that the activity of glucose-6-phosphate dehydrogenase is extremely high in the liver of fish. This enzyme mediates the initial step in the pentose phosphate shunt which is implicated in generating NADPH for fatty acid formation. Fried *et al.* (1969) suggested that in fish having no well-developed adipose tissue, liver may be a major site of fatty acid biosynthesis. Mayerle and Butler (1971) found that a 6-month starvation of the American eel, *A. rostrata*, leads to a significant increase in glycerol levels in blood and muscle but shows no effect on liver and kidney.

The lipolytic enzymes from fish muscle have been the subject of several investigations. It is known from the earlier studies that fish muscle contains a lipase able to catalyze the hydrolysis of short chain triglycerides (for review see Bilinski, 1969). Bilinski and Lau (1969) studied the lipolytic activity toward long chain triglycerides in the lateral line muscle of rainbow trout, *S. gairdneri*. Emulsification of substrates with phospholipids was necessary to demonstrate lipolysis which showed maximal activity at pH 7.3. An acid lipase, active toward tripalmitin and having the characteristics of a lysosomal enzyme, was also shown to occur in this tissue (Bilinski *et al.*, 1971). The enzyme had maximum activity at pH 4–4.5. Triton X-100 strongly stimulated the activity of acid lipase but it inhibited markedly the lipolytic activity occuring above pH 7. Palmitic acid and dipalmitin were the two main products of hydrolysis of tripalmitin. The physiological function of lysosomal enzymes is thought to consist of intracellular digestion of cell constituents, which may take place under normal physiological conditions or may have a character of pathological response in cells subjected to some kind of metabolic stress.

V. Buoyancy Control

A. Neutral Buoyancy

The locomotion of fish in the aquatic environment presents problems which are not encountered by terrestrial animals. Since two main body constituents, the proteins and the skeletal material, are denser than water, fish either have to evolve devices to lower their body density or have to provide an additional muscular effort to overcome sinking. Fish, having the same density as the ambient water, are said to have neutral buoyancy. Neutral buoyancy is particularly important to pelagic fish living permanently independent of any support other than sea water, and may be approached through economy in skeletal material and musculature, storage of compounds less dense than water (mainly lipids), inclusion of air and other gases in part of the body (swim bladder), and high water content. The following discussion is concerned with some biochemical aspects of buoyancy control in fish, more particularly with those bearing a relation to energy metabolism.

B. Lipid Deposits

The main classes of lipids which have been implicated in buoyancy control in fish are triglycerides, hydrocarbons (mainly squalene), diacyl glyceryl ethers (alkyl diacyl glycerols) and wax esters (see Fig. 6). Lewis (1970) compared the density of marine lipids from various sources and showed that squalene (sp. gr. 0.8562) and wax esters (sp. gr. 0.8578) are nearly equal in effectiveness as hydrostatic agents and greatly surpass the diacyl glyceryl ethers (sp. gr. 0.8907) and triglycerides (sp. gr. 0.9154–0.9168,). The hydrocarbon, pristane (sp. gr. 0.78), appears to be the most effective of the naturally occurring buoyant agents but it is rarely found in significant amounts in fish.

The elasmobranchs have no swim bladder, and lipid deposits often located in the liver appear to play an important role in buoyancy control in

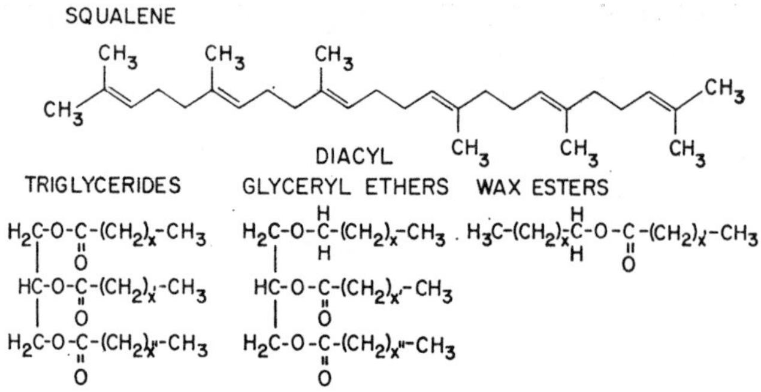

SQUALENE

DIACYL
TRIGLYCERIDES   GLYCERYL ETHERS   WAX ESTERS

H_2C-O-$\overset{\text{O}}{\underset{\text{}}{C}}$-$(CH_2)_x$-$CH_3$ H_2C-O-$\overset{\text{H}}{\underset{\text{H}}{C}}$-$(CH_2)_x$-$CH_3$. H_3C-$(CH_2)_x$-$\overset{\text{H}}{\underset{\text{H}}{C}}$-O-$\overset{\text{O}}{\underset{\text{}}{C}}$-$(CH_2)_x$-$CH_3$

HC-O-$\overset{\text{O}}{\underset{\text{}}{C}}$-$(CH_2)_x$-$CH_3$ HC-O-$\overset{\text{O}}{\underset{\text{}}{C}}$-$(CH_2)_x$-$CH_3$

H_2C-O-$\overset{\text{O}}{\underset{\text{}}{C}}$-$(CH_2)_{x''}$-$CH_3$ H_2C-O-$\overset{\text{O}}{\underset{\text{}}{C}}$-$(CH_2)_{x''}$-$CH_3$

FIG. 6. Structures of squalene, triglycerides, diacyl glyceryl ethers and wax esters. (Reproduced with permission of Malins, 1967).

these fish. The hydrocarbon, squalene, is found in considerable amounts in certain deep sea sharks of the family Squalidae (Malins and Wekell, 1970). These fish have large oily livers which may account for up to 25% of the total body weight and in some species up to 90% of the liver oil is composed of squalene. Corner et al. (1969) studied buoyancy control in several species of deep sea sharks and found that the lift provided by the liver almost exactly compensated for the weight in seawater of the rest of the animal. They estimated that in seawater squalene is 80% more effective per unit weight as a flotation agent than cod liver oil which is composed virtually entirely of triglycerides. Consequently squalene appears very economical in terms of the metabolic energy which has to be used by fish to provide lipid deposits responsible for buoyancy. According to Corner et al. squalene is laid down by deep sea sharks as a means of obtaining buoyancy and probably it has no role as a metabolic reserve. They also suggest that the fine regulation of buoyancy would more likely depend on lipids other than squalene. This interpretation is consistent with the observation that in dogfish, Scyliorhinus caniculas, the turnover rate of liver squalene is low (Sargent et al., 1970; Malins and Varanasi, 1972) although in this fish only very small amounts of squalene are present. Acetate and mevalonate are known to serve as precursors for squalene biosynthesis in fish (Diplock and Haslewood, 1965; Sargent et al., 1970; Kayama et al., 1971; Kayama and Shimada, 1972).

The diacyl glyceryl ethers represent another class of lipids implicated in the buoyancy control of elasmobranchs. A review covering the chemistry and metabolism of ether-linked lipids from fish and other marine animals is available (Malins and Varanasi, 1972). The role played by diacyl glyceryl ethers in buoyancy regulation in fish received particular attention in the case of spur dogfish, *S. acanthias*. Spur dogfish have relatively large livers containing approximately 60% oil composed largely of triglycerides and diacyl glyceryl ethers (Malins *et al.*, 1965; Malins, 1968). Malins and Barone (1970) postulated that the buoyancy of the spur dogfish is controlled by the ratio of diacyl glyceryl ethers to triglycerides occurring in the liver. In support of this hypothesis they found that this ratio was significantly increased in livers from dogfish maintained with lead weights in seawater. These findings suggest that the spur dogfish has at its disposal a regulatory mechanism involving the selective biosynthesis and catabolism of diacyl glyceryl ethers and triglycerides which allows it to adjust the body density to near neutral buoyancy. The metabolism of these two classes of lipids has been the object of thorough investigations in the spur dogfish. Malins and Sargent (1971) determined the relative rates of formation of ester and ether bonds in cell preparations from dogfish liver. The biosynthesis of triglycerides greatly exceeded that of diacyl glyceryl ethers. Oleyl alcohol was extensively oxidized to oleic acid in this system. Fatty alcohols were the important precursor in biosynthesis of the ether bond, but both fatty acids and fatty alcohols served as precursors of the ester linkages. Experiments with live dogfish indicated that it is unlikely that 2-acyl glycerols of dietary origin are required as precursors in the biosynthesis of diacyl glyceryl ethers (Malins and Robish, 1971) and triglycerides (Malins and Robish, 1972). Sargent *et al.* (1971) pointed out that the biosynthesis of low density lipids calls for high levels of reduced pyridine nucleotides since it generally involves the conversion of a carbonyl group to an alcohol group catalyzed by a reductase in the presence of NADPH. The oxygenation of the tissues of dogfish seems to be relatively poor and Sargent *et al.* suggested that the biosynthesis of low density lipids could provide an efficient mechanism for the oxidation of pyridine nucleotides and compensate for relatively inefficient respiratory system of these fish. There is strong evidence that the physiological function of triglyceride deposits in dogfish liver is not confined to buoyancy control. *In vivo* studies showed that the liver triglycerides serve as metabolic energy reserves; liver appears to export both free fatty acids and triglycerides to serum and thence to the muscle for utilization (Sargent *et al.*, 1972).

Wax esters are not found in significant amounts in elasmobranch fish but they are the major constituents of muscle lipids in some teleosts. Nevenzel (1970) reviewed the function and biosynthesis of wax esters in marine organisms. The occurrence of wax esters in the muscles of teleosts correlates better with the mesopelagic habitat and vertical migration patterns than with taxonomy. In some lantern fishes and gempylidae wax esters account for 80–90% of the total muscle lipids. According to Nevenzel the massive deposits of these lipids probably play no role as energy reserves but they serve as a means of approaching neutral buoyancy. By having a much lower density than triglycerides, wax esters are evidently very economic as flotation agents. Nevenzel suggests that a divorce between the mechanism of buoyancy control and the energy-supplying metabolism might be advantageous to fish by stabilizing buoyancy against short-term fluctuations. In this connection he points out that it seems likely that wax esters are not subject to the hormone-controlled mobilization of tri- glycerides and that glycerol, a key intermediate in pathways supplying energy, is not involved directly in the metabolism of wax esters. The diet does not appear to be the source of wax esters occurring in fish. Various tissues of marine teleosts were shown to have the ability to synthetize wax esters from fatty acids and alcohols and it is thought that the latter play a key role in regulating the deposition of wax esters in fish. Sargent *et al.* (1971, 1972) showed that preparations from the liver of spur dogfish catalyze the formation of wax esters. The rate of formation of the ester bond in wax esters was comparable to the rate of synthesis of the ester bonds in triglycerides. Wax esters occur only in trace amounts in dogfish liver but they account for about 25% of the total lipids in the blood serum, suggesting that they may play an important role in lipid metabolism in this species (Sargent *et al.*, 1971). The total amount of wax esters occurring in the body of dogfish is, however, quite limited so it seems unlikely that this class of lipids plays a significant role in buoyancy control.

C. Swim Bladder

Many fish control their buoyancy by maintaining an enclosure of free gases in a special organ, the swim bladder. The swim bladder is found in fish living at various depths; it occurs most frequently among fish living in the upper 200 m but it is also found in fish from greater depths (Marshall, 1962, 1970). The elasmobranchs, as well as some bottom-dwelling and active pelagic teleosts, have no swim bladder. The role of the swim bladder in buoyancy control consists of its ability to store certain volumes of gases and to vary its volume in order to preserve neutral buoyancy with changing depth. In this connection the organ has a unique property to concentrate gases from the bloodstream. Steen (1970) reviewed the present understanding of gas secretion and the role of the swim bladder as a hydrostatic organ. Additional information concerning swim bladder function and physiology may be found in the articles of Scholander (1954), Denton (1961), Kuhn *et al.* (1963), Alexander (1966) and Fänge (1966). The following discussion is confined to the peculiar aspects of lactic acid metabolism in the gas-secreting tissue.

The postulated mechanism for the formation of gases in the swim bladder implies that the gas-secreting tissues contain an enzyme system capable of producing significant amounts of lactic acid. Lactic acid has been implicated in unloading oxygen and other gases from the bloodstream into the swim bladders. Essentially, the increase in lactic acid concentration lowers the blood pH and decreases the capacity of oxyhemoglobin to bind oxygen which is in turn secreted into the swim bladder. This effect appears to be greatly amplified by a countercurrent mechanism in the rete structure of the swim bladder. In contrast to most other cells where anaerobic glycolysis is inhibited by O_2 (Pasteur effect), in the gas-secreting tissue of fish the production of lactic acid takes place even under extremely high partial pressures of O_2. D'Aoust (1970) conducted studies with the gas gland from vermillon rockfish, *Sebastodes miniatus*, and found that lactic acid production could occur under an O_2 pressure of up to 50 atm. Since glycogen reserves of the gas gland are quite limited, D'Aoust suggested that, *in vivo*, glucose from blood is the main source of lactic acid for gas secretion. Gesser and Fänge (1971) determined the levels of activity of lactate dehydrogenase and cytochrome oxidase in gas glands, white skeletal muscle and heart from several species of marine teleosts. The gas gland showed a high level of activity of lactate dehydrogenase which was predominantly of the anaerobic type also found in the white muscle. The activity of cytochrome oxidase was very low in the gas gland, being much below that observed in the white muscle. Gesser and Fänge pointed out that the positive correlation between the supply of O_2 and the cytochrome oxidase activity demonstrated in other tissues does not exist in the gas gland, because in this tissue the prime physiological function of O_2

TABLE III: Activities of enzymes in various metabolic pathways in gas gland tissue of the swim bladder and in the white muscle of the cod, *Gadus morhua*.

Enzymes	Gas Gland	White Muscle
Hexokinase	$1·45 \pm 0·11$	$0·10 \pm 0·002$
Glucose 6-phosphate dehydrogenase	$1·99 \pm 0·11$	$0·10 \pm 0·05$
6-Phosphogluconate dehydrogenase	$0·92 \pm 0·07$	$0·09 \pm 0·01$
Phosphofructokinase	$0·71 \pm 0·05$	$3·75 \pm 0·16$
Glycerophosphate dehydrogenase	$4·80 \pm 0·48$	$11·2 \pm 1·2$
Pyruvate kinase	345 ± 25	330 ± 23
Lactate dehydrogenase	1154 ± 46	268 ± 46
Malate dehydrogenase	163 ± 5	$90·7 \pm 7·2$
Fumarase	$1·47 \pm 0·08$	$2·55 \pm 0·18$
Cytochrome oxidase	$21·5 \pm 3·1$	110 ± 9

Enzyme activities are expressed as U/g wet wt. The results are given as Means $\pm$S.E.M. of six animals. (Reproduced with the permission of Boström *et al.* 1972).

is to supply gas for the control of buoyancy rather than to serve for the generation of energy by aerobic metabolism. Boström et al. (1972) found that in cod, *Gadus morhua*, the key glycolytic enzyme, phosphofructokinase, and the oxidative enzymes (fumarase and cytochrome oxidase) showed many times lower activities in gas gland tissue than in white muscle (see Table III). The activities of hexokinase, pentose phosphate shunt enzymes (glucose 6-phosphate dehydrogenase and 6-phosphogluconate dehydrogenase), as well as lactate dehydrogenase were considerably higher in gas gland tissue than in white muscle. These results suggest that major differences exist between the white muscle and the gas gland of fish in the pathways participating in the formation of lactic acid. In the gas gland the glucose 6-phosphate is largely diverted to the pentose shunt. The formed pentose phosphate may enter glycolysis after being converted to glyceraldehyde 3-phosphate. This mechanism is consistent with the absence of a Pasteur effect in gas gland tissue since it bypassed the step mediated by phosphofructokinase which is considered to be mainly responsible for the Pasteur effect.

D. Other Means of Buoyancy Control

Denton and Marshall (1958) found that some bathypelagic fish without a swim bladder derive their neutral buoyancy from the fact that they contain considerable amounts of body fluids which are less dense than sea water. These fish also have lightly ossified skeletons associated with reduced muscular systems, particularly along the trunk and tail. Chemical analysis of two bathypelagic species, *Gonostoma elongatum* and *Xenodermichthys copei*, showed that they contain only 12.6% and 10% of dry matter compared with 28% for a typical coastal marine fish, *Ctenolabrus rupestris*. *Gonostoma* and *Xenodermichthys* are not particularly rich in fat (3% wet weight) but their protein content amounting to 4–7% of the wet weight is very low when compared with 16% for *Ctenolabrus*. Blaxter et al. (1971) have extended Denton and Marshall's observations on the water content and body composition of marine fish as related to buoyancy control. They found that certain mesopelagic fish without swim bladders, living at a depth of 100–1000 m, have soft bodies with a very high water content (88–95%), low haemotocrits (5–9%), small hearts, a low proportion of red muscle and large lymph duct. Species possessing swim bladders have a lower water content (70–83%), higher haematocrits (14–35%), larger hearts, more prevalent red muscle, and smaller lymph ducts. Some active surface species have still less water (64–74%) and higher haematocrits (48–57%). Blaxter et al. (1971) indicate that their data are consistent with the idea that soft, watery fish use skeletal reduction and high water content as a buoyancy mechanism, but they found no evidence that they maintain an extra low ionic content in body fluids to assist buoyancy. According to Blaxter *et al.* the lower blood oxygen-carrying capacity of the watery fish is partially compensated for by a high proportion of anaerobic white muscle. Despite the fact that these fish would appear to be best adapted for short bursts of activity, some species make extensive daily vertical migrations calling for a prolonged muscular effort (Badcock, 1970). There is, however, evidence indicating that they pause during migration, presumably in order to repay the oxygen debt (Blaxter, 1970).

Although several species of elasmobranchs gain most of their buoyancy from fatty livers, in many other fish in this group a significant amount of lift is provided by tissues other than liver. In some species the low density tissues consist of a gelatinous layer or muscular tissue rich in fat, but the water content of the white muscle of elasmobranch fish is relatively constant (Bone and Roberts, 1969). The electric rays, *Torpedo nobiliana*, are almost neutrally buoyant due to the presence of considerable amounts of electrical tissue of comparative low density occurring concurrently with a fatty liver (Roberts, 1969).

Some scombroid fish live permanently in the pelagic environment despite the fact that they are denser than seawater. The following five species of scombroids are known to overcome their negative buoyancy through continuous swimming: Atlantic mackerel, *Scomber scombrus*, Pacific bonito, *Sarda chiliensis*, skipjack tuna, *Katsuwonus pelamis*, wavyback skipjack, *Euthynnus affinis*, and yellowfin tuna, *Thunnus albacares* (Magnuson, 1970). Magnuson (1970) made studies of the hydrostatic equilibrium of *Euthynnus affinis* and found that the minimum speed required by this fish to produce hydrodynamic lift is close to the maximum endurance speeds of other species. It appears that the locomotory system of these scombroids is better adapted for continuous swimming than that of many other fish. The ability of scombroid fish to furnish an uninterrupted muscular effort during their lifetime may be related to a metabolic specialization reflected by high blood hemoglobin, a muscle temperature higher than the environment, a cutaneous circulatory system, and the organization of the dark muscle (Magnuson, 1970; Zharov, 1967).

E. Energy Cost of Buoyancy Regulation

Alexander (1970, 1972) made approximate calculations of the energy costs of buoyancy regulation by marine teleosts. He estimated the energy expenditure in terms of oxygen consumption during swimming at constant depth and during vertical migration in the following three hypothetical cases: (1) fish having no neutral buoyancy, (2) fish with neutral buoyancy due to lipid deposits, (3) fish with neutral buoyancy due to a gas-filled swim bladder. In the first case the fish are denser than water and they must generate an upward hydrodynamic force to prevent themselves from sinking when they swim at a constant depth. According to Alexander's estimates the muscular effort necessary to achieve the required lift is in the order of 25 $cm^3/O_2/kg/h$. In the case of fish having enough lipid deposits to give a neutral buoyancy, no energy is required to counteract the sinking but the presence of lipids increases the bulk of the fish and so the drag it has to overcome when it swims. The extent of drag will depend on the density of lipids. Alexander calculates that, in order to give a neutral buoyancy, the lipids would increase the volume of the body of a typical marine fish by 54% in the case of triglycerides and by 32% in the case of wax esters. The amount of energy needed to overcome the extra drag is likely to be 10–17 $cm^3/O_2/kg/h$ depending on the type of lipids for a fish of average swimming activity. This energy cost will increase with swimming speed. A gas-filled swim bladder can provide the neutral buoyancy without a pronounced increase in body volume, and the energy cost due to the increase in drag would be only around 2 $cm^3/O_2/kg/h$. The neutral buoyancy, regulated by the swim bladder, remains in a rather unstable equilibrium due to the loss of gases through diffusion, and some additional energy is required to replace the gas lost. It appears however that for fish living at constant depth a swim bladder can provide neutral buoyancy for a smaller energy cost than lipids. Neutral buoyancy, provided by lipid deposits, seems more appropriate for sluggish species than for active ones.

Regarding the energy cost of vertical migration, there are also marked differences between fish having different mechanisms for buoyancy regulation (Alexander, 1972). The density of fish without swim bladder remains unaffected by the changes in depth. Fish denser than water must evidently provide an additional muscular effort when they move toward the surface but they can save almost comparable energy as they descend. Fish provided with neutral buoyancy due to the presence of incompressible lipid deposits require the same muscular effort for swimming upward and downward as for moving horizontally. The situation is more complex for fish having a swim bladder. For instance, during descent the volume of the swim bladder decreases due to the compression of gases. This results in an increased body density which upsets the neutral buoyancy. One may consider that fish may restore the neutral buoyancy through secretion of gases to the swim bladder, or that they will compensate for the lack of neutral buoyancy by an additional muscular effort. Alexander estimated the energy expenditure involved in these two alternatives. It would appear that fish which make small vertical migrations near the surface will use less energy if they maintain neutral buoyancy by keeping the volume of the swim bladder constant through secretion and resorption of gases. In more extensive vertical migrations gas secretions appear to be a much more costly process than hydrodynamic compensation through muscular work. Alexander points out that his estimates are too approximate to allow firm conclusions, but they do suggest a possible saving of energy during vertical migrations of some species by a partial replacement of the swim bladder by lipids as source of buoyancy. This interpretation is supported by the observations showing adaptive changes in buoyancy control with increasing vertical migration. For instance, Butler and Pearcy (1972) found that, in myctophids, lipids assume the primary buoyancy function as the gas-filled bladder regresses with age and that adults make more extensive diurnal migrations than juveniles.

VI. References

Ackman, R. G. (1964). *J. Fish. Res. Bd Can.* **21**, 247–254.

Alexander, R. McN. (1966). *Biol. Rev.* **41**, 141–176.

Alexander, R. McN. (1970). *In* "Proc. International Symposium on Biological Sound Scattering in the Ocean" (Farquhar, G. B., Ed.), pp. 74–85. Maury Center for Ocean Science, Dept. Navy, Washington, D.C.

Alexander, R. McN. (1972). *Symp. Soc. exp. Biol.* **26**, 273–294.

Andersen, P., Jansen, J. K. S. and Loyning, Y. (1963). *Acta Physiol. Scand.* **57**, 167–169.

Badcock, J. (1970). *J. mar. biol. Ass. U.K.* **50**, 1001–1044.

Barany, M. (1967). *J. gen. Physiol.* **50**, 197–216.

Barets, A. (1961). *Archs. Anat. miscrosc. Morph. exp.* **50**, 91–187.

Barrett, I. and Robertson Connor, A. (1964). *Bull. inter-Am. trop. Tuna Commn.* **9**, 219–268.

Bass, A., Brdiczka, D., Eyer, P., Hofer, S. and Pette, D. (1969). *Eur. J. Biochem.* **10**, 198–206.

Beamish, F. W. (1966). *J. Fish. Res. Bd Can.* **23**, 1507–1521.

Beamish, F. W. (1968). *J. Fish. Res. Bd Can.* **25**, 837–851.

Beamish, F. W. H. and Dickie, L. M. (1967). *In* "The Biological Basis of Fresh Water Fish Production" (Gerking, S. D., Ed.), pp. 215–242. Blackwell Scientific Publications, Oxford and Edinburgh.

Bilinski, E. (1963). *Can. J. Biochem. Physiol.* **41**, 107–112.

Bilinski, E. (1969). *In* "Fish in Research" (Neuhaus, O. W. and Halver, J. E., Eds), pp. 135–151. Academic Press, New York and London.

Bilinski, E. and Gardner, L. J. (1968). *J. Fish. Res. Bd Can.* **25**, 1555–1560.

Bilinski, E. and Jonas, R. E. E. (1964). *Can. J. Biochem.* **42**, 345–352.

Bilinski, E. and Jonas, R. E. E. (1970). *J. Fish. Res. Bd Can.* **27**, 857–864.

Bilinski, E. and Jonas, R. E. E. (1972). *J. Fish. Res. Bd Can.* **29**, 1467–1471.

Bilinski, E., Jonas, R. E. E. and Lau, Y. C. (1971). *J. Fish. Res. Bd Can.* **28**, 1015–1018.

282 E. BILINSKI

Bilinski, E. and Lau, Y. C. (1969). *J. Fish. Res. Bd Can.* **26**, 1857–1866.
Bing, R. J. (1965). *Physiol. Rev.* **45**, 171–213.
Bishop, C. M. and Odense, P. H. (1967). *J. Fish. Res. Bd Can.* **24**, 2549–2553.
Black, E. C., Connor, A. R., Lam, K. C. and Chiu, W. G. (1962). *J. Fish. Res. Bd Can.* **19**, 409–436.
Black, E. C., Robertson, A. C. and Parker, R. R. (1961). *In* "Comparative Physiology of Carbohydrate Metabolism in Heterothermic Animals" (Martin, A. S., Ed.), pp. 89–121. University of Washington Press, Seattle, Washington.
Blaxter, J. H. S. (1969). *FAO Fish Rpt.* **62**, 69–100.
Blaxter, J. H. S. (1970). *In* "Marine Ecology" (Kinne, O., Ed.), Vol. I, pp 213–285. Wiley-Interscience, London.
Blaxter, J. H. S., Wardle, C. S. and Roberts, B. L. (1971). *J. mar. biol. Ass. U.K.* **51**, 991–1006.
Boddeke, R., Slijper, E. J. and Van Der Shelt, A. (1959). *Proc. K. ned. Akad. Wet. C.* **62**, 576–588.
Bokdawala, F. D. (1967). *J. Anim. Morph. Physiol.* **14**, 231–241.
Bokdawala, F. D. and George, J. C. (1967a). *J. Anim. Morph. Physiol.* **14**, 60–68.
Bokdawala, F. D. and George, J. C. (1967b). *J. Anim. Morph. Physiol.* **14**, 223–230.
Bone, Q. (1966). *J. mar. biol. Ass. U.K.* **46**, 321–349.
Bone, Q. and Roberts, B. L. (1969). *J. mar. biol. Ass. U.K.* **49**, 913–937.
Boström, S. L., Fänge, R. and Johansson, R. G. (1972). *Comp. Biochem. Physiol.* **43B**, 473–478.
Boström, S. L. and Johansson, G. R. (1972). *Comp. Biochem. Physiol.* **42B**, 533–542.
Bosund, I. and Ganrot, B. (1969). *J. Fd. Sci.* **34**, 13–17.
Braekkan, O. R. (1956). *Nature, Lond.* **178**, 747–748.
Braekkan, O. R. (1959). *Fisk. Dir. Skr. (Tekn. undersok.)* **3**, 1–42.
Brainbridge, R. (1958). *J. exp. Biol.* **35**, 109–133.
Brainbridge, R. (1962). *J. exp. Biol.* **39**, 537–556.
Bremer, J. (1968). *In* "Cellular Compartmentation and Control of Fatty Acid Metabolism" (Gran, F. C., Ed.), pp. 65–88. Academic Press, New York and London.
Brett, J. R. (1962). *J. Fish. Res. Bd Can.* **19**, 1025–1038.
Brett, J. R. (1970). *In* "Marine Aquiculture" (McNeil, W. J., Ed.), pp. 37–52. Oregon State University Press, Corvallis.
Brett, J. R. (9172). *Resp. Physiol.* **14**, 151–170.
Brown, W. D. and Tappel, A. (1959). *Arch. Biochem. Biophys.* **85**, 149–158.
Burton, D. T. and Spehar, A. M. (1971). *Comp. Biochem. Physiol.* **40A**, 945–954.
Butler, D. G. (1968). *Gen. comp. Endocrinol.* **10**, 85–91.
Butler, J. L. and Pearcy, W. G. (1972). *J. Fish. Res. Bd Can.* **29**, 1145–1150.
Buttkus, H. (1963). *J. Fish. Res. Bd Can.* **20**, 45–58.
Buttkus, H. and Tomlinson, N. (1966). *In* "Physiology and Biochemistry of Muscle as Food" (Briskey, E. J., Cassens, R. G. and Trautman, J. C., Eds), pp. 197–203. The University of Wisconsin Press, Madison, Milwaukee and London.

Chang, V. M. and Idler, D. R. (1960). *Can. J. Biochem. Physiol.* **38**, 553–558.
Connell, J. J. (1970). *In* "Proteins as Human Food" (Lawrie, R. A., Ed.), pp. 200–212. The Avi Publishing Co., Westport, Conn.
Connor, A. R., Elling, C. H., Black, E. C., Collins, G. B., Gauley, J. R. and Trevor-Smith, E. (1964). *J. Fish. Res. Bd Can.* **21**, 255–290.
Corner, E. D. S., Denton, E. J. and Forster, G. R. (1969). *Proc. R. Soc. Ser. B* **171**, 415–429.
Crabtree, B. and Newsholme, E. A. (1972). *Biochem. J.* **126**, 49–58.
Dando, P. R. (1969). *J. mar. biol. Ass. U.K.* **49**, 209–223.
D'Aoust, B. G. (1970). *Comp. Biochem. Physiol.* **32**, 637–668.
Dawson, D. M., Goodfriend, T. L. and Kaplan, N. O. (1964). *Science N.Y.* **143**, 929–933.
Dean, J. M. (1969). *Comp. Biochem. Physiol.* **29**, 185–196.
Dean, J. M. and Goodnight, C. J. (1964). *Physiol. Zool.* **37**, 280–299.
Denton, E. J. (1961). *Prog. Biophys. biophys. Chem.* **11**, 178–234.
Denton, E. J. and Marshall, N. B. (1958). *J. mar. biol. Ass. U.K.* **37**, 753–767.
Diplock, A. T. and Haslewood, G. A. D. (1965). *Biochem. J.* **97**, 36P–37P.
Doudoroff, P. and Shumway, D. L. (1970). *FAO Fisheries Tech.* Paper No. 86, 291p.
Drummond, G. I. (1967). *Fortschr. Zool.* **18**, 360–429.
Drummond, G. I. (1971). *Am. Zool.* **11**, 83–97.
Drummond, G. I. and Black, E. C. (1960). *Ann. Rev. Physiol.* **22**, 169–190.
Dyer, W. J. and Dingle, J. R. (1961). *In* "Fish as Food" (Borgstrom, G., Ed.), pp. 275–327. Academic Press, New York and London.
Fänge, R. (1966). *Physiol. Rev.* **46**, 299–322.
Farkas, T. (1967a). *Progr. Biochem. Pharmacol.* **3**, 314–319.
Farkas, T. (1967b). *Amnls Inst. biol. Tihany* **34**, 129–138.
Farkas, T. (1969). *Acta biochim. biophys. Acad. Sci. Hung.* **4**, 237–249.
Fontaine, M. and Hatey, J. (1952). *Physiologia comp. Oecol.* **3**, 37–52.
Franzini-Armstrong, C. and Porter, K. R. (1964). *J. Cell Biol.* **22**, 675–696.
Fraser, D. I., Dyer, W. J., Weinstein, H. M., Dingle, J. R. and Hines, J. A. (1966). *Can. J. Biochem.* **44**, 1015–1033.
Fried, G. H., Schreibman, M. P. and Kallman, K. D. (1969). *Comp. Biochem. Physiol.* **28**, 771–776.
Fritz, I. B. (1968). *In* "Cellular Compartmentation and Control of Fatty Acid Metabolism" (Gran, F. C., Ed.), pp. 39–63. Academic Press, New York and London.
Fry, F. E. J. (1957). *In* "The Physiology of Fishes" (Brown, M. E., Ed.), pp 1–63. Academic Press, New York and London.
Fry, F. E. J. and Hochachka. (1970). *In* "Comparative Physiology of Thermoregulation" (Whittow, G. C., Ed.), pp. 79–134. Academic Press, NewYork and London.
Fukuda, H. (1958). *Bull. Jap. Soc. scient. Fish.* **24**, 24–28.
George, J. C. (1962). *Am. Midl. Nat.* **68**, 487–494.
George, J. C. and Bokdawala, F. D. (1964). *J. Anim. Morph. Physiol.* **11**, 124–132.
Gesser, H. and Fänge, R. (1971). *Int. J. Biochem.* **2**, 163–166.

Gordon, M. S. (1972a). *Mar. Biol.* **13**, 222–237.
Gordon, M. S. (1972b). *Mar. Biol.* **15**, 246–250.
Greene, C. W. (1926). *Physiol. Rev.* **6**, 201–241.
Greer Walker, M. (1971). *J. Cons. perm. int. Explor. Mer* **33**, 421–427.
Greer Walker, M. and Pull, G. (1973). *Comp. Biochem. Physiol.* **44A**, 495–501.
Halver, J. E. (1972). "Fish Nutrition". Academic Press, New York and London.
Hammond, B. R. and Hickman, Jr., C. P. (1965). *J. Fish. Res. Bd Can.* **23**, 65–83.
Hamoir, G. (1955). *Adv. Protein Chem.* **10**, 227–288.
Hamoir, G., Focant, B. and Destèche, M. (1972). *Comp. Biochem. Physiol.* **41B**, 665–674.
Heath, A. G. and Pritchard, A. W. (1962). *Physiol. Zool.* **35**, 323–329.
Herodek, S. (1969). *Annls Inst. biol. Tihany* **36**, 179–184.
Ho, C. H. and Vanstone, W. E. (1961). *J. Fish. Res. Bd Can.* **18**, 858–864.
Hoch, F. (1971). "Energy Transformations in Mammals: Regulatory Mechanism" W. B. Saunders Co., Philadelphia–London–Toronto.
Hochachka, P. W. (1961). *Can. J. Zool.* **39**, 767–776.
Hochachka, P. W. (1969). *In* "Fish Physiology" (Hoar, W. S. and Randall, D. J., Eds), Vol. 1, pp. 351–389. Academic Press, New York and London.
Idler, D. R. and Bitners, I. (1958). *Can. J. Biochem. Physiol.* **36**, 793–798.
Idler, D. R. and Bitners, I. (1959). *J. Fish. Res. Bd Can.* **16**, 235–241.
Idler, D. R. and Bitners, I. (1960). *J. Fish. Res. Bd Can.* **17**, 113–122.
Idler, D. R. and Clemens, W. A. (1959). International Pacific Salmon Commission Progress Report, pp. 1–80.
Idler, D. R. and Tsuyuki, H. (1958). *Can. J. Biochem. Physiol.* **36**, 783–791.
Ingram, P. (1970). *Int. J. Biochem.* **1**, 263–273.
Iverson, J. L. (1972). *J. Ass. Off. Anal. Chem.* **55**, 1187–1190.
Johnston, J. A., Freason, N. and Goldspink, G. (1972). *Experientia* **28**, 713–714.
Johnston, J. A. and Goldspink, G. (1973a). *J. mar. biol. Ass. U.K.* **53**, 17–26.
Johnston, J. A. and Goldspink, G. (1973b). *J. Fish Biol.* **5**, 249–260.
Jonas, R. E. E. and Bilinski, E. (1964). *J. Fish. Res. Bd Can.* **21**, 653–656.
Jonas, R. E. E. and MacLeod, R. A. (1960). *J. Fish. Res. Bd Can.* **17**, 125–126.
Jones, N. R. (1962). *In* "Recent Advances in Food Science" (Hawthorn, J. and Letch, M., Eds), pp. 151–166. Butterworths, Washington, D.C. and London.
Karlsson, J. (1971). *In* "Muscle Metabolism During Exercise" (Pernow, P. and Saltin, B., Eds), pp. 383–393. Plenum Press, New York and London.
Katz, A. M. (1967). *In* "Factors Influencing Myocardial Contractility" (Tanz, R. D., Kavaler, F. and Robert, J., Eds), pp. 401–415. Academic Press, New York and London.
Katz, A. M. (1970). *Physiol. Rev.* **50**, 63–158.
Kayama, M., Rizvi, S. Z. W. and Asakawa, S. (1971). *J. Fac. Fish. Anim. Husb. (Hiroshima Univ.)* **10**, 1–10.
Kayama, M. and Shimada, H. (1972). *Bull. Jap. Soc. sci. Fish.* **38**, 741–751.
Krueger, H. M., Saddler, J. B., Chapman, G. A., Tinsley, I. J. and Lowry, R. R. (1968). *Am. Zool.* **8**, 119–129.
Kuhn, W., Ramel, A., Kuhn, H. J. and Marti, E. (1963). *Experientia* **19**, 497–511.

Kutty, M. N. (1968). *J. Fish. Res. Bd Can.* **25**, 1689–1728.
Kutty, M. N. (1972). *Mar. Biol.* **16**, 126–133.
Larsson, A. (1972). *Gen. comp. Endocr.* **20**, 155–167.
Larsson, A. and Lewander, K. (1972). *Comp. Biochem. Physiol.* **43A**, 831–836.
Larsson, A. and Lewander, K. (1973). *Comp. Biochem. Physiol.* **44A**, 367–374.
Lauter, C. S., Brown, E. A. and Trams, E. G. (1968). *Comp. Biochem. Physiol.* **24**, 243–247.
Lawrie, R. A. (1953a). *Biochem. J.* **55**, 298–305.
Lawrie, R. A. (1953b). *Biochem. J.* **55**, 305–309.
Lech, J. J. (1970). *Comp. Biochem. Physiol.* **34**, 117–124.
Lee, R. F. and Puppione, D. L. (1972). *Biochim. biophys. Acta* **270**, 272–278.
Lee, D. J. and Sinnhuber, R. O. (1972). *In* "Fish Nutrition" (Halver, J. E., Ed.), pp. 145–180. Academic Press, New York and London.
Legge, J. W. (1971). *In* "Biochemistry and Methodology of Lipids" (Johnson, A. R. and Deveport, J. B., Eds), pp. 515–532. Wiley–Interscience, New York, London, Sydney and Toronto.
Leibson, L. G., Plisetskaya, E. M. and Mazina, T. I. (1968). *Z. Evol. Biokhim. Fiziol.* **4**, 121–127 (in Russian). English Translation: *Fish. Res. Bd. Can.* Translation Series No. 1253 (Vancouver Laboratory).
Lewis, R. W. (1970). *Lipids* **5**, 151–153.
Lipshaw, L. A., Patent, G. J. and Foa, P. P. (1972). *Horm. Metab. Res.* **4**, 34–38.
Lobanov-Rostovsky, I. (1969). *Ocean. Mag.* **1**, 5–17.
Love, R. M. (1970). "The Chemical Biology of Fishes". Academic Press, London and New York.
Magnuson, J. J. (1970). *Copeia* **1970**, 56–85.
Malessa, P. (1969). *Mar. Biol.* **3**, 143–158.
Malins, D. C. (1968). *J. Lipid Res.* **9**, 687–692.
Malins, D. C. and Barone, A. (1970). *Science* **167**, 79–80.
Malins, D. C. and Robisch, P. A. (1971). *Biochim. biophys. Acta* **248**, 430–433.
Malins, D. C. and Robisch, P. A. (1972). *Comp. Biochem. Physiol.* **43B**, 125–127.
Malins, D. C. and Sargent, J. R. (1971). *Biochemistry* **10**, 1107–1110.
Malins, D. C. and Varanasi, U. (1972). *In* "Ether Lipids: Chemistry and Biology" (Snyder, F., Ed.), pp. 297–312. Academic Press, New York and London.
Malins, D. C. and Wekell, J. C. (1970). *Prog. in Chem. Fats and Other Lipids* **4**, 339–363.
Malins, D. C., Wekell, J. C. and Houle, C. R. (1965). *J. Lipid Res.* **6**, 100–111.
Marshall, N. B. (1962). *Discovery Rept.* **31**, 1–122.
Marshall, N. B. (1970). *In* "Proc. International Symposium on Biological Sound Scattering in the Ocean" (Farquhar, G. B., Ed), pp. 69–73. Maury Center for Ocean Science, Dept. Navy, Washington, D.C.
Masoro, E. J. (1968). "Physiological Chemistry of Lipids in Mammals". W. B. Saunders Co., Philadelphia–London–Toronto.
Mayerle, J. A. and Butler, D. C. (1971). *Comp. Biochem. Physiol.* **40A**, 1087–1095.
Mazeaud, F. (1969a). *C.r. Séanc. Soc. Biol.* **163**, 24–28.
Mazeaud, F. (1969b). *C.r. Séanc. Soc. Biol.* **163**, 58–61.

Mazeaud, F. (1971). *C.r. Séanc. Soc. Biol.* **165,** 539–544.

Mazeaud, F. (1973). "Recherches sur la regulation des acides gras libres plasmatiques et de la glycemie chez les poissons". Thesis, Faculté de Sciences de Paris.

Mercy Bai, V. V. (1970). "Studies on the Food Requirements and Rate of Metabolism of Tropical Fish". Thesis, Madurai University, Madurai, cited by Kutty (1972).

Mills, G. L. and Taylaur, C. E. (1971). *Comp. Biochem. Physiol.* **40B,** 489–501.

Minick, M. C. and Chavin, W. (1970). *Am. Zool.* **10,** 500.

Minick, M. C. and Chavin, W. (1972a). *Comp. Biochem. Physiol.* **41A,** 791–804.

Minick, M. C. and Chavin, W. (1972b). *Comp. Biochem. Physiol.* **42B,** 367–376.

Mori, K., Hashimoto, Y. and Komata, Y. (1956). *Bull. Jap. Soc. sci. Fish.* **21.** 1233–1235.

Nag, A. C. (1972a). *J. cell Biol.* **55,** 42–57.

Nag, A. C. (1972b). *Am. Zool.* **12**(3), *Dissert. Abstr. Nr.* 427.

Nagai, M. and Ikeda, S. (1972). *Bull. Jap. Soc. sci. Fish.* **38,** 137–143.

Nakai, T., Shibata, T. and Saito, T. (1970a). *Bull. Fac. Fish. (Hokkaido Univ.)* **21,** 234–239.

Nakai, T., Shibata, T. and Saito, T. (1970b). *Bull. Fac. Fish. (Hokkaido Univ.)* **21,** 240–245.

Nakai, T., Shibata, T. and Saito, T. (1970c). *Bull. Fac. Fish. (Hokkaido Univ.)* **21,** 246–251.

Nakajima, Y. (1969). *Tissue and Cell* **1,** 229–246.

Nakano, T. and Tomlinson, N. (1967). *J. Fish. Res. Bd Can.* **24,** 1701–1715.

Nevenzel, J. C. (1970). *Lipids* **5,** 308–319.

Nishihara, H. (1967). *Arch. Histol. Jap.* **28,** 425–447.

Norum, K. R. and Bremer, J. (1966). *Comp. Biochem. Physiol.* **19,** 483–487.

Pande, S. V. and Blanchaer, M. C. (1971). *Am. J. Physiol.* **220,** 549–553.

Patton, S., Crozier, G. F. and Benson, A. A. (1970). *Nature, Lond.* **225,** 754–755.

Pernow, B. and Saltin, B. (1971). "Muscle Metabolism During Exercise". Plenum Press, New York and London.

Perrier, H., Perrier, C., Gudefin, Y. and Gras, J. (1972). *Comp. Biochem. Physiol.* **43A,** 341–347.

Pette, D. (1971). *In* "Muscle Metabolism During Exercise" (Pernow, B. and Saltin, B., Eds), pp. 33–49. Plenum Press, New York and London.

Phillips, Jr., A. M. (1969). *In* "Fish Physiology" (Hoar, W. S. and Randall, D. J., Eds), Vol. 1, pp. 391–432. Academic Press, New York and London.

Philpott, C. W. and Copeland, D. E. (1963). *J. Cell Biol.* **18,** 389–404.

Phleger, C. F. (1971). *Lipids* **6,** 347–349.

Plack, P. A. and Fraser, N. W. (1971). *Biochem. J.* **121,** 857–862.

Plack, P. A., Pritchard, D. J. and Fraser, N. W. (1971). *Biochem. J.* **121,** 847–856.

Plisetskaya, E. M. and Mazina, T. I. (1969). *Z. Evol. Biokhim. Fiziol.* **5,** 457–463 (in Russian). English Translation: *Fish. Res. Bd. Can.* Translation Series No. 1478 (Halifax Laboratory).

Pritchard, W., Hunter, J. R. and Lasker, R. (1971). *Fishery Bull. Fish Wildl. Serv. U.S.* **69,** 379–386.

Randall, D. J. (1970). *In* "Fish Physiology" (Hoar, W. S. and Randall, D. J., Eds) Vol. IV, pp. 253–292. Academic Press, New York and London.
Rayner, M. D. and Keenam, M. J. (1967). *Nature, Lond.* **214**, 392–393.
Roberts, B. L. (1969). *J. mar. biol. Ass. U.K.* **49**, 621–640.
Robinson, J. S. and Mead, J. F. (1970). *Can. J. Biochem.* **48**, 837–840.
Robinson, J. S. and Mead, J. F. (1973). *Can. J. Biochem.* **51**, 1050–1058.
Romanul, F. C. A. (1971). *In* "Muscle Metabolism During Exercise" (Pernow, B. and Saltin, B., Eds.), pp. 21–32. Plenum Press, New York and London.
Saddler, J. B. and Cardwell, R. (1971). *Comp. Biochem. Physiol.* **39A**, 709–721.
Sargent, J. R., Gatten, R. R. and McIntosh, R. (1971). *Mar. Biol.* **10**, 346–355.
Sargent, J. R., Gatten, R. R. and McIntosh, R. (1972). *Lipids* **7**, 240–245.
Sargent, J. R., Williamson, I. P. and Towse, J. P. (1970). *Biochem. J.* **117**, 26p.
Schmidt-Nielsen, K. (1972). *Science* **177**, 222–227.
Scholander, P. F. (1954). *Biol. Bull.* **107**, 260–277.
Smit, H., Amelink-Koutstaal, J. M., Vijverberg, J. and Von Vaupel-Klein, J. C. (1971). *Comp. Biochem. Physiol.* **39A**, 1–28.
Stansby, M. E. (1967). "Fish Oils". The Avi Publishing Co., Westport, Conn.
Staudte, H. W. and Pette, D. (1972). *Comp. Biochem. Physiol.* **41B**, 533–540.
Steen, J. B. (1970). *In* "Fish Physiology" (Hoar, W. S. and Randall, D. J., Eds), Vol. IV, pp. 413–443. Academic Press, New York and London.
Syrovy, I., Gaspar-Godfroid, A. and Hamoir, G. (1970). *Arch. int. Physiol.* **50**, 197–218.
Takashima, F., Hibiya, T., Watanabe, T. and Hara, T. (1971). *Bull. Jap. Soc. scient. Fish.* **37**, 307–312.
Takashima, F., Hibiya, T., Phan-Van Ngan and Aida, K. (1972). *Bull. Jap. Soc. scient. Fish.* **38**, 43–49.
Tarr, H. L. A. (1969). *In* "Fish in Research" (Neuhaus, O. W. and Halver, J. E., Eds), pp. 155–174. Academic Press, New York and London.
Tarr, H. L. A. (1972). *In* "Fish Nutrition" (Halver, J. E., Ed.), pp. 255–326. Academic Press, New York and London.
Tashima, L. and Cahill, Jr., G. F. (1965). *In* "Adipose Tissue" (Renold, A. E. and Cahill, G. F., Jr., Eds), pp. 55–58. American Physiological Society, Washington, D.C.
Tomlinson, N. and Geiger, S. E. (1962). *J. Fish. Res. Bd Can.* **19**, 997–1003.
Trams, E. G. (1969). *Mar. Biol.* **4**, 1–3.
Tsuyuki, H. (1973). *In* "Chemical Zoology" (Florkin, M., Ed.), Vol. 8, pp. 287–305. Academic Press, New York and London.
Vague, J. and Fennasse, R. (1965). *In* "Adipose Tissue" (Renold, A. E. and Cahill, G. F., Jr., Eds), pp. 25–36. American Physiological Society, Washington, D.C.
Waterman, R. E. (1969). *Am. J. Anat.* **125**, 457–494.
Wittenberger, C. (1967). *Revue roum. Biol. Zool.* **12**, 139–144.
Wittenberger, C. (1968). *Mar. Biol.* **2**, 1–4.
Wittenberger, C. (1972a). *Mar. Biol.* **16**, 279–280.

Wittenberger, C. (1972b). *St. Si Cerc. Biol. Seria Zoologie* **24,** 69–77 (in Roumanian).

Wittenberger, C., Coro, A., Suarez, G. and Potrilla, N. (1969). *Mar. Biol.* **3,** 24–27.

Wittenberger, C. and Diaciuc, I. V. (1965). *J. Fish. Res. Bd Can.* **22,** 1397–1406.

Yamamoto, M. (1968). *Can. J. Biochem.* **46,** 423–432.

Zharov, V. L. (1967). *Vop. Ikhtiol.* **7,** 209–224.

Survival at Freezing Temperatures

ARTHUR L. DEVRIES

Physiological Research Laboratory
Scripps Institution of Oceanography
University of California
La Jolla, California 92037

I. Introduction

The narrow range of temperatures within which most organisms carry on their life processes is called the biokinetic zone and lies between 4° and 45°C. Outside of this zone the rates of biological reactions of organisms tend to slow to such an extent that their life processes are seriously affected. As a consequence, at low temperatures cellular processes including growth and reproduction proceed only slowly or stop altogether. Despite the rate depressing effects of low temperatures on biochemical reactions, numerous poikilothermic (cold blooded) animals are known to carry on their life processes at temperatures outside of the biokinetic zone, that is, they have managed to adjust their biochemical machinery to compensate for this temperature effect. The ways in which these biochemical adjustments can occur have recently been reviewed by Hochachka and Somero (1973). The changes include evolution of enzyme variants which have high activities at low temperatures, increases in the amounts of enzymes present, and alterations in the control of the activities of enzymes.

As animals depart further from the biokinetic zone into colder waters, the next major obstacle they face is that of freezing. Freezing is nearly always lethal because it deprives biological reactions of the aqueous medium they require, concentrates solutes, denatures macromolecules and ruptures cell membranes. It is this latter problem, that of avoiding freezing in the marine environment, that is the subject of this chapter.

II. Low temperature marine environments: variation of temperature with depth and season

The oceans and seas of the world encompass more than two-thirds of the surface of the earth. Their average surface temperature varies between −2° and 30°C depending upon latitude. Even in tropical regions waters near 0°C occur in the deep trenches. Although organisms inhabiting these deep waters are not confronted with freezing because of the effect of pressure, they do have to contend with the rate depressing effect of temperature on their biological processes. Because the deep sea is relatively inaccessible and the animals retrieved from it usually do not survive long confinements in nets, little is known about the nature of the adaptive processes which permit their survival.

A. Temperate oceans

Although the temperate oceans as a whole appear to be relatively warm, a great deal of temperature variation occurs both with depth and season in certain parts of the world, and at times oceanic temperatures may drop to freezing. The major temperature patterns of the temperate oceans appear to be influenced considerably by ocean currents. A relatively warm mass of water moves eastward across the north pacific and eventually splits in the vicinity of British Columbia into the south-flowing California Current and the north-flowing Aleutian Current. Thus in winter, sea water does not usually freeze along the West Coast of North America until latitudes as far north as Alaska (54°N) are reached (Fig. 1). Along the East Coast of North America, however, freezing conditions in winter are found at more southerly latitudes (45°N) due to the combined effect of the cold south-flowing Labrador Current and the cold, east moving, continental air masses. Winter storms usually cause sufficient mixing of the offshore waters so that little ice formation occurs there. However, in the many shallow inlets and sheltered bays, thermal stratification occurs to the extent that ice does form, and in some areas it may reach a meter in thickness. Ice formation is especially common where fresh water streams empty into bays and form a fresh-water "lens", which in addition to being cooled by the air is also cooled by the sea water beneath it. The result is a thick layer of fresh-water ice. In such sheltered regions of the Atlantic Coast, ice formation has been observed to occur as far south as 40°N during cold winters. The temperate regions can thus be characterized by low winter temperatures with some ice formation, and by water temperatures some 15° to 20°C warmer in the summer.

B. Polar Oceans

In contrast to the temperate regions, the waters of the polar regions are much cooler and have much more ice. Because of the inaccessibility of the polar regions one often thinks of them mainly as large areas of snow and ice characterized by freezing temperatures throughout the year. However, even between and within the polar regions large differences exist. The Arctic Ocean can be thought of as a land-locked ocean and, therefore, many of its characteristics are controlled by the climate of the surrounding land masses. The Antarctic in contrast is an ice covered continent surrounded by an unrestricted Southern Ocean. As a consequence, the Arctic region experiences more seasonal changes in temperature than the Antarctic. Thus some regions of the Arctic are ice-free in the summer and water temperatures range well above freezing (4° to 7°C) (MacGinitie, 1955; Lewis and Weeks, 1970). The climate in Antarctica on the other hand is more severe and south of the Antarctic Convergence year-round water temperatures near −2°C are common and ice is abundant (Fig. 1). But even here, in some areas small seasonal differences exist in water temperatures and ice cover. In the northern part of the Antarctic Peninsula water

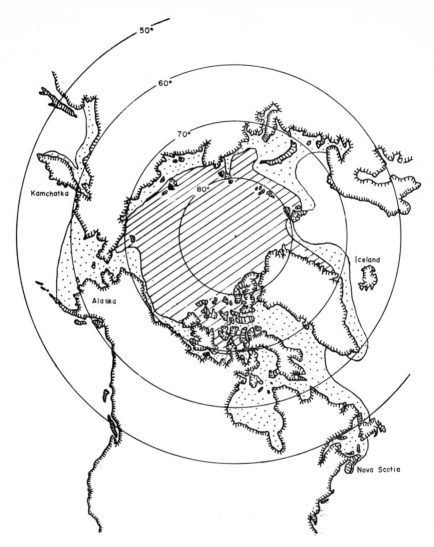

Fig. 1.

temperatures rise to $+2°C$ and the intertidal region is ice free for the entire summer. Other regions at more southerly latitudes are ice covered for all but a small part of the summer. In McMurdo Sound in the Ross Sea (79°S), for example, water temperature varies by only 0.2°C about the annual mean of $-1.86°C$ with both season and depth (Tressler and Ommundsen, 1962; Littlepage, 1965). During the early part of the austral

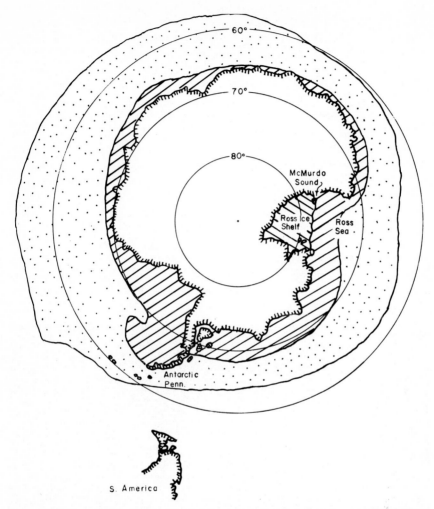

FIG. 1. Maps showing approximate range of pack ice in northern and southern hemispheres in summer ▨, and winter ▥. Extent of ice varies greatly from year to year depending on weather conditions. (Data assembled from U.S.N. Oceanographic Office, 1957; Bogdanov, 1963; Weyl, 1970).

winter, the cold dry air which originates on the polar plateau flows down the many glaciers onto the Ross Ice Shelf, and slowly moves toward the ocean. Part of this flow of cold air moves across McMurdo Sound, rapidly cooling the open water. This results in rapid formation of solid sea ice. By early summer, an ice thickness of 3 m is common.

Along with rapid sea ice growth, an ice platelet layer forms beneath the solid ice. This platelet layer is a loose matrix composed of many ice platelets or large crystals (2 mm in thickness by 10 cm in diameter) which form in the upper 33 m of the water column and float to the under side of the solid ice, where they freeze together in a random fashion (Dayton *et al.*, 1969). In the shallow waters, these same ice platelets are found frozen to the bottom in large aggregates which have been appropriately termed anchor ice (Pearse, 1962; Dayton *et al.*, 1969). During the winter periods of sub-ice platelet formation, the upper 33 m of the water column is also usually saturated with minute ice crystals which appear as thousands of tiny reflective "needles" (Littlepage, 1965).

Despite the abundance of ice and the low temperatures, a relatively rich fish fauna, capable of avoiding freezing, is present in the coastal waters of Antarctica (Norman, 1940). Many fishes belonging to the family Notortheniidae spend much of their lives in close association with the submarine ice environment of the Antarctic Ocean. For example, the fingerlings of *Trematomus nicolai*, a common antarctic coast fish, have been observed in large numbers "clinging" to the vertical faces of tabular icebergs by forming a sort of suction device with their large pelvic fins. It is suggested by

FIG. 2. *Trematomus bernacchii* resting on the ice platelets of a patch of anchor ice which forms on the bottom to depths of 30 m during the winter in McMurdo Sound, Antarctica. (Photo, P. K. Dayton).

Andriashev (1970) that the adaptive significance of this phenomenon is that the vertical face of the iceberg provides a habitat which is rich in attached diatoms and small crustacean prey, and is also a safe refuge from predators. Both young and adults of *Trematomus borchgrevinki* spend much of their time swimming in the surface waters beneath the annual ice of McMurdo Sound and often swim up into the holes and tunnels within the sub-ice platelet layer, presumably to forage for food and to escape their predator, the Weddell seal. *Trematomus bernacchii* and *T. hansoni* are both sluggish benthic fishes, and they have frequently been observed resting on or foraging for food among the masses of anchor ice on the bottom of McMurdo Sound (Fig. 2.)

III. Avoidance of freezing through habitat selection

A. Invertebrates

Most marine invertebrates do not have to worry about freezing since they are iso-osmotic or slightly hyper-osmotic to seawater (Potts and Parry, 1964). Intertidal invertebrates, however, have to contend with air temperatures well below their freezing points as well as the freezing of intertidal waters. Accordingly they have developed various types of behavior to avoid these conditions. On New England shores, for example, barnacle-eating snails move offshore in the winter (Kanwisher, 1955). Similarly, in Antarctica the limpet, *Patinigera polaris*, and the viviparous clam, *Lasaea consanguinea*, can be found in the intertidal region only during the summer. During the winter they can be found only below the low tidemark because of the large amount of intertidal ice (Stout and Shabica, 1970; Hedgpeth, 1969). Not all invertebrates leave the intertidal region during winter, however. Some clams like *Venus mercenaria* avoid freezing by living in the mud (Williams, 1970), and others like *Mytilus edulis* (to be discussed later) actually freeze.

B. Fishes

The body fluids of most marine teleosts contain only about one third as much salt as does seawater (Prosser and Brown, 1961; Potts and Parry, 1964) and, as a result, their body fluids freeze between $-0.7°$ and $-0.9°C$ (Black, 1951). During the winter there are large parts of the polar oceans and coastal north temperate oceans at temperatures of $-1.7°$ to $-1.9°C$, so that many fishes living in these waters are supercooled by approximately 1°C. At these temperatures ice is always present in the shallow sea water and thus it seems most unlikely that supercooling could exist in shallow water fishes unless their integuments were able to act as a barrier to ice. Studies by Scholander and co-workers (1957) have shown that the integuments of fishes such as the killifish, toad fish and tautog do not act as barriers to ice propagation when the body fluids are supercooled by 1° or 2°C. Thus, unless the body fluids of fishes possess some compound which serves to stabilize the supercooled state, or unless their membranes are altered to prevent ice propagation, they must escape to ice-free water to survive the winter.

Some shallow-water fishes of the Arctic Ocean avoid freezing by migrating to fresh water where the temperature can never fall below 0°C. The arctic char, *Salvelinus alpinus*, for example, usually leaves the ocean in the autumn and spends the winter in fresh water lakes and streams (Andrews and Lear, 1956).

The absence of ice in the depths can be explained by the facts that (1) heat is removed from sea water at the surface, (2) ice floats, and (3) increased hydrostatic pressure depresses the freezing point of sea water (0.0075°C for each 10 m increment in depth). Thus, it is not uncommon for supercooled fish to avoid contact with ice by either living in or migrating to deeper water. The long horn sculpin, *Myoxocephalus octodecemspinosus*, for example, is common in the shallow waters of the Atlantic coast of North America in summer, but during the winter it can only be found in water deeper than 20 m (Leim and Scott, 1966). This fish shows no resistance to freezing in the winter (Duman and DeVries, 1973).

In the fjords of northern Labrador Scholander *et al.* (1957) described a number of fishes which experience water temperatures of −1.7°C throughout the year. This group of fishes, consisting of *Boreogadus saida*, *Lycodes turneri*, *Liparis koefoedi*, *Gymnacanthus tricuspis* and *Icelus spatula*, inhabit only the deep bottom water of the fjords (200–300 m). The blood serum freezing points of this deep-water group range from −0.9° in *L. koefoedi* to −1.0°C in *B. saida*. Since the water temperature is a constant −1.7°C throughout the year, these fishes spend their entire lives with their body fluids supercooled by about 0.8°C. That they are indeed supercooled can be demonstrated by bringing them to the surface where contact with ice crystals in the water initiates freezing. From the results of this simple experiment, it can be argued that these fishes lack factors which stabilize the supercooled state. Interestingly, Scholander *et al.* (1957) noted that a few specimens of the tomcod, *B. saida*, taken from the bottom of the fjord could survive contact with ice at −1.7°C despite the fact that their blood reportedly froze at −1.0°C. The basis of freezing resistance in this species was not explained, but it seems likely that their body fluids possess "antifreeze" compounds which lower the freezing point, or stablize the supercooled state. The reported freezing point of −1.0°C did not take into consideration the unusual freezing point depression which results from the presence of "antifreeze" compounds (see section IVB).

In the Antarctic Ocean a number of fishes exist in a supercooled state throughout their life span. The extremes in supercooling can be observed in McMurdo Sound where the water at 500–600 m is uniformly $-1.9°C$ throughout the year (Littlepage, 1965). The most extreme case is the liparid fish, *Liparis sp.*, whose blood serum freezes at $-0.9°C$ and which lives only in the deep water. When brought to the surface and exposed to ice this fish always freezes (DeVries, 1968). The zoarcid fish, *Rhigophila dearborni*, freezes at $-1.5°C$ and is thus supercooled by only $0.4°C$. In the winter it almost always freezes when raised through the ice-laden surface waters of McMurdo Sound. In the summer, however, when the water temperature is about $0.1°C$ warmer and relatively free from ice crystals, this fish can survive in the surface waters. In the aquarium they survive at $-1.8°C$ as long as no ice is present. If a handful of finely chopped ice is added, however, they freeze immediately. Another deep-water benthic fish, *Trematomus loennbergii*, whose serum freezes at $-1.8°C$, has also been observed to freeze occasionally when raised through the ice-laden surface waters in the winter. At $-1.86°C$, the freezing point of seawater in McMurdo Sound, this fish survives poorly in the aquarium. However at $-1.7°C$, a temperature at which no ice formation occurs, it survives with few problems (Wohlschlag, 1964).

As long as the deep-water arctic and antarctic fishes remain in their natural ice-free habitats, their supercooled state is stable enough to permit survival. These fishes have never been captured in shallow water near shore in the Antarctic. The fact that the deep-water zoarcids and liparids of both the Arctic and Antarctic have not evolved a mechanism for complete protection from freezing, but rather survive by inhabiting only the ice-free deep water, indicates that their adaptation to the cold is incomplete. Their susceptibility to freezing is not entirely unexpected if one considers that these species belong to families whose distributions are primarily in the warmer waters of the northern hemisphere. In addition, their adaptation to the cold antarctic water at the metabolic level, at least in the case of the zoarcid fish, has been shown to be slight, despite the fact that this fish is an endemic species (Wohlschlag, 1964).

In some studies of supercooling and freezing resistance in fishes, certain investigators have put forth the view point that the supercooled state in the absence of ice is unstable, and that fishes must have elaborated compounds which serve to stabilize supercooling. Notably, Umminger (1969a) argues that the large amounts of the reducing sugar glucose, which build up in the blood of the killifish, *Fundulus heteroclitus*, during cold acclimation serves this purpose. The high glucose level, however, does not afford protection beyond its small contribution to the lowering of the freezing point when these fish are exposed to ice. Smith (1970, 1972) also argues that the high levels of reducing sugars present in some antarctic fishes, such as *Notothenia rossi* and *N. neglecta coriiceps*, stablize the supercooled state. These views, however, seem unwarranted when one considers that vials of water can be supercooled by 8° to 10°C for 10 months without the occurrence of freezing (Dorsey, 1948). In addition, fishes in ice-free, deep water appear to survive equally well whether they possess antifreeze compounds or not.

IV. Survival in the presence of ice

A. Invertebrates

1. FREEZING TOLERANCE

In the north Atlantic intertidal regions, invertebrates are exposed to winter air temperatures of −20° to −30°C (Kanwisher, 1965). Sessile invertebrates such as mussel, *Mytilus edulis*, and the snail, *Littorina rudis*, cannot escape into the relatively warm water and therefore quickly freeze (Kanwisher, 1966); upon thawing they show no ill effects from being frozen. Often, freezing occurs only during the period of low tide, but in some areas in the Arctic the mussels remain frozen for up to six months a year (Kanwisher, 1955).

In studies of *M. edulis* near Woods Hole, Massachusetts, Kanwisher (1955) has shown that upon exposure to air temperatures of −15°C when the tide is out for 2 h, the water within the soft tissues of the mussel freezes, and the internal temperature is within a few tenths of a degree of the air temperature. The same observations probably apply to the smaller snails and barnacles.

On the basis of calorimetric measurements, it was found that at −20°C about 70% of the body water of both *M. edulis* and *L. rudis* was in the form of ice (Kanwisher, 1955). Microscopic studies of the tissues of such preparations revealed that the cells were shrunken and distorted and that ice was present only in the extracellular spaces. Upon thawing the cells resumed normal shape and appearance (Kanwisher, 1959).

Williams (1970) has also studied freezing tolerance in intertidal molluscs. In this case *M. edulis* collected from Woods Hole, Massachusetts, was found to tolerate freezing at temperatures only as low as −10°C. At this temperature 66% of the body water is frozen. The difference in the lower lethal temperatures observed in *M. edulis* by Kanwisher and Williams may well have been caused by seasonal variation, since specimens for both studies were collected in the same locality, but at different times of the year. As is the case with several insects, some intertidal molluscs have indeed been shown to have seasonal variations in tolerance to freezing.

While the common oyster, *Crassostrea virginicus*, is tolerant to freezing regardless of season, the periwinkle, *Littorina littorea*, is tolerant to freezing only in the winter season (Kanwisher, 1966).

Freezing tolerance in intertidal invertebrates appears to be closely associated with the ability to resist dehydration, as in some insects. Both *M. edulis* and *L. littorea* can withstand several days out of water on the east coast of North America, whereas in the more extreme polar regions, the intertidal fauna (consisting mostly of mussels, snails and limpets) is frozen for six months of the year (MacGinitie, 1955; Hedgepeth, 1969).

Williams (1970) found that acclimation to a salinity greater than normal (150% sea water) resulted in a lower tolerance temperature of −15°C in *M. edulis*. However, changes in freezing tolerance following adaptation to higher and lower salinity did not appear to be the result of a change in the basic freezing tolerance. That is, the change in freezing tolerance noted with adaptation to a high salinity resulted only from an increase in the concentration of small cellular solutes such as sodium chloride and the free amino acids. Thus the same degree of dehydration was noted (64% of water frozen) at the limit of freezing tolerance for *Mytilus* regardless of whether it was adapted to high or low salinities.

In certain areas of the polar regions, ice formation is so heavy that it not only converts the intertidal region into lifeless blocks of solid ice, but it also invests the shallow waters to depths up to 30 m and affects the survival of the many invertebrates living there. For example in McMurdo Sound, Antarctica, invertebrates such as the asteroid, *Odontaster validus*, the nemertean worm, *Lineus corrugatus*, the isopod, *Glyptonotus antarcticus* and various pycnogonids are associated with the shallow-water anchor ice aggregations (Dayton *et al.*, 1969). While foraging about for diatoms and detritus which collect on the platelets of the anchor ice, these invertebrates are in little danger of freezing. However, as the aggregations of anchor ice increase in size during the course of the winter, their buoyancy increases to the extent that small disturbances in the water column cause them to break free from the bottom and float to the under side of the solid sea ice, where solidification of the entire mass eventually occurs. If the invertebrates, which are transported to the underside of the solid sea ice in this fashion, are unable to escape from the interstices of the uplifted anchor ice formations before solidification occurs (a matter of days), then they are eventually frozen solid. On the permanent ice around the Dailey Islands which are located on the western edge of McMurdo Sound, many invertebrates and fishes can be found lying on the surface of the ice, where the processes of ablation at the surface and accretion at the bottom of the ice have transported the organisms entrapped by uplift of anchor ice from the bottom to the surface. In some areas around these islands the density of invertebrates on the surface reaches 10 per m² (DeVries, unpublished observations). The presence of anchor ice in the shallow waters of McMurdo Sound thus becomes an important factor in determining the distribution of organisms, which is evidenced by the lack of sessile invertebrates such as sponges and bryozoans in waters less than 30 m deep (Dayton *et al.*, 1969). For motile invertebrates, the risk of becoming entrapped and frozen must be offset by a substantial gain in the increased food and protection available in the anchor ice zone.

2. PHYSIOLOGICAL AND CHEMICAL RESPONSES TO FREEZING ENVIRONMENTS

For years man has attempted to freeze organisms in such a way that, after thawing, they will survive. So far this has been accomplished only for spermatozoa, tumor cells and red blood cells which are all single cells, and always with the aid of cryoprotective agents such as glycerol, dimethyl-sulfoxide or polyvinylpyrolidone (see review by Mazur, 1970). Successful cryopreservation has never been possible in the laboratory for biological materials showing a level of organization higher than the cellular level, yet in nature examples can be found in the arthropod and molluscan phyla where biological organization is indeed complex. In the arthropods, both larval and adult insects can tolerate freezing for long periods at temperatures as low as −40°C (Asahina, 1969; Miller, 1969). In most of the insects freezing tolerance is generally attributed to the protection afforded by the large amounts of glycerol which accumulate with the onset of winter (Asahina, 1969). However there are some insects, both larval and adult, which survive wintertime freezing in the absence of glycerol (Asahina, 1969).

Since glycerol has more often than not been implicated in freezing tolerance of insects, it would seem likely that it might be present in inter-tidal invertebrates and be responsible for their freezing tolerance. Examination of the tissues and body fluids of *Mytilus edulis*, *Littorina littorea* and *Crassostrea virginicus* revealed that glycerol as well as dimethylsulfoxide was absent (Kanwisher, 1966; Williams, 1970).

Calorimeteric data on *M. edulis* and *V. mercenaria*, the latter a form less tolerant to low temperatures, led Williams (1970) to postulate that in the frost tolerant *M. edulis*, 20% of the body water is osmotically inactive and is possibly "bound water". That is, this fraction of the intracellular water is unavailable for ice formation in the extracellular space, despite a lower vapor pressure there relative to the intracellular space. This "tying up" of cellular water does not occur in the frost intolerant *V. mercenaria*. The nature of the solutes responsible for tying up or binding cellular water is not known, but inorganic ions and low molecular weight organic molecules such as amino acids and sugars are unlikely candidates as the vapor pressures of their solutions at high concentrations do not significantly deviate from what one would predict on the basis of colligative relationships. It is possible that some unusual compound might be present which exerts an osmotic influence out of proportion to the number of particles in solution. However it seems more likely that the osmotically inactive water or "bound water" arises from the structural properties of the cell itself rather than from the presence of a soluble compound.

B. Fishes

1. INCREASED LEVELS OF INORGANIC IONS

In most fresh water and marine teleosts sodium chloride is the principal electrolyte present in the body fluids, and it is responsible for 80 to 90% of their body fluid osmolality. Potassium and calcium ions, urea and the free amino acids account for much of the remainder. When temperate and boreal marine teleosts encounter low water temperatures, the concentration of sodium chloride in the blood serum increases (Eliassen *et al.*, 1960; Pearcy, 1961; Woodhead and Woodhead, 1959; Gordon *et al.*, 1962; Raschack, 1969; Umminger, 1969*a*). The extent of the increase in this electrolyte varies among species (Table I). For instance when *Fundulus heteroclitus* is transferred from 20° to −1.5°C water, the concentration of sodium chloride in the plasma increases by only 13% (Umminger, 1969*a*), whereas in the winter flounder, *Pseudopleuronectes americanus*, it increases by 18% (Pearcy, 1961). In *Myoxocephalus scorpius* taken from Kiel Bay on the Baltic Sea where the winter water temperature is −0.5°C, the electrolyte content of the plasma is 20% over that found in the summer when water temperatures are around 10°C (Raschack, 1969). In the boreal cod, *Gadus callarias*, the chloride level in plasma taken from specimens captured at −1.5°C is 15% over that of those captured at 15°C (Eliassen *et al.*, 1960; Leivestad, 1965). In the antarctic fish *Notothenia rossii*, the serum concentration of sodium chloride in the winter is 15% over that of the summer (Smith, 1970, 1972).

With most temperate fishes the depression of the freezing point of the blood associated with low temperatures is only partially due to an increase in sodium chloride. Pearcy (1961) reported that in *P. americanus* sodium chloride accounts for 83% of the depression of the plasma freezing point in the summer, whereas in the winter it accounts for only 57%. In other words when the freezing point depression of the plasma is increased from 0.63° to 1.10°C, about 0.4°C of the increase is due to solutes other than sodium chloride. In *M. scorpius* taken from the brackish water of Kiel Bay in the Baltic Sea the decrease in plasma freezing point (−0.64° to −0.86°C) associated with low temperature (−0.5°C) is also only partially due to

TABLE I: Effect of temperature on electrolyte levels in blood of temperate, arctic and antarctic marine teleosts.

Species and Season	Water Temp. (°C)	Sodium (mequiv/l)	Chloride (mequiv/l)	References
TEMPERATE				
Fundulus heteroclitus				
warm	+20·0	182	145	Umminger (1969*a*)
cold	−1·5	208	171	
Pseudopleuronectes americanus				
summer	+16·0	194	147	Duman and DeVries (1973)
winter	−1·0	250	178	
ARCTIC				
Myoxocephalus scorpius				
summer	+4·0 to +7·0	—	200	Scholander *et al.* (1957)
winter	−1·7	216	234	Gordon *et al.* (1962)
ANTARCTIC				
Trematomus borchgrevinki	−1·9	274	235	
Trematomus hansoni	−1·9	270	254	DeVries (1971*b*)
Notothenia gibberifrons	0·0 to +1·0	243	220	
Notothenia neglecta corriiceps	0·0 to +1·0	243	220	
Trematomus bernacchii				
summer	+0·5	248	231	DeVries (1971*b*)
winter	−1·9	—	254	

increases in electrolytes, and the remainder is attributed to non-dissociated organic compounds. In *Taurulus bubalis*, a long spined sea scorpion, the decrease in plasma freezing point at low temperatures is not as great as that of *M. scorpius* and is due exclusively to inorganic electrolytes (Raschack, 1969).

Increases in the serum levels of sodium chloride in response to low temperatures in the laboratory and in the environment have often been attributed to the breakdown of osmoregulatory ability in many temperate and boreal fishes (Woodhead, 1964; Doudoroff, 1945). However, with many arctic and antarctic fishes living permanently in near-freezing habitats, the levels of sodium chloride in the blood are higher than those found in temperate fishes and in fact show either no variation (Table I) (DeVries, 1968), or only a little variation with season (Smith, 1970). In addition, the levels of sodium chloride in the blood of the arctic fishes, *M. scorpius* and *Gadus ogac*, show a natural seasonal variation, increases being observed in the levels of sodium chloride as well as in non-dissociated organic compounds during the winter when water temperatures are low (Scholander *et al.*, 1957; Raschack, 1969). With these arctic fishes sodium chloride accounts for 87% of the depression of the freezing point of the blood in the summer, while in the winter it accounts for only 62% in the sculpin and 79% in the fjord cod (Gordon *et al.*, 1962). In addition, analyses for potassium during the winter indicate that this ion is not present at concentrations much higher than those found in the plasma of temperate marine fishes which inhabit warmer waters.

Studies of fishes in the Antarctic indicate that sodium chloride accounts for slightly less than half of the depression of the serum freezing point in those fishes showing the greatest resistance to freezing, while most of the remainder results from glycoproteins which lower the freezing point. For instance, in *T. borchgrevinki*, which lives in the coldest water where ice is most abundant, sodium chloride accounts for only 42% of the depression of the freezing point. In the cases of *T. bernacchii* and *T. hansoni* this electrolyte accounts for 44% when they inhabit the ice-laden shallow waters and 46% when inhabiting the deep waters where ice is absent. In *Chaenocephalus aceratus*, a hemoglobin-free fish whose serum freezes at $-1.3°C$, sodium chloride accounts for 55% of the freezing point depression.

The concentrations of potassium, magnesium and calcium ions have not been determined in the blood of the antarctic fishes, except in the serums of *Notothenia neglecta* and *N. rossii* where potassium levels are not exceptionally high (Smith, 1970). In the arctic sculpin and fjord cod the concentration of potassium ion is about the same as that found in teleosts living in warmer temperate waters (Gordon *et al.*, 1962).

In contrast to marine teleosts, fresh water fishes tend to lose electrolytes from their body fluids as temperature is decreased (Prosser *et al.*, 1970; Umminger, 1971). In view of this fact it seems likely that the gain and loss of sodium and chloride ions in teleosts results from the effect of temperature upon the enzymes involved in ion transport. The fact that K^+/Na^+ ratios in the muscle of the cod, *G. callarias*, decrease as the temperature of exposure is lowered does, however, suggest that the temperature effect varies for different electrolyte transport systems (Leivestad, 1965). The effect of temperature on the permeability of membranes must also be important, as recent studies have shown that a few of the antarctic fishes are extremely impermeable to ions and water (Read, 1973). Whether this will be the case for other cold-water fishes is not known.

Although the levels of electrolyte are increased in marine teleosts as temperatures decrease, it is not clear that this increase has much significance in freezing avoidance. It can be argued that increased salt levels do depress the freezing point of the body fluids. However an increase in sodium chloride from 150 mM/l to 250 mM/l would lower the freezing point by only 0.35°C, which is quite small compared to the difference between the freezing point of a temperate fish (−0.7°C) and that of sea water (−1.86°C). Thus, since the temperature effect on electrolyte levels can be observed in species of fishes which experience freezing conditions (ice and low temperatures) and also in those that experience only low temperatures, it would appear that it is a general effect and not a mechanism which fishes have evolved to avoid freezing.

2. INCREASED LEVELS OF SMALL ORGANIC SOLUTES

Since inorganic ions have been found to account for so little of the depression of the freezing point of plasma in some of the cold adapted fishes of the temperate and polar regions, the presence of high concentrations of osmotically active organic solutes has been investigated (DeVries, 1968; Umminger, 1969a, and Raschack, 1969). Even with the serums from cold-acclimated *F. heteroclitus*, where most of the increase in plasma freezing point depression is due to inorganic ions, part of it results from elevated levels of free glucose, in which increases of 430% have been observed.

Umminger (1969b) speculates that the primary role of the increase in the level of glucose is the prevention of spontaneous nucleation in the absence of ice. However, the high levels of glucose cannot afford protection against nucleation if external ice is present (Umminger, personal communication). It has been suggested that the increase in serum freezing point depression not due to sodium chloride in *P. americanus* during the winter, results from the presence of some organic solute (Pearcy, 1961), which recently has been shown to be large molecular weight proteins (Duman and

DeVries, 1973). In the plasma of arctic sculpin and fjord cod the concentrations of organic solutes commonly found in the blood of teleosts are not extraordinarily high. However, the levels of non-protein nitrogen (NPN) are two and four times higher in the sculpin and cod, respectively, than in temperate marine teleosts. Since the NPN in the body fluids of most organisms can be attributed to small nitrogen containing compounds, it was reasonable for Gordon *et al.* (1962) to postulate that there would be more than enough solute to account for the low freezing points if all of the NPN were present in molecules containing only one nitrogen atom per molecule. However, despite their systematic analyses of the serum, the high level of NPN could not be correlated with high serum levels of small nitrogen-containing compounds such as urea, free amino acids, purines, pyrimidines and amines. Concentrations of other solutes such as reducing sugars, alcohols and lipids were not significantly high either.

3. MACROMOLECULES AS BIOLOGICAL ANTIFREEZE AGENTS IN FISHES

Examination of the osmotic role played by salts and low molecular weight organic compounds identified in the serum revealed that they could supply enough solute to account for only 68% of the winter serum freezing point depression in the sculpin. It was speculated that the remaining 30% was due to an "antifreeze" compound whose presence was associated with the high NPN level (Gordon *et al.*, 1962). However, despite exhaustive analyses of the serum, no compound with antifreeze properties was identified. As with the populations of *M. scorpius* in the fjords of northern Labrador, the concentrations of electrolytes in the winter populations inhabiting Kiel Bay in the Baltic Sea accounts for only 73% of the plasma freezing point depression, in contrast to 78% in the summer. This winter increase in freezing point depression was thus interpreted as being partly due to an increase in non-dissociated organic compounds which, however, were not identified (Raschack, 1969).

In early studies (Gordon *et al.*, 1962), the quest for the nature of the solutes which imparted freezing resistance to northern fishes involved examination of levels of electrolytes in the blood. When it was found that the latter were not significantly increased in the cold, efforts were then turned towards identifying common low molecular weight solutes such as glucose, urea and the free amino acids. When none of these solutes was found to exist in high concentration, they next screened for low molecular weight NPN compounds which is easily done by precipitating the proteins with trichloroacetic acid (TCA), sodium tungstate or zinc hydroxide. Examination of the supernatants revealed that one of the fish had high levels of NPN and the other did not, even though they were both resistant to freezing.

In the antarctic fishes high levels of NPN have also been found. As in the case of the arctic fishes, these high levels are not due to elevated concentrations of urea or free amino acids. In the antarctic fishes the majority of the NPN has been shown by dialysis experiments to be associated with macromolecular solutes (DeVries and Wohlschlag, 1969; DeVries, 1970). The high NPN was subsequently shown to be due to the presence of a group of glycoproteins which are responsible for the low blood serum freezing points in polar fishes (DeVries and Wohlschlag, 1969; Duman and DeVries, 1973).

As an example, when the serum of the temperate black perch, *Embiotoca jacksoni*, which has a freezing point of −0.7°C, was dialyzed against dis-

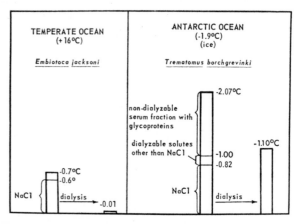

FIG. 3. Freezing points of blood serums from the temperate marine fish, *Embiotoca jacksoni* and the antarctic fish *Trematomus borchgrevinki* before and after dialysis against distilled water. The pore size of the dialysis membrane is approximately 24Å.

tilled water to remove the small solutes, the freezing point rose to −0.01°C, indicating that none of the depression of the freezing point was due to macromolecular solutes (Fig. 3). In contrast to the perch, the serum of the antarctic cod, *T. borchgrevinki*, has a freezing point of −2.1°C which after dialysis froze at −1.1°C, a value greater than 50% of the total freezing point depression. By partitioning the blood solutes with a dialysis membrane Scholander and Maggert (1971) have been able to show that the low freezing point of the blood plasma (−2.0°C) of the arctic sculpin, *Myoxocephalus scorpioides*, is also associated with the colloidal fraction. In the temperate winter flounder, *P. americanus*, 0.8°C of the total freezing point depression of −1.4°C remains with the non-dialyzable fraction of the blood (Duman and DeVries, 1973). Recent unpublished studies in our laboratory have shown that for fishes with freezing points above −1°C, the depression results exclusively from small solutes while for those that have freezing points lower than −1.0°C, a non-dialyzable macromolecule always gives rise to part of the depression.

Recently the blood of the shorthorn sculpin, *Myoxocephalus scorpius*, the tomcod, *Microgadus tomcod*, and the winter flounder, *Pseudopleuronectes americanus*, has been shown to have macromolecules which possess antifreeze properties. All these macromolecules are soluble in TCA. An exception to this trend can be found in the antarctic zoarcid fish, *R. dearborni*, whose macromolecular antifreeze is insoluble in 10% TCA but soluble in 95% ethanol. It is possible that eventually the high level of NPN in the polar cod, *G. ogac*, may also be attributed to the presence of macromolecules.

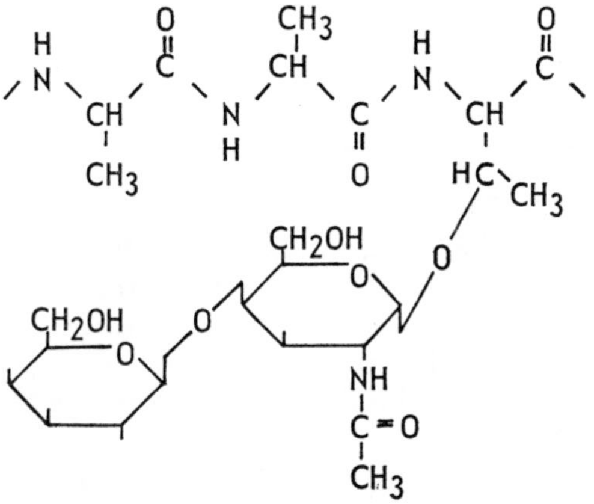

FIG. 4. Basic repeating unit of glycoproteins with antifreeze properties. The protein backbone is composed of two amino acids, alanine and threonine in the sequence of -ala-ala-thr- with the disaccharide attached to every threonine.

The glycoproteins have been isolated and purified by ion exchange chromatography from the colloidal fraction of the blood of several antarctic fishes (DeVries *et al.*, 1970). They are composed of repeating units of glycotripeptides in which the disaccharide, β-D-galactopyranosyl-(1→4)-2-acetamido-2-deoxy-α-D-galactopyranose is linked to the threonine residue of the tripeptide alanyl-threonyl-alanine (Fig. 4) (DeVries, *et al.*, 1971; Lin *et al.*, 1972; Shier *et al.*, 1972). A total of eight distinct glycoproteins has been isolated by ion exchange chromatography and identified on the basis of analytical acrylamide gel electrophoresis. These glycoproteins have been assigned numbers 1 to 8 beginning from the cathode and proceeding towards the anode. Most of the glycoproteins recovered after purification procedures are mixtures of glycoproteins 3, 4 and 5 and mixtures of 7 and 8. These mixtures have been further purified to yield the individual glycoproteins. The molecular weight of the smallest glyco-protein is 2600 daltons while that of the largest is approximately 33 000 daltons (DeVries *et al.*, 1970; Lin *et al.*, 1972). The molecular weights of the intermediate sizes are given in Table II. The most common forms are

TABLE II: Molecular weights of glycoproteins isolated from the serum of the antarctic fish, *Trematomus borchgrevinki*.

Glycoproteins	MW (daltons)
1	33 700
2	28 800
3	21 500
4	17 000
5	10 500
6	7900
7	3500
8	2600

glycoproteins 3, 4, 5 ,7 and 8, glycoproteins 1 and 2 being present only in trance amounts (DeVries *et al.*, 1970). The three smallest glycoproteins (6, 7 and 8) differ from the larger ones (3, 4 and 5) not only in size but also in that beginning at the seventh amino acid position, proline residues replace some of the alanine residues in the glycotripeptide repeating sequence until the C-terminal residue is reached (Lin *et al.*, 1972) (Fig. 5).

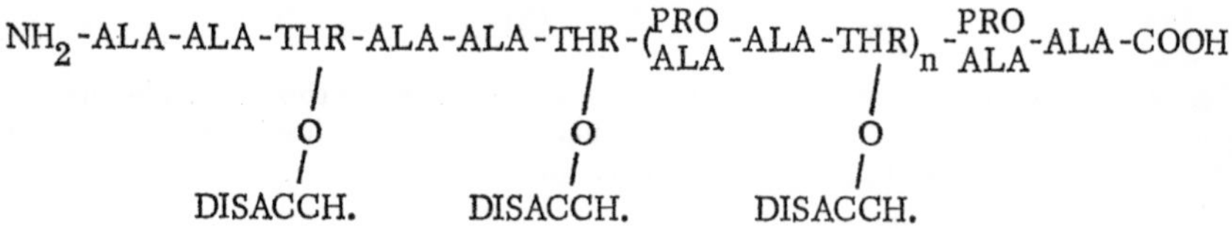

$$NH_2-ALA-ALA-THR-ALA-ALA-THR-\binom{PRO}{ALA}-ALA-THR)_n-\binom{PRO}{ALA}-ALA-COOH$$

DISACCH.   DISACCH.   DISACCH.

DISACCH. : β-D-galactosyl (1→4) α-N-acetylgalactosamine

Glycoprotein 7, n=4; glycoprotein 8, n=2.

FIG. 5. Complete structure of glycoproteins 7 and 8. Starting from the seventh amino acid from the amino terminus, proline is occassionally found to replace alanine. (From Lin *et al.*, 1972).

Despite knowledge of the primary structures of the glycoproteins their exact secondary structure is not known. However, studies involving dialysis, viscosity, and circular dichroism all indicate that they are expanded structures (DeVries *et al.*, 1970). Comparisons of the spectra obtained from circular dichroism measurements of polyglutamic acid and the glyco-proteins show they are similar. Tiffany and Krimm (1969) argue that polyglutamic acid exists as an extended helix. Whether or not the glyco-proteins are random coils or extended helices is not known, but it is clear that they are expanded structures, a feature which is undoubtedly of great significance when considering their function.

To most people "freezing point" is the temperature at which a solid is in equilibrium with its liquid, and therefore freezing point and melting point are assumed to be the same. The glycoproteins, however, lower the temperature at which ice will form, but, remarkably, do not also lower the melting point of the ice. For example, a 2% glycoprotein solution freezes at approximately −0.9°C, but it will not melt until the temperature is raised to −0.02°C, the melting point determined by the solution's normal colli-gative properties. Thus the glycoproteins can be characterized as having "antifreeze" properties, where it is stressed that only the freezing process, and not the melting process, is affected. The difference between the freezing point and melting point has been referred to as a thermal hysteresis in the freezing-melting behavior (DeVries, 1971a). The magnitude of this hysteresis in solutions of both the large and small glycoproteins at several concentrations is illustrated in Fig. 6.

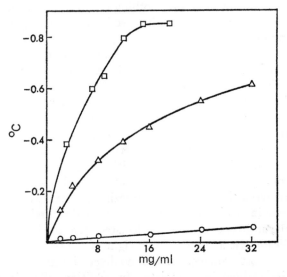

FIG. 6. Freezing points and melting points of aqueous solutions of glycoproteins as a function of concentration. Freezing points of glycoproteins 1 to 5, —□—□—; glycoproteins 7 and 8, —△—△—. The melting points, —O—O—, of all the glycoprotein solutions are nearly the same. (From Lin *et al.*, 1972.)

A thermal hysteresis has been detected in the blood of all the antarctic fishes examined so far (DeVries, 1971a; Hargens, 1972) when the freezing points and melting points are measured directly (see section VI). Fishes living in the coldest ice-laden water have the lowest freezing point and the greatest difference between the freezing and melting points. This obviously is a reflection of greater amounts of glycoproteins in those fishes associated with colder waters (DeVries, 1971a; DeVries and Somero, 1971). For example, among the different members of the *Trematomus* fishes of McMurdo Sound, the magnitude of hysteresis is greatest for *T. borchgrevinki* which encounters ice crystals in its habitat, whereas it is smallest in *T. loennbergii* which remains in the ice-free deep water (Table III).

In the Arctic, the melting-freezing point technique has been used to demonstrate the presence of a thermal hysteresis in the blood of several shallow-water fishes (Scholander and Maggert, 1971). During the winter the shallow-water shore cod, *Eleginus gracilus*, and the sculpin, *M. scorpioides*, captured in freezing waters have melting-freezing point differences of 0.8° and 1.0°C, respectively, while the deep-water cod, *Theragra chalcogramma*, and the flounder, *Limanda aspera*, have a difference of only 0.1°C or less. The hysteresis in *E. gracilus* was shown to result from the presence of an antifreeze in the colloidal fraction of the blood by preparing an ultrafiltrate of the plasma (Scholander and Maggert, 1971). Other than its relatively large size, nothing is known about the antifreeze in this arctic cod or in any of the other arctic fishes having unusually low blood freezing points.

Hystereses in the freezing-melting behaviors of the blood have also been observed in temperate fishes which experience freezing conditions during the winter. The freezing points of the blood serums of several northern fishes such as the atlantic smelt, *Osmerus mordax*, atlantic tomcod, *Microgadus tomcod*, grubby sculpin, *M. aeneus*, shorthorn sculpin, *M. scorpius*, and the winter flounder, *P. americanus*, are between −1.4° and −1.7°C while the melting points are between −0.7° and −0.9°C (Table III). When the blood serums of these fishes are dialyzed across cellulose membranes of approximately 24Å pore size, the solutes responsible for the differences between the melting and freezing points always remain inside the dialysis bag indicating they are macromolecules. Preliminary chemical analyses of the antifreeze compounds isolated from the blood of the winter flounder indicate that it contains only eight of the twenty common amino acids found in most proteins, and that no detectable amount of galactose is present in its carbohydrate moiety. In addition, dialysis experiments indicate that the minimum molecular weight is greater than 5000 daltons. Thus, although a macromolecule, it differs from the glycoprotein antifreezes of the antarctic fishes both in size and composition.

TABLE III: Freezing and melting points and their differences in antarctic, arctic and north-temperate fishes.

	Water Temperature (°C)	Freezing point (°C)[1]	Melting point (°C)[2]	Freezing point Melting point Difference (°C)
ANTARCTIC				
Trematomus borchgrevinki	−1·9	−2·34	−1·07	1·27
Notothenia gibberifrons	+0·5	−2·24	−1·07	1·17
Notothenia neglecta coriiceps	+0·5	−2·06	−1·00	1·06
Rhigophila dearborni	−1·9	−1·70	−0·94	0·76
Chaenocephalus aceratus	+0·5	−1·45	−0·98	0·47
ARCTIC				
Myoxocephalus scorpioides	−1·7	−2·0	−1·0	1·0
Eleginus gracilus	−1·7	−2·0	−1·2	0·8
NORTH-TEMPERATE				
Myoxocephalus scorpius	−1·2	−1·69	−0·81	0·88
Myoxocephalus aeneus	−1·2	−1·60	−0·84	0·76
Microgadus tomcod	−1·2	−1·51	−0·81	0·70
Pseudopleuronectes americanus	−1·2	−1·47	−0·71	0·76
Osmerus mordax	−1·2	−1·22	−0·82	0·40

[1] Freezing point was taken as the temperature of initial ice crystal growth in the presence of a seed crystal.
[2] Melting point was taken as the temperature at which the last ice crystal melted.

At first sight it might seem unusual that rather different molecules possess the same antifreeze capability. However, in the antarctic fishes the antifreeze activity appears to depend upon a specific hydrogen bonding of the hydroxyl groups of the carbohydrate moiety to the oxygens in the ice lattice, in which case the positions of the hydroxyls would have to correspond to those of the oxygens (see VB). It is not unreasonable to expect that other configurations of amino acids and sugars could exist that would have a similar correspondence. After all, other macromolecules such as myoglobin and hemocyanin have similar basic functions but differ greatly in size and composition.

4. SEASONAL OCCURRENCE OF ANTIFREEZE COMPOUNDS

Compared to antarctic fishes, northern fishes experience large seasonal temperature changes (10° to 20°C) in their environments and as a consequence they have need for and possess antifreeze compounds only during the winter. Exposure to temperatures of 10° to 12°C causes the winter flounder, *P. americanus*, to lose all of its antifreeze within three weeks (Table IV), while similar acclimation causes the shorthorn sculpin, *M. scorpius* and the tomcod, *M. tomcod* to lose only about half of their antifreeze. During the summer when water temperatures are 12° to 15°C, none of these fishes possesses antifreeze as evidenced by the lack of hystereses in the freezing-melting behavior of their blood serums.

Changes in the levels of antifreeze in the winter flounder appear to be associated with the photoperiod as well as the temperature, but the interplay between these environmental factors is not entirely understood at this time. It seems unlikely that temperature alone could be responsible because sudden changes in the water temperature in the shallow bays and inlets of the North Atlantic Coast during early spring or late autumn could cause these fishes to degrade their antifreeze compounds prematurely. Such a loss would endanger them if cold weather caused renewed freezing in the surface water before the antifreeze compounds could be resynthesized.

For the antarctic fishes, extended periods of warm acclimation cause only a small rise in the blood serum freezing point (Table IV). Most of the change is due to decreases in the levels of sodium chloride and, therefore, the levels of glycoprotein antifreeze cannot be modulated by temperature. It appears that antarctic fishes have lost the capability for adapting to higher temperatures since warm acclimation does not cause the upper incipient lethal temperature of +6°C to rise in *T. borchgrevinki* and *R. dearborni*, even after 60 days at +4°C (DeVries, 1973).

The seasonal change in antifreeze level in the northern fishes is striking. Although the factors responsible for the initiation of antifreeze synthesis

TABLE IV: The effect of warm acclimation on the freezing and melting point of blood serums of antarctic and north-temperate fishes.

Species	Acclimation period (days)	Environmental and Acclimation Temperature (°C)		Freezing point[1] (°C)		Melting point[1] (°C)		Freezing-melting point difference (°C)	
		Env.	Acclim.	Before	After	Before	After	Before	After
ANTARCTIC									
Trematomus borchgrevinki	30	−1·9	+2·0	−2·34	−2·14	−1·07	−0·94	1·27	1·1
Rhigophila dearborni	30	−1·9	+2·0	−1·70	−1·50	−0·94	−0·84	0·76	0·66
NORTH-TEMPERATE									
Myoxocephalus scorpius	36	−1·2	+12	−1·69	−0·68	−0·81	−0·65	0·88	0·03
Pseudopleuronectes americanus	24	−1·2	+12	−1·47	−0·58	−0·71	−0·58	0·76	0·0
Microgadus tomcod	34	−1·2	+12	−1·51	−0·60	−0·81	−0·58	0·70	0·02

[1] Freezing and melting points are taken respectively as the temperatures at which ice crystal formation occurs in the presence of a small seed crystal, and at which the last ice crystal melts.

in the autumn and for antifreeze degradation in the spring remain to be elucidated in detail, it is clear that temperature must play a prominent role. Perhaps low temperatures weaken the binding of a repressor protein to the operator region of the gene to the extent that transcription is allowed to proceed. In bacteria repression and induction occur within a few minutes or hours. If such a control mechanism is involved in the bio-synthesis of fish antifreezes, it apparently takes much longer because acclimation-induced changes, such as the appearance of new isozymes in fishes, takes at least a couple of weeks.

The loss in ability of the antarctic fishes to alter the level of glycoprotein antifreeze is as striking as the ability of northern fishes to alter theirs seasonally. In the constant low temperature environment of the Antarctic Ocean the need for enzymes and membranes which function at higher temperatures no longer exists. Perhaps as a consequence, over evolutionary time the genetic variability may have been reduced compared to a temper-ate fish which experiences large temperature changes. The fact that neither warm nor cold acclimation alters the glycoprotein antifreeze levels is consistent with this hypothesis.

V. Glycoprotein antifreezes in antarctic fishes

The properties of a solution that depend upon the number of particles in solution and not on the kinds of particles are called colligative properties. These properties are a manifestation of a common phenomena, i.e., solute molecules decrease the tendency of water molecules to escape from one phase to another or from one solution to another. The colligative properties of a solution are: (1) the osmotic pressure, (2) the vapor pressure lowering, (3) the boiling point elevation and (4) the freezing point depression. The relationship that exists between the freezing point depression and molal concentration is

$$\Delta T = \frac{RT_0^2 M}{1000\,L}\,m$$

where R is the gas constant, T_0 is the normal freezing point, M is the molecular weight of the solvent, L is the molar latent heat of fusion and m is the molality of the solute (Bent, 1965). The freezing point depressions for galactose and sodium chloride are shown in Fig. 7. The fact that the curve shown for sodium chloride has a slope almost twice that of galactose clearly illustrates that the colligative properties are dependent upon the number and not upon the kind of particles in solution. The fact that sodium chloride ionizes into two particles at the same molal concentration as galactose explains why the former has twice the effect on the freezing point of water as the latter which is several times larger than sodium chloride.

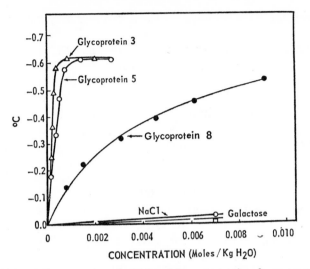

FIG. 7. Freezing points as a function of molal concentration for aqueous solutions of galactose, sodium chloride and glycoproteins 3, 5 and 8. Freezing points were determined with a Fiske Osmometer. Freezing points of glycoprotein 8 were determined by direct observation. (From DeVries, 1971*b*).

The freezing point data present in Fig. 7 clearly show that the manner in which the glycoproteins lower the freezing point of water is not a colligative effect. This is supported by the hysteresis in the freezing-melting behavior, as well as by the fact that the small glycoproteins (6, 7 and 8) depress the freezing point of water only half as much as the much larger glycoproteins 3, 4 and 5.

A. Mechanism: Water structuring vs. water binding theory

The arrangement of molecules in a liquid has neither the simplicity of long range order as found in a solid, nor the simplicity of independent motion as in a gas. Liquid structure is thus caught somewhere in between these two extremes and is accordingly the most poorly understood. In the case of intracellular water, this uncertainty in structure has caused something of a bottleneck in several aspects of cellular physiology, and thus much research is currently being directed towards it. Mostly on the basis of NMR and absorption isotherm studies, there is evidence that intracellular water has a higher degree of order than that found in pure bulk water. As an example of the importance of this concept, it has been suggested (Tait and Franks, 1971) that bound water may be responsible for the large K^+ and Na^+ gradients found across cellular membranes, thus relieving the cell of the huge work load involved in operating membrane ion pumps.

Water structuring in cells probably originates at the surface of macromolecules, and of these molecules fibrous and extended coil configurations tend to bind much greater amounts of water than globular configurations (Ling, 1972). One might expect this since in order to influence large volumes of water, the architecture of a macromolecule must be such that a maximum amount of interaction can occur between the molecule and water. From this point of view then, the antifreeze glycoproteins are good candidates for influencing water structures. To date all of the physcial data indicate that they are expanded molecules. Circular dichroism studies indicate that they may in fact be extended helical coils. In addition the larger glycoproteins have a more pronounced effect on the depression of the freezing point (Raymond and DeVries, 1972) which suggests that they interact with larger volumes of water than do the smaller glycoproteins.

In an attempt to gain some insight into how expanded molecules can intereact with water, it is worthwhile to examine the non-ideal behavior of solutions of certain extended long chain, flexible polymers composed of repeating residues. The anomalous colligative effects observed with these polymers could be explained on the basis that each repeating residue acts independently of its neighbor (Brey, 1958). Alternatively this anomalous behavior might result from the fact that such polymers contain considerable solvent within the domain of their coils. Solvent within the coils is trapped and is moved wherever the molecule moves (Tanford, 1963). Such water could conceivably be immobilized by means of hydrogen bonding, thus effectively removing it from the bulk solvent and making it unavailable for ice formation. With the glycoproteins, it is possible that water is likewise immobilized within their structure. There are a few instances in which the freezing points of structurally immobilized water are lower than expected. Solvent trapped within synthetic gels consisting of polyvinyl alcohol and polyacrylic acid has a freezing point 1° to 2°C lower than the swelling buffer (Bloch et al., 1963). A similar structurally based mechanism may explain why the glycoproteins lower the freezing point of water.

A water structuring hypothesis, however, seems inconsistent with the hysteresis in the freezing-melting behavior, and an alternative and perhaps more attractive hypothesis is that the glycoproteins exert their antifreeze effect by interfering with freezing at the water-ice interface by some shielding or coating process (DeVries, 1971a). Perhaps the hydroxyls of the carbohydrate moiety are positioned at regular intervals along the polypeptide in such a manner that they can hydrogen bond to the oxygens in the ice lattice. The presence of a thermal hysteresis does suggest that the glycoproteins bind to the surface of ice crystals, effectively lowering their potential for nucleation (DeVries, 1971a). Evidence that specific binding is involved comes from a number of experiments. One is that during freezing the glycoproteins are always equally partitioned between the liquid and solid phases (Duman and DeVries, 1972). In other words they are not excluded from the solid phase during freezing as are solutes such as sodium chloride, galactose and large molecular weight dextrans (MW = 20 000 and 40 000 daltons). However, glycoproteins which have been inactivated by converting them to carboxyglycoproteins lose this binding property.

Another observation which supports the binding theory is that the freezing points obtained for the serum of the antarctic cod show a great deal of variability from one determination to another (DeVries, 1971a). In the case of *T. borchgrevinki*, this variability amounts to 0.7°C and recent observations show that it is related to the size and surface characteristics of the seed crystal. A large, irregular crystal will begin growing at higher temperatures than a smaller, smooth one.

Two possible ways in which the binding of the glycoproteins to ice surfaces may prevent ice from growing and which are consistent with the above observations are (1) by increasing the surface free energy of the ice (Raymond and DeVries, 1972) and (2) by coating the ice with a barrier of unfreezable water. Although we have no evidence that either of these mechanisms is employed by the glycoproteins, it may be instructive to briefly describe them to show how a melting-point freezing point hysteresis might arise.

Because there is a surface tension associated with an ice-water interface, an irregular surface, such as a corrugated surface, would have a higher free energy than a smooth one. Thus, for a surface which is forced to grow in an irregular manner, a greater driving force, or lower temperature would be required before freezing occurs. The quantity which determines this temperature is dS/dV, the derivative of the surface area with respect to the volume of the crystal (Kuhn, 1956):

$$\Delta T = \frac{\sigma T_o M}{L \rho} \frac{dS}{dV}$$

where σ is the coefficient of surface tension, T_o is the normal freezing point, M is the molecular weight of the solvent, L is the latent heat of fusion, and ρ is the density of ice. It is possible that the glycoproteins cause such a freezing point depression by interfering with the growth of ice, forcing it to grow in an irregular manner. An equilibrium would be maintained between the solid and liquid phases over a finite temperature range by a sort of negative feedback. For example, a decrease in temperature would cause some freezing to occur. This would happen in such a way as to increase the irregularity, or dS/dV, of the crystal until the freezing point is lowered to the new temperature, at which point equilibrium is again reached. Bulk freezing would occur only when the temperature is dropped below the maximum freezing point depression, which is determined by the maximum irregularity or maximum value of dS/dV allowed. Bulk melting would occur when the temperature is raised above the minimum freezing point depression, which is determined by the minimum value of dS/dV (which is approximately zero for a smooth crystal).

A second possibility is that the glycoproteins are surrounded with a layer of bound water which is reportedly very resistant to freezing (Kuntz et al., 1969). If the glycoproteins then bind to ice, a coating of bound, unfreezable water might effectively remove the crystal as a nucleation site. The freezing point would be the temperature at which ice could penetrate the bound water barrier, and the melting point would be the same as that for ordinary ice.

B. Effect of chemical and enzymatic modifications on glycoprotein activity

To determine how the structure of a protein is related to its function, protein chemists often alter the structure of a protein by the use of enzymatic or chemical modifications and then study the effects on function.

A number of chemical and enzymatic modifications of the purified glycoproteins have been done in the past several years and they all strongly implicate the hyroxyl-rich carbohydrate moiety in an important functional role (DeVries et al., 1970; DeVries 1971a; Shier et al., 1972). For example, acetylation of as few as 30% of the hydroxyl groups results in loss of all activity, which can be restored upon deacetylation (Komatsu et al., 1970). If the galactose ring is opened by treatment with sodium periodate or removed by Smith degradation, activity is again lost (Komatsu et al., 1970). More specific modifications such as acetonation of the glycoproteins results in an isopropylidene derivative which is also inactive (Shier et al., 1972). Mild acid hydrolysis of this derivative removes the isopropylidene group and regenerates the original activity. Protective blockage of the various hydroxyl groups of the sugar moiety followed by acetylation and removal of the protective group has allowed us to alter selectively each hydroxyl group and test the effect on activity (Table V) (Shier et al., 1972). The only modification which did not destroy activity was the oxidation of the hydroxyl of carbon 6 on the galactose and N-acetyl-galactosamine residues to an aldehyde group (DeVries, 1971a). If this

TABLE V: Freezing point depressing activity of chemically modified glycoproteins (Shier *et al.*, 1972).

Structure*	Freezing point depressing activity†
$R_1 = R_3 = H$ $R_2 = CH_2OH$	100
$(R_1)_2 = >C(CH_3)_2$ $R_2 = CH_2OH$, $R_3 = H$	<5
$(R_1)_2 = >C(CH_3)_2$, $R_2 = CH_3OAc$, $R_3 = Ac*$	<5
$R_1 = H$, $R_2 = CH_2OAc$, $R_3 = Ac$	<5
$R_1 = R_3 = H$, $R_2 = CHO$	95
$R_1 = R_3 = H$, $R_2 = CO_2H$	<5
$(R_1)_2 = >C(CH_3)_2$, $R_2 = CHO$, $R_3 = H$	<5
$(R_1)_2 = >C(CH_3)_2$, $R_2 = CHO$, $R_3 = Av$	<5
$R_1 = H$, $R_2 = CHO$, $R_3 = Ac$	<5

*Ac is the acetyl group.
†The freezing point depressing activity of the native glycoprotein was assigned as 100. The activities of other compounds were expressed relative to the native glycoprotein.

aldehyde group is then oxidized to a carboxyl group, the product is an inactive carboxyglycoprotein (Shier *et al.*, 1972).

From these studies, it is clear that the hydroxyl groups of the carbohydrate moiety are important for the glycoprotein's function. The relationship of size to function is also important. For example, if a few of the peptide bonds in the polypeptide chain are cleaved using a proteolytic enzyme, then some of the activity is sacrificed. In fact only two bonds need to be cleaved before glycoprotein 5 begins to lose activity (Komatsu *et al.*, 1970; Raymond and DeVries, 1972). Figure 8 shows how the activity of the cleaved glycoproteins falls off with decreasing molecular weight. The pattern is similar to that of the uncleaved small glycoproteins.

Regardless of whether the glycoproteins work by a water structuring mechanism or a surface binding one, the studies described above clearly show that the specific effect of the glycoproteins on the freezing point is a

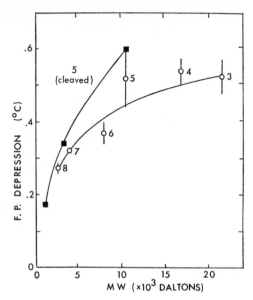

Fig. 8. Freezing point depression of glycoproteins vs. molecular weight. —O—O—, uncleaved glycoproteins 3 through 8; —■—■—, glycoprotein 5 before and after cleavage. Freezing points were determined from freezing curves. All concentrations were 6 mg/ml. (From Raymond and DeVries, 1972).

very specific one, in which the architectural features of the glycoprotein molecules are of great importance.

C. Distribution of glycoproteins in tissues and body fluids of fishes

The glycoproteins present in the antarctic fishes exist in a number of forms which vary slightly in composition and size (DeVries, 1971a) and they have moderately different activities (Lin et al., 1972). All eight sizes (see Table II) are present in the blood, pericardial fluid and coelomic fluid. On the basis of freezing-melting point differences, they are absent from the vitreous humor, bile and urine. The reason for so many different sizes with different activities is not known. The explanation that the small glycoproteins (6, 7 and 8) could be precursors of the large glycoproteins does not seem very likely because to replace the internal prolines of the small glycoproteins with alanines would require a very complex enzyme system.

Preliminary freezing and melting point determinations of extracts of the muscle tissues indicate that they contain more glycoprotein than can be accounted for by their presence in the extracellular space. Thus, intracellular glycoprotein exists. If one postulates that glycoproteins are synthesized outside the muscle cell (perhaps in the liver like serum proteins such as the albumins and globulins (Spiro, 1970)), then on the basis of size considerations, glycoproteins 7 and 8 (MW = 3500 and 2600 daltons) would be the most likely intracellular glycoproteins, because they would be more likely to penetrate cells than the larger ones. In this case penetration must be selective for certain membranes because the glyco-proteins do not cross those barriers which separate the urine, bile and vitreous humor from the blood stream. On the other hand, if the glyco-proteins are synthesized by each cell, then they could be both large and small.

D. Role of membranes in preventing ice propagation

Studies of the freezing behavior of pure glycoprotein solutions suggest they prevent freezing by binding to the surface of ice crystals and thus prevent them from growing (DeVries, 1971a; Duman and DeVries, 1972). However at first glance such a mechanism is difficult to reconcile with what one observes with fishes in McMurdo Sound. At the environmental temperature of $-1.9°C$ there is no evidence that ice crystals ever enter the antarctic fishes even though they are continually in contact with ice. However, in the laboratory at $-2.0°C$ ice is propagated across the integu-ments of such fish and macroscopic crystals appear in their body fluids and tissues (Smith, 1972). Thus under environmental conditions ice crystals are either prevented from gaining entry into the fish or, if they do so they are microscopic and do not increase in size once they are within the bodies of the fish. If one assumes that small ice crystals do enter the fish, then it is unclear how these ice crystals are disposed of because the melting point of ice in the blood of the fish is around $-1.0°C$ which is warmer than the summer water temperature ($-1.5°C$) of McMurdo Sound. Of course the crystals could be melted if a "thermal site" were present in the core of the fish, although so far no one has been able to demonstrate deep body temperatures which are more than $0.05°C$ warmer than the surrounding water (DeVries, 1969; Smith, 1972). Small ice crystals do, however, have higher surface free energies than large crystals and hence will melt at lower temperatures. It is possible that the glycoproteins may influence the surface free energies of the small crystals causing them to melt at much lower temperatures than in the absence of the glycoproteins.

Perhaps a more attractive hypothesis is that the membranes of cold-water fishes are more impervious to ice at $-1.9°C$ than those of temperate water fishes. In this case the primary function of the glycoproteins would be stabilization of the supercooled state of the body fluids. That membranes are able to form barriers to ice propagation is illustrated by the presence of vitreous humor and urine which are both supercooled by 1°C in several of the antarctic fishes. A comparative study of the abilities of membranes to resist ice propagation in temperate water fishes and in those which possess antifreeze glycoprotein would be of great interest. Perhaps northern fishes such as the sculpin would show a seasonal resistance to ice propagation across their membranes.

Another possibility is that the glycoproteins are incorporated into the membranes in such a manner that they are made less permeable to ice. If they do function in this capacity then it is not difficult to visualize how they could prevent ice from propagating across the body wall into the super-cooled urine of the bladder, or across the cornea into the vitreous humor.

E. Conservation of glycoproteins in antarctic fishes

Some of the antarctic nototheniids are the most stenothermal fishes known to exist. The *Trematomus* fishes inhabiting McMurdo Sound live at −1.9°C throughout the year and will die of heat if they are exposed to temperatures above 6°C (Somero and DeVries, 1967). Even if these fishes are warm acclimated to temperatures of +2° to +3°C for 60 days, their upper lethal temperature remains unchanged (DeVries, 1973). After lengthy warm acclimation, the level of the glycoprotein is also unchanged indicating that it is a fixed trait, a finding which contrasts with the seasonal variation found in the levels of antifreeze compounds in the blood of northern fishes.

In view of the fact that the levels of glycoprotein in the blood of the antarctic fishes cannot be altered by temperature and that the concentrations in the blood are high (3% w/v), one can predict that these fishes must possess adaptations which permit them to conserve their glycoproteins. Thus, the rate of synthesis and degradation (turnover) of the glycoproteins should be slow. When liver and other tissue homogenates are examined for galactosyl-transferase activity using glycoprotein which lacks the terminal galactose as an acceptor, very little transferase activity can be detected (DeVries, 1973). Such a finding suggests that the turnover rate may in fact be very slow. When a labelled glycoprotein (either large or small) is prepared and injected into the antarctic fish it takes approximately three weeks for half of the labelled glycoprotein to disappear, indicating a slow turnover rate.

It would seem that the kidneys of the antarctic fishes must also have some unusual modifications which prevent the loss of the small glycoproteins. That this is the case is apparent from the fact that even though all of the antarctic fishes have high levels of glycoproteins 7 and 8 (MW 3500 and 2600 daltons) in their blood, none of these glycoproteins can be found in their urine. In vertebrates possessing glomerular kidneys, globular proteins with a molecular weight of less than 40 000 daltons pass through the pores in the capillaries of the glomerulus and appear in the urine (Bloom and Fawcett, 1969). Inulin having a molecular weight of 5000 daltons appears in the urine of glomerular fishes soon after it is introduced into the blood stream. If the antarctic fishes have glomerular kidneys then the two small glycoproteins should appear in their urine as both are smaller than inulin, unless some mechanism exists for recovering them from the formative urine. From a consideration of the energetics involved, recovery of the glycoproteins from the formative urine would be very expensive. The fact that no glycoproteins have been found in the urine of these fishes suggest they are aglomerular. Histological examinations of the kidneys of the two antarctic nototheniids, *Dissostichus mawsoni* and *T. bernacchii* indicate their kindeys indeed do lack glomeruli (Dobbs and DeVries, 1974). Whether all of the antarctic nototheniids are aglomerular is not known at this time.

It is unlikely that aglomerularism evolved as a mechanism for conserving the small glycoproteins because nototheniids which live in the warm water ($+5°C$) around New Zealand are also aglomerular. Most likely ancestoral aglomerularism permitted evolution of small glycoproteins which could not be conserved if the fish had glomerular kidneys.

Not all of the cold water fishes are aglomerular, however. The kidney of the zoarcid, *R. dearborni*, is glomerular but its antifreeze is larger in size than that of the nototheniids which live in the same waters (DeVries, 1973). Similarly, all of the northern fishes which we have examined are glomerular and preliminary observations suggest that their antifreezes are also much larger than the smallest glycoproteins found in the nototheniids (DeVries, 1973). More detailed studies of the size and composition of the northern antifreeze compounds are in progress and should give a clearer picture of the relationship between the size of the antifreeze compounds and the structure of the kidney.

VI. Relationship between freezing point and freezing resistance

To determine whether body fluid freezing points are good estimates of freezing resistance, physiologists generally compare the freezing point with the incipient lower lethal temperature in the presence of ice. For a number of temperate marine teleosts which do not tolerate temperatures below $-1.0°C$, the temperatures of freezing of the fish and its blood are the same. Chum salmon, *Oncorhynchus keta*, and sockeye salmon, *O. nerka*, freeze at any temperature below the freezing point of their blood (Brett and Alderdice, 1958). The winter flounder, *Pseudopleuronectes americanus*, however, tolerates $-1.5°C$ water in the presence of ice, but its blood freezes at −0.78°C when measured with a vapor pressure osmometer (Umminger, 1970). Blood from the same fish has a freezing point of −1.15° C (Table VI) when measured with a freezing point osmometer (Pearcy, 1961). When the freezing point of blood is observed in a capillary the freezing point is −1.37°C (Duman and DeVries, 1973). The freezing point obtained by the capillary technique indicates that no supercooling exists,

TABLE VI: Comparison of the freezing points of blood from winter flounder, *Pseudopleuronectes americanus*, obtained by various methods.

Technique	Freezing point of Blood (°C)	Freezing point of Organism (°C)	References
Vapor pressure lowering	−0·71	−1·5	Umminger (1970)
Melting point	−0·75	−1·34	Duman and DeVries (1973)
Freezing point osmometer	−1·15	−1·0 to −1·5	Pearcy (1961)
Incipient temperature of ice formation	−1·37	−1·34	Duman and DeVries (1973)

while that obtained by vapor pressure osmometry indicates that super-cooling is a prominent feature of this fish. In the antarctic fishes *N. rossi* and *N. neglecta coriiceps*, freezing points obtained by the melting point technique (Ramsay and Brown, 1955) indicate that, when living in −1.7°C ice-laden water, these fishes are supercooled by 0.7°C (Smith, 1972). For the same fish the blood freezing point obtained with a freezing point osmometer was −1.8°C while the temperature of incipient ice crystal growth was −2.15°C (DeVries, 1971a). Thus, depending on how the freezing point is determined a fish may or may not appear to be super-cooled. These examples indicate the confusion which presently exists in the literature on lower lethal temperatures, body fluid freezing point and their relation to supercooling. Thus at this time it would seem appropriate to review some of the freezing point measurement techniques and their relation to the actual freezing process.

The freezing point of a liquid is the temperature at which the vapor pressure of the liquid and solid phases are equal. When a solute which causes a freezing point depression is added, it lowers the vapor pressure of the liquid with respect to that of the solid. A well known thermo-dynamic relationship exists between the difference in vapor pressure and the freezing point, so if the difference in vapor pressure can be measured, the freezing point can be calculated. Normally (that is, for an ideal solution having normal colligative properties) the difference is caused only by the lowering of the vapor pressure of the liquid. This can be accurately measured (Ehrmantraut, 1966) and, therefore, the freezing point can be calculated. However, if the solute causes a difference in vapor pressure by *raising* the vapor pressure of the solid, then measuring the vapor pressure lowering of the liquid will say nothing about the difference and hence nothing about the freezing point. The pitfall in the vapor pressure method of determining the freezing point, therefore, is the assumption that all solutes interact only with the liquid phase. A more accurate freezing point could be obtained if the vapor pressure of both liquid and solid phases were measured.

In any case, for the present, measuring freezing points directly is the most reliable method. However, there are several ways of doing this and all methods will not always give the same result.

In most studies of fishes, freezing points have been obtained by determining the melting point of the last small ice crystal in the blood. Here it has been assumed that the freezing point and the melting point are the same as they are in all ideal solutions. As has already been mentioned, however, the antifreeze solutions are not ideal (i.e., their freezing point and melting point are not the same) so this method can also be misleading.

We have found that the most accurate way of measuring the freezing point is to measure the temperature at which a small crystal in a capillary tube in a bath begins to grow as the bath temperature is slowly dropped. This is a modification of a technique devised by Ramsay and Brown (1955). An alternative method which gives similar results, the freezing curve method, measures sample temperatures vs. time as the bath temperature is slowly dropped through the sample freezing point (Raymond and DeVries, 1972).

Another technique which is similar to this latter method, except for its speed, measures the temperature rise that occurs on freezing a supercooled solution. The latent heat released on freezing raises the temperature to its freezing point and the temperature can be directly measured. This is the method used in a freezing point osmometer. Capillary freezing points of solution of glycoproteins 3, 4 and 5 have been found to be approximately the same as those obtained using a freezing point osmometer, but freezing points of glycoproteins 6, 7 and 8 obtained by these two methods differ greatly (Fig. 9).

An explanation for this can be found in the facts that glycoproteins 6, 7 and 8 are sensitive to the rate of freezing (Fig. 10) (Raymond and DeVries, 1972) and that the freezing point measured with the osmometer is obtained by rapid freezing. If solutions of the small glycoproteins are cooled slowly to their freezing point then they show significant depressions of the freezing point (Figs. 6, 9). As a result, if the blood serum is cooled slowly

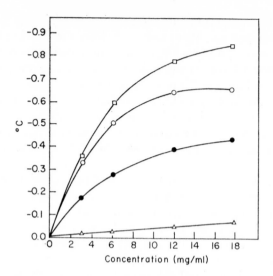

FIG. 9. Comparison of freezing points of solutions of glycoproteins determined with Fiske Osmometer (—O—O—, glycoproteins 3 to 5; —△—△—, glycoproteins 7 and 8) and the temperatures of the incipient ice crystal growth (—□—□—, glycoproteins 3 to 5; —●—●—, glycoproteins 7 and 8).

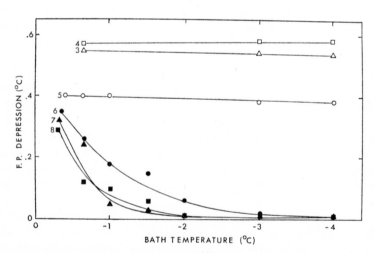

FIG. 10. Freezing point depression of glycoproteins 3 to 8 vs. freezing rate (bath temperature). Samples were supercooled to various bath temperatures and seeded. The temperature then rose to a plateau which was taken as the freezing point. All concentrations were 6 mg/ml. (From Raymond and DeVries, 1972).

toward its freezing point then ice formation occurs at a lower temperature than if cooled quickly.

From such experimental results it is clear that very different freezing points can be obtained depending on the technique used. The most meaningful freezing points, in a biological sense, are those obtained with the capillary technique where the temperature is slowly lowered until freezing can actually be observed, because this is the temperature at which ice will propagate through the body fluids of the fish. The freezing points we have emphasized as important are always the same or lower than the incipient lower lethal temperature of fishes which exhibit extreme resistance to freezing and, therefore, in this context supercooling is nonexistent. If the melting points are taken as the freezing points, then the temperature of initial ice propagation should be measured and considered the supercooling point. In this context then it could be said that the glycoprotein antifreezes serve to stabilize the supercooled state. When the manner in which the antifreezes exert their effect is better understood, much of the confusion in freezing terminology should disappear. For the present, it would seem that the problem of describing freezing resistance in fishes is greatly simplified by equating the freezing point to the temperature of ice formation.

VII. Acknowledgments

I thank James A. Raymond and Yuan L. DeVries for their assistance in the preparation of this article. They were also the source of some of the ideas set forth.

VIII. References

Andrews, C. W. and Lear, E. (1956). *J. Fish. Res. Bd Can.* **13**, 843–860.
Andriashev, A. P. (1970). *In* "Antarctic Ecology" (Holdgate, M. W., Ed.), Vol. 1, pp. 297–304. Academic Press, London and New York.
Asahina, E. (1969). *Adv. in Insect Physiol.* **6**, 1–47.
Bent, H. A. (1965). "The Second Law", p. 237. Oxford University Press, New York.
Black, V. S. (1951). *Univ. Toronto Biol. ser.* **59**, No. 71, 53–89.
Bloch, R., Walter, D. H. and Kuhn, W. (1963). *J. Gen. Physiol.* **46**, 605–615.
Bloom, W. and Fawcett, D. (1969). "A Textbook of Histology", pp. 669–673. W. B. Saunders Co., Philadelphia.
Bogdanov, D. V. (1963). *Deep Sea Research* **10**, 520–523.
Brett, J. R. and Alderdice, D. F. (1958). *J. Fish. Res. Bd Can.* **15**, 805–813.
Brey, W. S., Jr. (1968). "Principles of Physical Chemistry", pp. 306–308. Appleton-Century-Crofts, New York.

Dayton, P. K., Robilliard, G. A. and DeVries, A. L. (1969). *Science, N.Y.* **163,** 273–274.

DeVries, A. L. (1968). "Freezing Resistance in some Antarctic Fishes". Stanford University, California.

DeVries, A. L. (1969). *Antarctic J. U. S.* **4,** 104–105.

DeVries, A. L. (1970). *In* "Antarctic Ecology" (Holdgate, M. W., Ed.), Vol. 2, pp. 320–328. Academic Press, London and New York.

DeVries, A. L. (1971*a*). *Science, N.Y.* **172,** 1152–1155.

DeVries, A. L. (1971*b*). *In* "Fish Physiology" (Hoar, W. S. and Randall, D. J., Eds.), Vol. 6, pp. 157–190. Academic Press, London and New York.

DeVries, A. L. (1973). Unpublished data.

DeVries, A. L., Komatsu, S. K. and Feeney, R. E. (1970). *J. biol. Chem.* **245** 2901–2908.

DeVries, A. L. and Somero, G. N. (1971). *In* "The Ocean World" (Uda, M., Ed.), pp. 101–113. Japan Society for the Promotion of Science, Tokyo.

DeVries, A. L., Vandenheede, J. and Feeney, R. E. (1971). *J. biol. Chem.* **246,** 305–308.

DeVries, A. L. and Wohlschlag, D. E. (1969). *Science, N.Y.* **163,** 1073–1075.

Dobbs, G. H. and DeVries, A. L. (1973). Unpublished data.

Dorsey, N. E. (1948). *Trans. Am. Phil. Soc.* **38,** 247–328.

Doudoroff, P. (1945). *Biol. Bull.* **88,** 194–206.

Duman, J. G. and DeVries, A. L. (1972). *Cryobiology* **9,** 469–472.

Duman, J. G. and DeVries, A. L. (1973). Submitted for publication.

Ehrmantraut, H. C. (1966). *In* "Clinical Pathology of the Serum Electrolytes" (Sunderman, F. W. and Sunderman, F. W., Jr., Eds., pp. 181–188). Charles C. Thomas, Springfield, Illinois.

Eliassen, E., Leivestad, H. and Moller, D. (1960). *Arbok. Univ. Bergen, Mat.-Nat., Ser.* **14,** 1–24.

Gordon, M. S., Amdur, B. N. and Scholander, P. F. (1962). *Biol. Bull.* **122** 52–62.

Hargens, A. R. (1972). *Science, N.Y.* **176,** 184–186.

Hedgpeth, J. W. (1969). *Ant. J. of U.S.* **12,** 106–107.

Hochachka, P. W. and Somero, G. N. (1973). "Strategies of Biochemical Adaptation." W. B. Saunders Company, Philadelphia, London and Toronto.

Kanwisher, J. W. (1955). *Biol. Bull.* **109,** 56–63.

Kanwisher, J. W. (1959). *Biol. Bull.* **116,** 258–264.

Kanwisher, J. W. (1966). *In* "Cryobiology; Freezing in Intertidal Animals" (Meryman, H. T., Ed.), pp. 487–494. Academic Press, New York.

Komatsu, S. K., DeVries, A. L. and Feeney, R. E. (1970). *J. biol. Chem.* **245,** 2909–2913.

Kuhn, W. (1956). *Helv. chim. Acta* **39,** 1071–1086.

Kuntz, I. D. Jr., Brassfield, T. S., Low, G. D. and Purcell, G. V. (1969). *Science* **163,** 1329–1331.

Leim, A. H. and Scott, W. B. (1966). "Fishes of the Atlantic Coast of Canada." Ottawa, Fish. Res. Bd Can.

Leivestad, H. (1965). *Spec. Publs. int. Commn NW. Atlant. Fish.* **6,** 747–752.

Lewis, E. L. and Weeks, W. F. (1970). *In* "Symposium on Antarctic Ice and Water Masses" (Deacon, G., Ed.), pp. 23–34. Tokyo.

Lin, Y., Duman, J. G. and DeVries, A. L. (1972). *Biochem. Biophys. Res. Commun.* **46,** 87–92.

Ling, G. N. (1972). *In* "Water Structure at the Water-Polymer Interface" (Jellinek, H. H. G., Ed.), pp. 4–13. Plenum Press, New York.

Littlepage, J. L. (1965). *In* "Biology of the Antarctic Seas" (Lee, M. O., Ed.), Vol. 2, pp. 1–37. Washington, D.C., American Geophysical Union.

MacGinitie, G. E. (1955). "Distribution and Ecology of the Marine Invertebrates of Point Barrow, Alaska." Smithsonian Institution, Washington, D.C.

Mazur, P. (1970). *Science, N.Y.* **168,** 939–949.

Miller, L. K. (1969). *Science, N.Y.* **166,** 105–106.

Norman, J. R. (1940). *Discovery Rept.* **18,** 3–104.

Pearcy, W. G. (1961). *Science, N.Y.* **139,** 193–194.

Pearse, J. S. (1962). Letter to editor. *Sci. Am.* **(5) 207,** 12.

Potts, W. T. W. and Parry, G. (1964). "Osmotic and Ionic Regulation in Animals", Vol. 19, p. 171. Pergamon Press, New York.

Prosser, C. L. and Brown, F. A., Jr. (1961). "Comparative Animal Physiology", 2nd ed., pp. 57–80. W. B. Saunders, Philadelphia.

Prosser, C. L., MacKay, W. and Kato, K. (1970). *Physiol. Zool.* **43,** 81–89.

Ramsay, J. A. and Brown, R. H. J. (1955). *J. Scientific Instruments* **32,** 372–375.

Raschack, M. (1969). *Intern. Rev. Ges. Hydrobiol. Hydrog.* **54,** 423–462.

Raymond, J. A. and DeVries, A. L. (1972). *Cryobiology* **9,** 541–547.

Read, L. (1973). Unpublished data.

Scholander, P. F. and Maggert, J. E. (1971). *Cryobiology* **8,** 371–374.

Scholander, P. F., van Dam, L., Kanwisher, J. W., Hammel, H. T. and Gordon, M. S. (1957). *J. cell. comp. Physiol.* **49,** 5–24.

Shier, W. T., Lin, Y. and DeVries, A. L. (1972). *Biochim. Biophys. Acta* **263,** 406–413.

Smith, R. N. (1970). *In* "Antarctic Ecology" (Holdgate, M. W., Ed.), Vol. 1, pp. 329–336. Academic Press, London and New York.

Smith, R. N. (1972). *Br. Antarct. Surv. Bull.* **28,** 1–10.

Somero, G. N. and DeVries, A. L. (1967). *Science* **156,** 257–258.

Spiro, R. C. (1970). *A. Rev. Biochem.* **39,** 599–638.

Stout, W. E. and Shabica, S. (1970). *Ant. J. of U.S.* **5,** 134–135.

Tait, M. J. and Franks, F. (1971). *Nature, Lond.* **230,** 91–94.

Tanford, C. (1963). "Physical Chemistry of Macromolecules", 710pp. John Wiley and Sons, New York.

Tiffany, M. L. and Krimm, S. (1969). *Biopolymers* **8,** 347–359.

Tressler, W. L. and Ommundsen, A. M. (1962). "Seasonal Oceanographic Studies in McMurdo Sound, Antarctica", Tech. Report TR-125, p. 141. U.S. Navy Hydrog. Office.

Umminger, B. L. (1969a). *J. exp. Zool.* **172,** 283–302.

Umminger, B. L. (1969b). *J. exp. Zool.* **172,** 409–424.

Umminger, B. L. (1970). *Biol. Bul.* **139,** 574–579.

Umminger, B. L. (1971). *Physiol. Zool.* **44,** 20–27.

U.S. Naval Oceanographic Office. (1957). Publ. 705.

Weyl, P. K. (1970). "Oceanography." pp. 180–182. Wiley, New York.

Williams, R. J. (1970). *Comp. Biochem. Physiol.* **35**, 145–161.

Wohlschlag, D. E. (1964). *In* "Biology of the Antarctic Seas" (Lee, M. O., Ed.), Vol. 1, pp. 33–62. Washington, D.C., American Geophysical Union.

Woodhead, P. M. J. (1964). *Helgo aender Wiss. Meeresuntersuch.* **10**, 283–300.

Woodhead, P. M. J. and Woodhead, A. D. (1959). *Proc. zool. Soc. Lond.* **133**, 181–199.

Subject Index

A

Adrenalectomy, 84, 107
Adrenalin, 148, 269, 270, 271
Aëdes aegypti, 20, 27, 31, 32, 33, 41, 48, 50, 57, 64, 90, 111, 142, 149
Aëdes detritus, 110, 111, 142, 143
Aglomerularism, 323
Air flow, 293
Alaska, 225, 227, 291
Aleutian Current, 291
Alleles, kinetic studies of, 228
Allelic variation, 221, 222
Ambystoma gracile, 22, 48, 50, 80, 82
  A. tigrinum, 22
Amino-acids
  effect on enzymes, 228
  effect on freezing tolerance, 299
  electrophoretic detection of, 234
  in generation of energy, 253
  in haemocyanins, 207
  non-electrolyte, 150
  sequencing data, 222
Ammonia
  endogenous, 32
  excretion of, 33–40
Ammonotelism, 33
Anal papillae
  in *Aëdes aegypti*, 20, 57, 64, 76, 90, 149
  electrolyte uptake, role in, 17
  water transfer, site of, 64
Anguilla, effects of hypophysectomy, 85
Anguilla anguilla, 26, 60, 61, 65, 80, 84, 114, 124, 134, 137, 144, 260, 262, 263, 270, 271
  A. rostrata, 55, 60, 137, 253
  A. vulgaris, 257
Antarctic Ocean, 314
Antarctica, 294
Anthraquinones, 193, 194

"Antifreeze", 296, 305, 306, 310, 312, 316, 322
  seasonal changes in, 312, 314
Artemia salina, 61, 92, 109, 110, 139, 140, 143, 148, 181, 201
Arctic sculpin, 303, 305
Astaxanthin, 176, 178, 179, 181, 183, 184, 188, 189

B

Balance
  acid-base, 49, 50, 51, 262
  ionic, 51
  mineral, 13, 14, 92, 104
  water, 64
Baltic Sea, 301
Bilichromes
  derivation of, 205
  in gastropod secretions, 206
  in skeletal structures, 205
  location of, 204
Biokinetic zone, 290
Blood
  freezing point of, 296, 306, 312, 317, 324
  glucose level in, 253
  sodium chloride, seasonal variations in, 303
Body
  fluid, 278, 296, 300
  fluid, electrolyte loss from, 304
  fluid, extracellular, 150
  fluid, freezing of, 295, 298, 323
  fluid, freezing point depression, 304
  fluid, glycoproteins in, 320
  fluid volume, 52
  size, 69
  size, gill surface relationship to, 69
  tissues, 300, 304

Water movement—*continued*
   transfer sites, 64
   uphill transfer, 6
   Van 't Hoff's law, 2
Wax esters, 275, 276
Weddell seal, 295
Weightlessness of fish, 266
Whale, blue, 179
White scattering, 172

X

X-rays, 127

Z

Zooanthellae, 187
Zooplankton, 172, 173, 179, 200